Textbook of GENERAL HORTICULTURE

by

Julian Claude Schilletter
Harry Wyatt Richey

BIOTECH
2025
Biotech Books

ISBN 978-81-7622-030-9

Published by: **BIOTECH BOOKS**®
4762-63/23, Ansari Road, Darya Ganj,
New Delhi - 110 002
Phone: 09899295259, 011-43003222
E-mail: biotechbooks@yahoo.co.in
Website: www.biotechbooks.in

Typeset at: **Classic Computer Services**
Delhi - 110 035

PRINTED IN INDIA

TO THE MEMORY OF

ROBERT RICHEY

PREFACE

The object of the present book is to set forth the essential principles of horticulture in a manner suitable for a general introductory college course.

Horticulture, although one of the oldest arts, is a young science. Ideas are changing so rapidly from year to year that no uniform practice has been established for the textbook treatment of subject matter. Difficulties arise in deciding the proper emphasis to be placed on the various phases of horticulture, and there is a natural tendency to overstress those particular fields in which the writer's or teacher's special interests lie.

A need for a textbook in general horticulture has existed for many years. Particularly is this true in the agricultural colleges that require such a course in many of the departmental curriculums. This general course in horticulture varies in the agricultural colleges. Some departments give the student a choice between a course in vegetable crops and a course in fruit crops. Some give plant propagation as the general course. Some term the course "general horticulture" and then proceed to devote most of the work to fruits or to vegetables, depending upon the personal interests of the instructor. Again, there are instances where the course is handled by several instructors during the semester or quarter. In the latter instance a fruit specialist will give the fruit work, and a vegetable specialist the vegetable work and a florist may give the ornamental phases.

The authors have had several years' experience in teaching a general course in horticulture, and the present text is an attempt on their part to accord each division of horticulture such treatment as will result in a well-balanced presentation of the subject matter as a whole.

Although this volume is written for the student who may desire a general knowledge of horticulture rather than for the one who intends to specialize in the subject, it is felt that the book offers a background of material that will be of value to the student who desires to take special work dealing with fruits, vegetables or ornamentals. This volume presupposes but little biological knowledge on the part of the student. Even if he has had elementary courses in such subject materials, it has been found advisable to refresh his memory on certain fundamental facts, particularly those concerning structures and functions of the plant. This general basic knowledge is necessary in order that he may have a clear understanding of the reasons for performing the various horticultural practices in a particular manner. The approach to the subject matter is from the viewpoint of fundamentals rather than from the culture of individual crops. Although the methods of performing certain practices are described in many instances, the emphasis has been placed on the reasons for the practices rather than on how they are performed.

Two sets of questions are presented at the end of each chapter. The first is designed to call the student's attention to certain facts that it is believed should be a part of his general knowledge. The second set is designed to stimulate his curiosity by presenting practical problems, to test his knowledge of facts and to train his judgment in selecting the essential facts and properly evaluating them in solving practical horticultural problems. No general horticultural text will be perfectly adapted to all sections of the United States. In some instances a particular section might require several volumes, and in other cases local conditions might necessitate specialized treatments for certain crops. The text material is widely applicable, but the instructor should be responsible for a knowledge of local horticultural practices, and the student responsible for a knowledge of the fundamental principles influencing such practices.

The first four chapters deal with rather broad subjects in an effort to present a picture of the field of horticulture—how it fits into the field of agriculture and how it enters into the economic, political and social life of the individual. The remaining chapters deal almost entirely with the fundamental plant processes and their application to horticultural practices.

Since free use has been made of much of the published material pertaining to the subject matter of the text, specific acknowledgment would be impossible, but the authors wish to acknowledge this valuable assistance. In particular are they indebted to Dr. E. S. Haber of the Department of Horticulture of Iowa State College, who contributed some of the material incorporated in the manuscript.

JULIAN CLAUDE SCHILLETTER,
HARRY WYATT RICHEY.

CONTENTS

TEXTBOOK OF GENERAL HORTICULTURE

CHAPTER I

GENERAL HORTICULTURE

Horticulture is that division of agriculture which relates to the culture of those plants commonly known as "fruits," "vegetables" and "ornamental plants." Etymologically the word "horticulture" (*hortus*, "garden," and *cultura*, "cultivation") means the culture of a garden or enclosure in contrast to the word "agriculture" (*agri*, "field," and *cultura*, "cultivation") which means the cultivation of a field. Obviously this distinction, although satisfactory during feudal times, is no longer valid, for many single horticultural crops are now produced in larger areas. The layman usually thinks of horticulture as being synonomous with gardening or the intensive culture of plants in small areas, but horticulture is much more than gardening. It is an industry and occupies important economic and cultural positions in the field of agriculture.

Agriculture generally divides itself into two broad groups—animals and plants. Subjects in agriculture dealing with plants include horticulture, agronomy and forestry. Each of these subjects is an applied science and naturally considers the practical application of its scientific endeavor to a much greater degree than does botany, which focuses its attention on the fundamental principles of plant life. Again, horticulture, agronomy and forestry are accepted as industries, and the industries look to these fields for the solution of practical problems. The facts that determine whether a particular plant belongs to the subject of horticulture, agronomy or forestry are (1) type of culture, (2) custom and (3) purpose for which grown.

Horticultural crops are often considered as those which require intensive culture and are grown in small areas in contrast to agronomic or field crops which require extensive culture and are grown in large areas. The foregoing distinction does not always hold true, as there are horticultural crops grown in large areas and field crops, such as tobacco, that require intensive culture. Crops receiving intensive culture, however, as contrasted with crops receiving extensive culture becomes the principal line of separation between agronomic crops and horticultural crops. Custom sometimes establishes a crop in one field or another. For example, Irish potatoes and sweet corn are generally classed as horticultural crops, but some sections of North America class them as field crops. Again, custom establishes tomatoes and melons as vegetable crops in North America but as fruit crops in Europe. The purpose for which the particular plant is grown may confine it to one field or another. Pecans and walnuts, when grown as orchard crops, are associated with horticulture; and when grown for wood production they are associated with forestry. Bluegrass, when grown as a lawn crop, is considered horticulture; but when grown as a pasture crop it becomes agronomic. Other examples might be cited, but sufficient evidence is presented to show that there are borderline cases and that the student rightfully becomes confused. In order that the beginner may become familiar with the materials that horticulture proposes to study and with the terminology of the field of horticulture, a classification of horticultural crops is given later in this chapter. The classification makes no attempt to list the names of all horticultural crops.

Horticulture has been, is now and will continue to be of great economic, political and social importance to mankind. The edible products of horticulture plants furnish essential vitamins and necessary food for man and often furnish the chief if not the sole diet for considerable groups of people. Approximately 25 per cent by weight of the food consumed by man in the United States is from horticultural crops. Although the figures given at the end of this chapter indicate the magnitude of the horticulture of the United States as based on commercial value, a second or amateur phase has affected more directly the lives of a greater number of people. It would be hard to estimate the number of people growing horticultural plants just for the sheer joy derived from the beauty of the plants and the skill required in growing them satisfactorily. In this group of amateur horticulturists will be found great contrasts, varying from the individual growing one geranium plant in a tin can in a crowded city tenement to the individual producing many different kinds of plants, with each kind receiving the special treatments best adapted to its particular culture.

CONCEPT

Horticulture might be defined as the culture of fruits, vegetables and ornamental plants. Such a definition, although correct, is so broad and vague in meaning that it does not present a clear picture of the term. Any general definition would be similarly vague, and a specific definition would become very involved, complex, and at times questionable, if not contradictory. It is much better, therefore, for one to have a concept or an understanding of the term "horticulture" rather than a dictionary definition. Horticulture is both an art and a science. It is an art because propagating, pruning, spraying, etc., require special techniques which must be mastered if the work is to be done satisfactorily. The skills attained through practice develop into an art. The scientific reasons underlying these techniques which explain why the various operations are performed in particular fashions constitute a true science.

The primary objective of horticulture is to find ways by which horticultural plants can be made to yield the optimum benefits to mankind. In order to attain this objective, the horticulturist obtains facts relative to the geographical distribution of horticultural plants; the sources and uses of such plants about the home; their structure and manner of growth; the influence of climate and soil on their development; the manner in which they are propagated or reproduced; the manner, time and degree of pruning the various plants; the pests that trouble the plants and the means of controlling them; and the manner of harvesting, storing, transporting and using the finished product.

DIVISIONS

Horticulture divides itself somewhat naturally into the culture of fruits, vegetables and ornamental plants and the production of nursery stock and seeds, which constitute a special type of horticultural farming. The horticulturist refers to these coordinated groups as pomology, olericulture, ornamental horticulture, nursery stock and seed production.

A classification of horticultural crops might be made upon the basis of botanical relationships, in which case they would be arranged in their evident places according to similarity of parts. The horticultural plants might be grouped according to use or according to temperature requirements. In other words, it is hardly possible to give a classification that will hold under all conditions. The one suggested here is based upon plant growth. This is done in order to bring about uniformity in a consideration of the crops studied in the various branches of horticulture. In the classification of horticultural plants an attempt is made to group most of the important fruit and vegetable crops according to the terms commonly used rather than on the strictest scientific basis. No attempt is made to list all the important ornamental plants, because there are hundreds of them widely scattered over the world.

Fruits

To the botanist a fruit is a ripened ovary, but the horticultural concept of the word is more difficult to define. To the horticulturist some botanical fruits are only seeds; in other cases the ovary may be but a small part of the fruit, and in still others the horticultural fruit may not possess seed. To a horticulturist a fruit is the edible product usually of a woody plant, which in its development is closely associated with a flower. The horticulturist recognizes two large groups of fruits: tree fruits and small fruits. Tree fruits are produced on trees, while small fruits, in general, are produced on shrubs or vines. Both tree fruits and small fruits are produced on deciduous and non-deciduous, or evergreen, plants. Deciduous plants are those which drop their leaves during the dormant season, but evergreen plants never become entirely denuded of functional leaves, although some of the older leaves drop each year. The horticulturist recognizes various types or kinds of fruits that in some instances are entirely different from those recognized by the botanist.

Horticulturally fruits may be true fruits or false fruits. True fruits are those which are developed wholly from tissues of a single ovary. Such fruits are also simple fruits, as the peach, plum and orange. An aggregate fruit is formed when several ovaries of a single flower develop independently but remain attached to their common stem, as in the case of the raspberry. A false fruit, as the apple, pear and strawberry, is one that is composed of tissues in addition to those of the ovary. A multiple fruit is a false fruit that is formed when several adjacent flowers and their adjacent tissues unite to form a single fruit, as is true of the pineapple.

The horticulturist often speaks of fleshy fruits as "drupe" fruits, "pome" fruits and "berries." A drupe fruit is a true simple fruit in which only a single ovary of the flower is involved in its composition, as in the peach, plum, apricot and other so-called stone fruits. The part of the ovary surrounding the seed of drupe fruit becomes hard and stony and is known as the "pit"; the outer layers usually become soft and succulent, as in the peach, or dry and leathery, as in the almond. A pome fruit, as the apple, is a false fruit in which the receptacle, calyx or other parts of the flower in addition to the ovary are involved in its development. A true botanical berry is a true or simple fleshy fruit that has seed embedded throughout the pulpy ovarian tissues. Most horticultural "berries" are not true botanical berries at all, but custom has established such fruits as the raspberry, blackberry, cranberry and strawberry as berries.

Fruit plants may be grouped as follows:

I. Tree fruits.
- *A.* Deciduous.
  1. Pome (false fruits).
     Apple, pear, quince, medlar, etc.
  2. Drupe (simple fruits).
     Peach, cherry, plum, apricot, etc.
- *B.* Evergreen.
  1. Citrus (simple fruit).
     Orange, grapefruit, lemon, lime, etc.
  2. Avocado (simple fruit).
  3. Mango (simple fruit).

II. Small fruits (including vine fruits).
- Grape (simple fruit—a true berry).
- Strawberry (false fruit).
- Currant (simple fruit).
- Gooseberry (simple fruit).
- Red raspberry (aggregate fruit).
- Black raspberry (aggregate fruit).

Vegetables

No convenient or accurate definition can be given for the great group of horticultural plants known as vegetables. Since the term is used widely and includes a large number of diverse plants, it is well to form a concept of it as used by horticulturists. This concept differs, since the line of demarcation between fruits and vegetables is not a clear one; for in the United States tomatoes and melons are considered as vegetables, and in Europe they are considered as fruits. As generally accepted, however, the term refers to two groups of commodities: first, those products of herbaceous plants which afford fresh material for culinary purposes or which are normally cooked before consumption; second, those products of herbaceous plants, such as lettuce and tomatoes, which are consumed without being cooked and which are known as "salad plants."

Vegetable crops may be grouped as follows:

Tuber.
- Potato, Jerusalem artichoke.

Root.
- Chinese artichoke, sweet potato, turnip, rutabaga, beet, carrot, parsnip, radish, horseradish, celeriac, salsify.

Bulb.
- Onion, shallot, garlic, chive.

Stem.
  Asparagus, kohlrabi.
Leaf.
  Cabbage, brussels sprouts, kale, spinach, chard, mustard, lettuce.
Petiole.
  Celery, sea kale, rhubarb.
Inflorescence.
  Globe artichoke, cauliflower, broccoli.
Seed.
  Pea, bean.
Fruit.
  Tomato, muskmelon, watermelon, cucumber.

Ornamentals

Ornamental horticulture confines itself to those plants which are grown for their aesthetic value. Horticulturally, a plant is classified as an ornamental when it is used to decorate the landscape or when it is grown because it or its product satisfies the desire for beauty. The purpose for which the plant is grown determines whether or not it is classed as an ornamental plant. For example, many people grow apple trees, and even vegetable plants, for their aesthetic value. If grown for this purpose, such plants would be classified as ornamental plants; however, custom generally establishes an apple tree as a fruit plant and a red cabbage as a vegetable regardless of the purpose for which they are grown. Ornamental plants may be grouped as follows:

I. Flowers.
  A. Annuals, or perennials treated as annuals.
    Zinnia, petunia, cosmos.
  B. Biennials.
    Columbine, hollyhock.
  C. Perennials.
    Dahlia, peony, tulip, iris.
II. Shrubs and vines.
  A. Deciduous.
    Spirea, honeysuckle, deutzia.
  B. Evergreen.
    1. Narrow leaf.
      Mugho pine, Pfitzer juniper.
    2. Broadleaf.
      Rhododendron, laurel, boxwood.
III. Trees.
  A. Deciduous.
    Elm, maple.
  B. Evergreen.
    1. Narrow leaf.
      Spruce, pine, cedar.
    2. Broadleaf.
      Magnolia, citrus.

SOURCES OF PLANTS

When the Norsemen discovered North America, they named the new continent "Vineland" because of the abundance of grapes that they found growing so profusely. Early Spanish, French and English explorers and settlers also mention many kinds of horticultural plants found on the new continent. The present horticultural industry in the United States, however, is largely dependent on plants that were introduced from other parts of the world. The introduced plants were crossed with native sorts, and the present list of horticultural plants is comprised of native species, imported species and hybrids between some of the native and imported species.

The fruit industry of the United States would rank very low in value today if it consisted of only the native species. This fact is indicated when one compares the native species of fruits with the introduced species. Many species of the raspberry, blackberry, dewberry, strawberry, cranberry, blueberry, huckleberry, gooseberry, currant and grape are indigenous to or hybrids with native species of the United States. Species of the plum and apple are native to various sections of this country, but their fruits are decidedly inferior to those of the introduced species. The chief commercial fruits of the United States, as the apple, pear, quince, peach, plum, cherry, apricot, orange, grapefruit, lemon, lime, kumquat, fig, pomegranate, mango, avocado, pineapple, date, European grape, Persian walnut and almond, have been imported from other lands.

Although the early settlers of North America found certain vegetables being grown by the Indians, it is significant to note that none of them, with the exception of Jerusalem artichoke, was indigenous to the United States. Several important vegetables, such as corn, peppers, tomatoes, Irish potatoes, sweet potatoes, and beans, are of New World origin, but most other vegetables of any importance are of Old World origin. A vegetable as important as the tomato, now producing an annual cash income of around 25 million dollars in the United States, did not come into general use until after 1840.

An investigation of the nativity of the ornamental plants listed in a representative nursery catalogue of the United States reveals that most of the herbaceous annual and perennial plants and about 70 per cent of the woody trees and shrubs are of foreign introduction.

Although hundreds of species and varieties of horticultural plants have been introduced into the United States, the search continues into all parts of the world for plant materials that may be valuable in horticulture.

COMMERCIAL IMPORTANCE

From the early beginnings of horticulture in the United States steady progress has been made, until today horticulture has become more than a billion-dollar industry. The importance of this industry can best be appraised by noting the value of the products that it sells and the extent to which these products are used.

TABLE 1.—AVERAGE CASH INCOME* IN MILLIONS OF DOLLARS FROM SALE OF FARM PRODUCTS, 1935–1937†

Commodity	1934	1935	1936	1937
Fruits and vegetables	870	913	1,050	1,165
Cotton and cottonseed	777	690	905	864
Grain	626	598	824	1,073
Other crops	614	612	683	780
All crops	2,887	2,813	3,462	3,882
Meat animals	1,199	1,679	2,001	2,013
Dairy products	1,133	1,289	1,417	1,475
Poultry and eggs	467	620	617	635
Other livestock and products	106	106	136	149
All livestock and products	2,905	3,694	4,173	4,272
Total crops and livestock	5,792	6,507	7,633	8,154

* Cash income relates to the value of quantities actually sold off the farms of the state where produced.

† GAY, M. C., Marketing Fruits and Vegetables Cooperatively, *Farm Credit Admin. Cir.* C-110, p. 2, 1938.

During the last decade fruits and vegetables combined ranked third among agricultural commodities as producers of cash income. In 1937 the farmers' cash income therefrom represented 14 per cent of the total cash income from agricultural products of that year. The exports of fruits and vegetables grew in importance from 16 per cent of the total value of food-products exports in 1926 to 46 per cent in 1936.

Unlike some other agricultural commodities, fruits and vegetables are used primarily for human food or beverages. Although hundreds of uses are found for cotton and its by-products, most fruits and vegetables must find their ways to the table to maintain places of economic importance. The uses of fruits and vegetables as raw materials for the manufacture of industrial products is insignificant.

Fruits and vegetables contain important vitamins and minerals and are prominent in the diets of many people. Although there is a popular belief that the per capita consumption of these products has greatly increased during the past thirty years, there are no official statistics to prove this to be a fact. Unfortunately official statistics for all fruits and for all vegetables were not available until 1918, and even these figures do not give the production of these crops which were

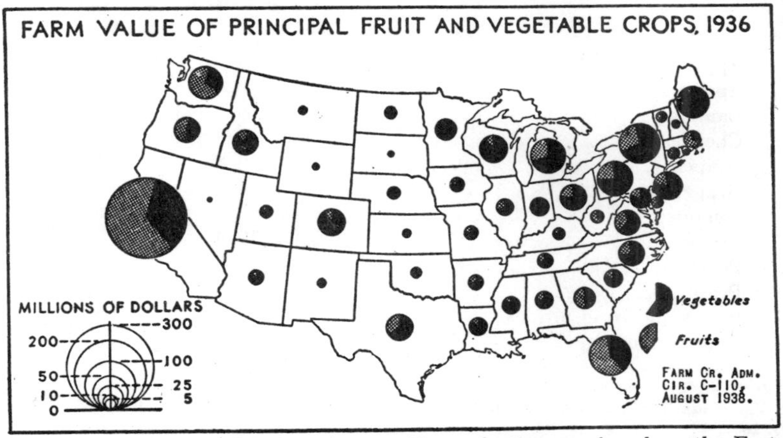

FIG. 1.—Principal areas of fruit and vegetable production are found on the Eastern seaboard and Pacific coast states.

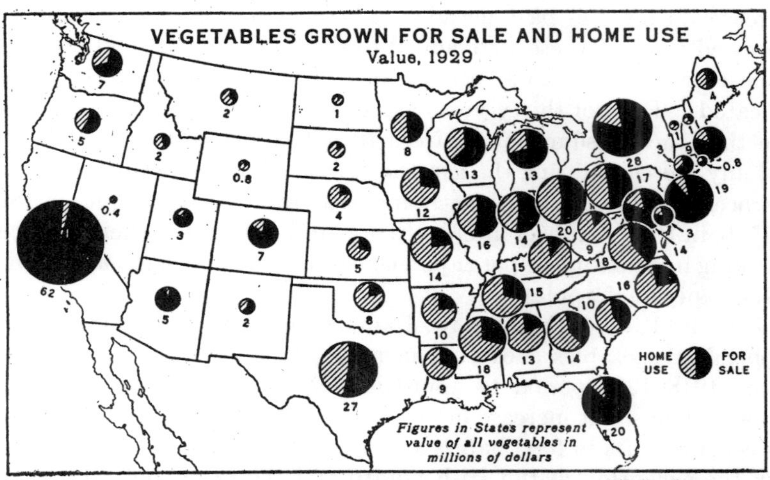

FIG. 2.—Value of vegetables grown for sale and home use. (*U.S. Department of Agriculture.*)

raised in home gardens and used locally. Although the figures shown in Tables 2 and 3 are of necessity only rough approximations of actual consumption, they do serve to show some interesting trends in per capita consumption. It will be noted in Table 2 that the apparent

TABLE 2.—APPARENT UNITED STATES CONSUMPTION OF IMPORTANT FRESH FRUITS*†

Commodity	Pounds per capita	
	1923–1925	1933–1935
Apples	60.2	46.3
Apricots	0.3	0.2
Bananas	18.5	17.1
Cherries	0.8	0.8
Grapes	9.8	8.8
Grapefruit	6.0	6.9
Lemons	3.7	3.6
Oranges	19.4	26.2
Peaches	14.3	11.6
Pears	5.9	4.8
Plums and fresh prunes	2.0	1.6
Strawberries	3.4	3.1

* Includes fruit that may have been wasted, canned in the home, etc.

† THOMPSON, J. M., The Orange Industry—An Economic Survey, *Calif. Agr. Exp. Sta. Bul.* 622, p. 23, 1938.

total per capita consumption of 12 principal fresh fruits shows a slight decrease during the past decade. These figures do not include the processed fruit; if these were added to the total, there would probably be no change in the total per capita consumption during the period indicated. Data of thirty years ago are rather meager, but they indicate that the consumption of fresh fruit has changed very little and that any increase during the last thirty years is perhaps due to the use of processed fruit, as there was an increase from 7 lb. per capita in 1899 to 25 lb. in 1927. It is of interest to note in Table 2 that there has been a striking increase in the per capita consumption of oranges, apparently at the expense of apples.

Note in Table 3 that the total annual per capita production of 16 important vegetables shows an increase from approximately 275 lb. in the 1918–1921 period to approximately 290 lb. in the 1932–1936 period. The slight increase in per capita production is not important, but certain shifts in food habits are of interest. It will be noted that, with the exception of the Irish potato and the sweet potato, the per capita consumption of each of the 16 vegetables listed shows an increase. Rather striking increases have occurred in the consumption of leafy and green vegetables. This is notably true of lettuce which increased approximately 150 per cent in per capita production from 1918–1921 to 1932–1936. Except in a few cases, official statistics are not available for comparing present vegetable consumption with that of thirty years ago. These figures indicate a decline in the per capita consumption of Irish potatoes from 190 lb. in the 1902–1906 period to 144 lb. in more recent years and a threefold increase in the per capita consumption of canned vegetables, or an increase from 9 lb. per capita in 1899 to 27 lb. in more recent years.

TABLE 3.—PER CAPITA PRODUCTION OF SIXTEEN VEGETABLES BY SPECIFIED PERIODS, 1918–1936*

Commodity	Pounds					
	Market		Manufacture		Total	
	1918–1921	1932–1936	1918–1921	1932–1936	1918–1921	1932–1936
Asparagus	0.5	1.8	0.4	0.8	0.9	2.6
Beans, snap	1.0	3.0	0.6	1.0	1.6	4.1
Cabbage	14.0	15.2	1.4	2.2	15.4	17.4
Muskmelons	5.6	6.5	....	....	5.6	6.5
Cauliflower	0.8	2.2	....	....	0.8	2.2
Celery	3.6	6.5	....	....	3.6	6.5
Corn, sweet	.....	.....	9.8	8.7		
Cucumbers	1.1	1.3	1.5	1.6	2.6	3.0
Lettuce	4.4	11.2	....	....	4.4	11.2
Onions	9.1	11.4	....	....	9.1	11.4
Peas, green	0.2	2.0	2.5	2.8	2.7	4.7
Spinach	0.6	1.7	0.4	0.7	1.1	2.3
Sweet potatoes	35.2	30.5	....	....	35.2	30.5
Tomatoes	5.4	8.1	20.0	23.4	25.4	31.5
Watermelons	11.4	12.3	....	....	11.4	12.3
Irish potatoes	155.0	144.0	....	....	155.0	144.0

* Per capita data obtained by dividing average production for the period from figures reported by the U.S. Department of Agriculture by the total average population for the period. Data on the Irish potato were taken from *U.S. Dept. Agr. Misc. Pub.* 267, p. 83.

TABLE 4.—RECEIPTS FROM SALES OF ORNAMENTAL CROPS*

Commodity	Millions of dollars	
	1919	1929
Flowers and flowering plants	62	98
Nursery stock	20	58
Flower and vegetable seeds	..	14
Bulbs	..	5

* Horticulture, Fifteenth Census of the United States, pp. 13, 57, 91, 109.

The data in Table 4 indicate a marked increase in the value of ornamental plants, but the total value remains insignificant when compared to that of the horticultural crops used for food.

During recent years there has been a marked increase in the percentage of nursery sales of ornamental plants, with a decrease in the percentage of sales of fruit plants. The data in Table 5, although for but one year, are representative of this trend. This is undoubtedly due to the more widespread interest in ornamental gardening. It has been reported, however, that the average home grounds are only 22 per cent planted when compared with an established standard for a well-landscaped home.

TABLE 5.—PERCENTAGE OF SALES OF VARIOUS NURSERY PRODUCTS, 1927*

Ornamentals:		Fruits:	
Shrubs	20.30	Apples	6.41
Evergreens	17.79	Plums	3.56
Ornamental trees	12.37	Peaches	4.95
Roses	11.21	Cherries	4.31
Perennials	7.07	Small fruits	5.84
Bulbs	2.09	Nuts	0.45
Forest seedlings	1.65	Total	25.52
Total	74.48		

* Survey of U.S. Nurseryman Association, 1927.

Review Questions

1. What factors place a particular plant in the field of horticulture?
2. What is the concept of horticulture?
3. What is the difference between intensive and extensive culture when used in reference to plants?
4. What is the primary objective of horticulture?
5. What are the divisions of horticulture?
6. Give a botanical definition of a fruit.
7. Give a horticultural definition of a fruit.
8. Distinguish between a fruit and a vegetable.
9. Define the following terms and give a horticultural example of each:
   a. Annual.
   b. Biennial.
   c. Perennial.
   d. Deciduous.
   e. Evergreen.
10. Are many of the present commercial fruits of the United States indigenous to the United States?
11. Are many of the present commercial vegetable crops of the United States indigenous to the New World?
12. Has the annual per capita consumption of fruits in the United States increased materially during the last decade?
13. What particular fruit in the United States shows a material increase in the annual per capita consumption during the last decade?

14. Has the annual per capita consumption of vegetables in the United States increased materially during the last decade?

15. What particular vegetable in the United States shows a material increase in the annual per capita consumption during the last decade?

Problems

1. State and discuss one specific horticultural problem of economic significance in your home community.

2. State and discuss one specific horticultural problem of political significance in your home community.

3. State and discuss one specific horticultural problem of social significance in your home community.

4. Make a block graph showing the percentage of the total cash income for each of the following during 1937: fruits and vegetables, cotton and cottonseed, grain, meat animals, dairy products and poultry and eggs.

Suggested Collateral Readings

1. BAILEY, L. H., "The Standard Cyclopedia of Horticulture," Vol. II, pp. 1501–1523, The Macmillan Company, New York, 1925.

2. CORBETT, L. C., *et al.*, "Fruits and Vegetables," pp. 107–124, "U.S. Department of Agriculture Yearbook," 1925.

3. KNOTT, J. E., "Vegetable Growing," pp. 29–36, Lea & Febiger, Philadelphia, 1935.

4. MARTIN, J. N., "Botany with Agricultural Applications," pp. 77–88, John Wiley & Sons, Inc., New York, 1920.

5. "U.S. Department of Agriculture Yearbook," 1937.

CHAPTER II

HORTICULTURAL ENTERPRISES OF THE WORLD

Horticultural crops have been important sources of human food since the dawn of history. They have been and are important as a group rather than as individual crops; and none, with the possible exception of the Irish potato, has attained the importance among world crops of wheat, which is used for bread, or of cotton, which is used for clothing. For example, according to government reports, the production of fruit reported by countries for which estimates were available averaged 66 million tons per year in the three-year period 1933–1936. During the same period, the world production of corn averaged 122 million tons per year. The world production of fruits was greater than that of either cotton or tobacco. The world figures for vegetables as a group are not nearly so complete as for other agricultural crops, but they indicate a larger total than even that for fruits. The world production of Irish potatoes alone was estimated at more than 216 million tons during each year of the period 1927–1931.

The staple crops of wheat, corn and cotton are of greater monetary importance as world crops than any of the individual horticultural crops, such as apples, onions or potatoes. Every student of agriculture should be familiar with those horticultural crops which are important sources of food to large populations in various parts of the world and with the important ones that enter into national and international trade. A study of world production of particular crops is most enlightening, not only from the standpoint of the adaptations of the particular crops to climatic and soil conditions but also from the standpoint of world markets.

GEOGRAPHICAL DISTRIBUTION

Climate, soil and markets are all important factors which combine to concentrate the commercial production of horticultural crops into limited areas in various parts of the world. In general, the ability of a region to produce a crop successfully is limited by climate and soil.

Climate

Temperature and moisture, the two chief agencies of climate, which is considered as the average condition of the atmosphere, must be favorable for the successful growing of horticultural crops, as either alone can limit production. The absence of adequate available water limits plant distribution on the earth more definitely than does temperature.

Temperature.—Of the climatic agencies influencing plant growth, temperature is one of the most important. Its effect may be modified by several factors, such as duration of high or low temperatures and the rapidity of temperature changes as well as the amount of moisture in the air, in the soil and in the plant.

Moisture.—Although many horticultural areas of the world depend upon rainfall to supply adequate moisture, there are also many areas where irrigation is being used to supplement the natural precipitation, and many others where irrigation is depended upon entirely for the water. Irrigation is generally a costly practice, but this fact does not seem so important when it is remembered that generally the profit per acre from horticultural crops is higher than that from agronomic crops.

Artificial climates are created for crops grown in glasshouses. The cost of maintaining artificial heat in these houses generally limits this type of enterprise to the production of special horticultural crops. It is of interest to note in Table 6 the area under glass and the kinds of crops that are grown in a few principal countries of the world.

TABLE 6.—AREA UNDER GLASS IN A FEW PRINCIPAL COUNTRIES OF THE WORLD, AND PERCENTAGE OF AREA DEVOTED TO FRUITS, VEGETABLES AND FLOWERS, 1930*

Country	Area, square feet	Crops: Percentage of total area		
		Fruits	Flowers	Vegetables
Netherlands	190,082,563	†	80.0	20.0
United States	173,371,536	....	46.0	45.2
Belgium	65,639,120	48.2	19.5	23.4
Denmark	13,982,760			
Sweden	13,858,908			
Norway	4,142,245			
Alaska	43,873			

* "International Statistics of Agriculture," consisting of separate reports by countries based on 1930 census.

† Some grapes grown under glass.

SOIL

The soil may be the limiting factor in the production of particular or all horticultural crops in any given area. It may be too infertile, too wet or too dry or too heavy or too light for profitable production of horticultural crops. Any one of these soil factors may be a limiting factor in the profitable production of a selected crop.

IMPORTANT HORTICULTURAL CROPS IN WORLD TRADE

The distribution of the important horticultural enterprises of the world will be briefly reviewed here. These enterprises will include (1) fruits, (2) vegetables, (3) ornamentals, (4) nursery stock and seeds.

FRUITS

Commerce in fruits on a large scale is of comparatively recent development but has become one of the important and widespread branches

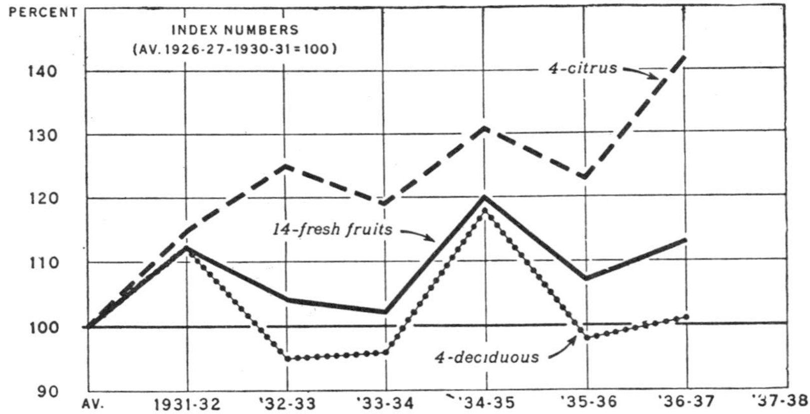

FIG. 3.—World production of 14 principal fruits; 4 principal citrus fruits; and 4 principal deciduous fruits. (*U.S. Department of Agriculture.*)

of national and international trade. Although fruit production is increasing in many of the chief importing countries, these countries have not attempted to stimulate the production of fruit to the same extent as they have stimulated the staple products such as wheat and rye. Moreover, many of the densely populated countries cannot spare the land required, even if climatic conditions were favorable for fruit production.

Fresh fruit can nowadays be brought to markets 5,000 miles away from producing areas within one month after harvest. As a result, areas in South Africa, Australia and South America have grown in importance for the production of fruits because they can ship their fruit products to markets at great distances from the producing areas.

The estimated total world production of fruits including grapes used for wine and raisins averaged 65 million tons yearly in the period

1932–1937. Of this total, grapes comprised 52 per cent; apples, 15 per cent; oranges, 9 per cent; pears, 6 per cent; plums, 5 per cent; bananas, 3 per cent; and all other fruits, 10 per cent.

The trend in the world production of 14 principal fruits, excluding grapes for wine and raisins, is upward. The increase, however, is principally the result of the rapid increase in the production of citrus fruits.

The principal fresh fruits entering into world trade are grapes, apples, oranges, pears, plums and bananas. The principal nut fruits

FIG. 4.—Covent Garden Market, London; horticultural products from all parts of the world are sold on this market. (*Courtesy Blue Anchor.*)

entering world trade include almonds, Brazil nuts, chestnuts, coconuts, filberts, pecans and walnuts. In addition to these fresh fruits and nuts, it should be remembered that great quantities of processed fruit, such as canned fruit, dried fruit and wines, enter world commerce.

Practically every country of the world produces fruit of one kind or another, but comparatively few export it in any great quantity. Some countries export the surplus of one kind of fruit and then import the kinds that they cannot produce. For example, Canada exports apples but imports citrus fruits. Chile, Argentina, South Africa and Australia depend largely upon the export trade for the sale of fruit, and their fruit areas have grown rapidly, mainly because of the demand from such countries as England, and Poland and Finland (until their recent conquest).

Fruits from all over the world are sold in the markets of European countries. Great Britain imports a greater variety and quantity than any other country in Europe. Covent Garden Market, situated in the heart of London, is known as the "hub of the produce world." Moreover, most of the best known firms in the fruit, flower and vegetable trade—firms of international reputation—make Covent Garden their headquarters. If any country, be it the United States, South Africa, Chile or Japan, has a new horticultural product or a new package to test, it consigns the experimental sample to Covent Garden, and the verdict from that market may determine if the product or the package will be of any value in international trade.

The fresh or processed fruit of the United States is known over most of the world. During recent years, fruit exports have maintained third rank among all agricultural exports, outranked only by cotton and tobacco and outranking grains and meats. On a value basis in 1935–1936 fruits made up 11 per cent of all agricultural products exported, as compared with 3 per cent for grains and grain products and 5 per cent for meats, including animal fats and oils.

Fruit crops may be classified geographically as follows: those of temperate regions, those of the subtropical regions and those of the tropical regions. There are more species of fruits in the tropics than in any other region, but most of them are of local importance and because of their highly perishable nature do not enter into world commerce.

Fruits of Temperate Regions.—Although many kinds of fruit are grown in temperate regions, the most important include the grape, apple, peach, pear, plum and strawberry. The distribution and production of the first three in the temperate regions of the world will be discussed briefly.

Grape.—More grapes are produced in the world than any other fruit. They are grown in nearly every country of the temperate region; and the total production, including grapes for table use, for wine and for drying, is almost as large as the total of all other fruits produced in the temperate regions combined. Many species of grapes are grown in various parts of the world; but from the viewpoint of world culture, one species is of great importance. This is known as the "European grape" (*Vitis vinifera*) and is used for fresh fruit, for dried fruit and for wine making. Many people in the United States refer to this type as the "California grape," because most of the acreage in the United States devoted to this species is in California. Its greatest commercial value is reached in those countries of the world noted for their wines, such as France, Italy and Spain.

FIG. 5.—Grape vines planted on terraces in Germany.

The wine grape is grown in many countries of the world, but the requirements of its culture for satisfactory wine are rather exacting. It must have a long summer; a moderately fertile, well-drained, warm

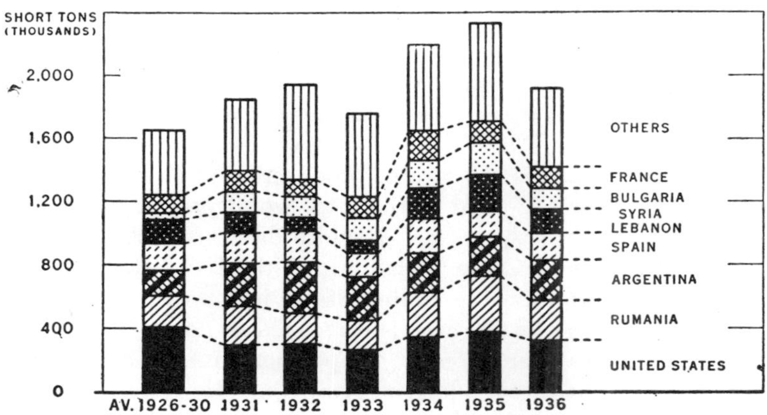

FIG. 6.—Table grapes: total production in leading countries reporting. (*U.S. Department of Agriculture.*)

soil; a relative low water supply during the growing months—even less with relatively high temperatures—and bright sunshine during the three months in which the fruit matures. In Europe a mean temperature of about 60°F. in the month of September is one of the reasons for production of high-quality wine.

Apple.—The apple is the most important tree fruit in the world. The tree itself is quite resistant to low temperature and adapted to a wide range of climatic and soil conditions. Most of the apples are produced in countries of temperate regions. The main groups include dessert and culinary, cider and crab apples. Dessert and culinary apples form the main bulk of the production in most countries, but the largest part of the crop in Germany, France and the United Kingdom consists of cider varieties which are too astringent for raw consumption.

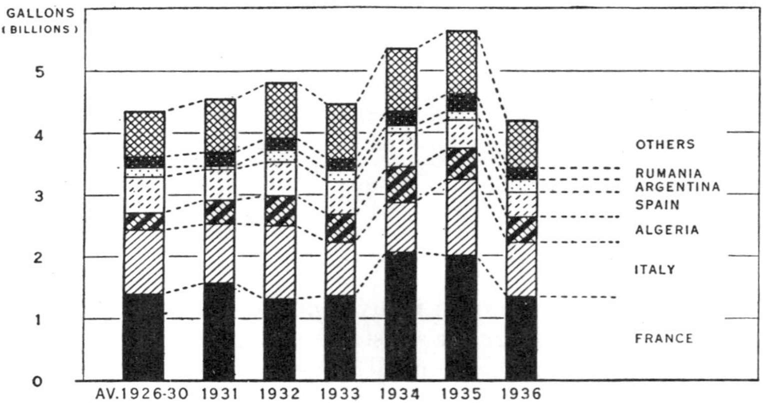

FIG. 7.—Wine: total production in leading countries reporting (excluding the United States). (*U.S. Department of Agriculture.*)

Apples are marketed as fresh, dried and canned fruit and are also used in making a long list of products such as vinegar, brandy, candy, pectin, cider and jelly.

The United States is the largest producer of apples in the world, responsible for over one-third of the total world production in 1931–1936. It also exports the most. Definite data are lacking for such important countries as China, Belgium and the U.S.S.R., but important countries other than the United States include France and Germany, where the bulk of the crop is consumed at home.

Canadian apples offer the most competition to American apples in foreign markets. Other countries competing for foreign markets before 1939 included Switzerland, Czechoslovakia, Italy and Austria. The major importing countries were the United Kingdom, France, Germany, the Netherlands and Belgium.

Peach.—Peaches are widely grown and rank seventh in world fruit production. Because of the fresh fruit's tender character, it plays a minor role in foreign trade, but considerable quantities of canned and dried peaches enter international trade. The peach tree, unlike the apple, yields well only in restricted localities and under special climatic

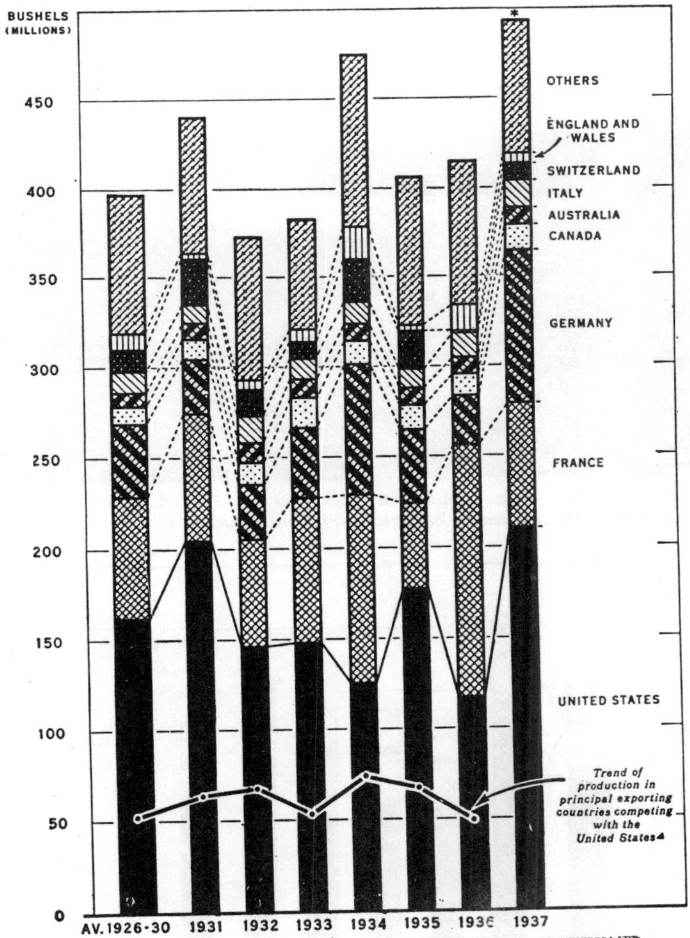

FIG. 8.—Apples: total production in the leading countries reporting. (*U.S. Department of Agriculture.*)

conditions. It is more subject to early spring frost and to winter killing of the buds than is the apple.

The United States is the world's largest producer of peaches, accounting for around 60 per cent of the total. European peach growing is confined to the warmer areas of Portugal, Spain, France and

Italy. In Germany, Holland, Belgium and the north of France and Great Britain the tree is grown under the artificial conditions of hot-houses or trained on the south side of walls.

FIG. 9.—General view of an orchard in Cape Province, South Africa, in September; pears, oranges with peaches in background in full bloom. (*Courtesy F. W. Allen.*)

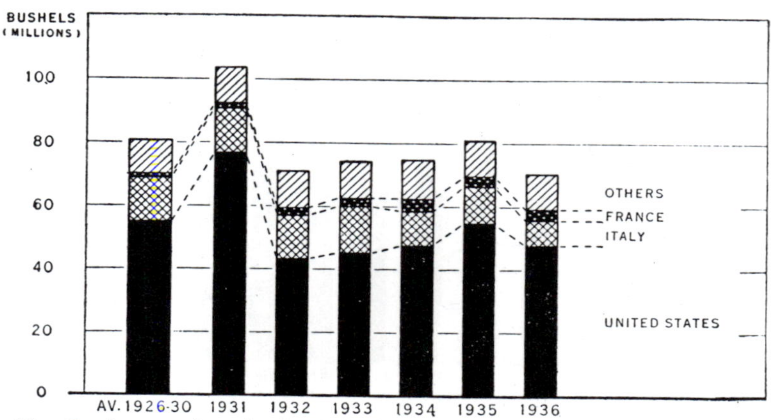

FIG. 10.—Peaches: total production in leading countries reporting. (*U.S. Department of Agriculture.*)

The peach also does well in the south temperate zone, as Chile, Australia, New Zealand and South Africa. Owing to the difference in seasons, peaches are shipped from these countries in February and March.

Fruits of Subtropical Zones.—The regions where subtropical fruits are produced are between the true tropics, where frost never occurs, and the temperate regions, where normally the temperature falls below freezing and stays below for a considerable part of the winter season. In these subtropical, or intermediate, zones the temperature occasionally goes below freezing but not as a rule below 25°F. The influence of large bodies of water and the protection of mountain ranges, or planting where topography gives good air drainage, sometimes extend this type of region as "islands" into territory beyond the usual subtropical regions.

The types of fruits grown in subtropical regions merge into those grown in the true tropics; and no hard-and-fast division can be drawn on the basis of fruit types except that forms possessing resistance to light freezing temperature are of major importance in the subtropics. Diverse types are cultivated, but the orange, grapefruit and lemon are the most commonly known. Dates, figs and olives are also important. Many true tropical fruits such as the avocado, mango, papaya and even the pineapple are also grown on a commercial scale in the subtropical regions.

Orange.—The orange is the second most important commercial tree fruit grown in the world and is widely distributed in both tropical and subtropical regions. The principal types include sweet, mandarin and sour; the sweet orange is the most important. According to government reports for 1927–1931, commercial production of all oranges was important in 27 countries of the world, with the United States the largest producer. Other principal producing countries were Spain, China, Japan, Brazil and Italy. Spain was the largest exporter of oranges followed by Palestine, Italy, the United States, Brazil and the Union of South Africa.

Date Palm.—The prophet Mohammed said, "Honor your maternal aunt the palm for it was created from the clay left over after the creation of Adam on whom be peace and the blessings of God." The foregoing statement indicates the importance of the date palm to the people inhabiting the Sahara Desert and the desert areas of Arabia which are made habitable because of the date palm. In fact, the palm is said to have 800 different uses for the people of desert areas. It might be classed as a shade crop for fruits, as it furnishes shade for the growth of such plants as the fig, apricot, pomegranate and peach. Beneath these smaller trees are grown beans and other vegetables—a three-story agriculture.

Its geographical distribution is throughout the hot, arid regions of the world, but the date palm requires much water at certain times and is really a tree of the oasis, some of which are said to have been in cultivation for over 2,000 years. The greatest date-growing section of the world is that around Basrah on the conjoined Tigris and Euphrates rivers. Although the plant will survive a range of tempera-

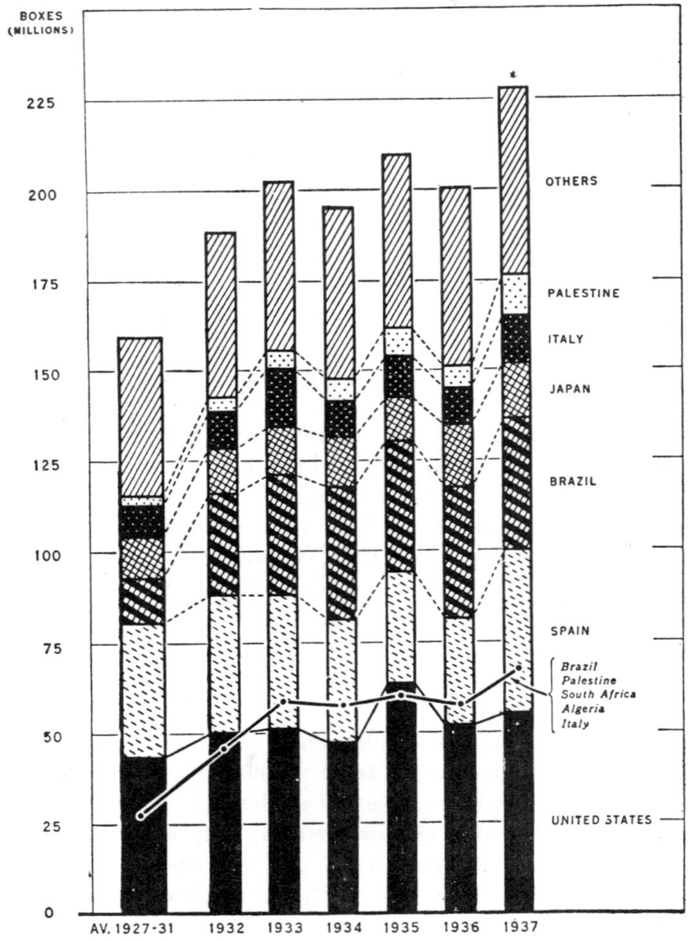

FIG. 11.—Oranges: total production in leading countries reporting. (*U.S. Department of Agriculture.*)

ture from 4 to 125°F., it will not fruit unless it has a hot summer. It is deep rooted and is tolerant of alkaline soils and even of salt water. The irrigation water of northern Africa in summer commonly has a temperature of 75°F.; occasionally, in the cases of hot springs, the water will be 95°F. In general, then, the expression "head in the fire and feet in the water" is a good summary of the climatic requirements of the date palm.

The countries growing the date palm include Arabia, Iraq, Egypt, Tripoli, Tunisia, Algeria, Morocco, northwest India, Baluchistan,

FIG. 12.—Fig trees growing under partial shade afforded by date palms, Algeria. (*U.S. Department of Agriculture.*)

Iran, southern Spain and Brazil and in the United States, California and Arizona.

Fruits of Tropical Regions.—There are hundreds of kinds of tropical fruits, but only four are of great commercial value—the banana, the pineapple, the coconut and the Brazil nut. Others, such as the mango, avocado, papaya, guava, sapodilla, cherimoya, soursop and sweetsop, would be of greater importance if they could be shipped more readily. Less important tropical fruits include the jujube, ceriman, cashew nut, tamarind, feijoa, and breadfruit.

Some of the fruits mentioned are of great importance to the people of specific areas. For example, the mango is as important to the people of India as the apple is to the inhabitants of temperate regions. The

FIG. 13.—Papaya orchard in Hawaii. (*U.S. Department of Agriculture.*)

people of Guatemala hold the avocado in great esteem; Guatemalans say that a cup of coffee, a tarantella and an avocado make a meal. Of the papaya Pope remarks, "Excepting the banana there is no fruit grown in the Hawaiian Islands that means more to the people of this territory than the papaya, if measured in terms of comfort and enjoyment furnished the people." This also applies to the papaya in other tropical countries. The Polynesians of the Pacific could hardly exist without the breadfruit.

Banana.—The banana is the great starch food of the tropics. Shipments go to all parts of the world. Bananas can be grown in subtropical zones, but the major production areas occur in tropical climates with considerable rainfall. Irrigation is practiced in certain districts; the water requirement of the banana plant is enormous, however, and consequently irrigation is feasible only where an abundant supply is available.

FIG. 14.—Harvesting the banana. (*United Fruit Co.*)

Bananas are produced in great quantities in Central and South America and the West Indies. They are grown in the tropics of Africa, Asia and Australia where the rainfall is plentiful; also in the Canary Islands, Hawaiian Islands, Philippine Islands, Malay Archipelago, Fiji Islands and various other islands of the Pacific within the torrid zone. The Atlantic coast of Central America offers ideal conditions for banana culture. A few miles back from the Carribean Sea, at an elevation of not more than 250 ft., is a region that has hot days, humid nights and an annual rainfall of 80 to 200 in. This region, which was almost uninhabited fifty years ago and is now a section of large banana farms, owes its present development to the banana.

VEGETABLES

On account of the large bulk and perishable nature of fresh vegetables, they do not enter world commerce so extensively as do some of the fruit and nut crops. Many vegetable crops, however, are grown for canning, and large quantities of the canned products enter world trade. Fresh vegetables important in world trade include potatoes, onions and tomatoes.

Large quantities of fresh vegetables are shipped into England from France, Spain and Italy. When business is normal, the whole region north of the Alps receives early vegetables from these southern regions. France has a considerable traffic from Algeria, which is sheltered from the cold north winds by the waters of the Mediterranean. Egypt ships tons of early onions between March and May to Liverpool, London, Hull, Hamburg, Trieste and even the United States. Even though vegetables do not enter world commerce extensively, many kinds are grown throughout the world, and the crops are important in the local commerce of all countries.

No attempt will be made to list all the vegetable crop plants of the world, as hundreds of them are cultivated, particularly in the case of all the tropical sorts. Many vegetable crops are not known outside certain small areas. The different kinds grown in the temperate regions are numerous, but only a few of them will be mentioned.

Vegetables can be grown commercially in localities great distances from the original habitats of the plants. This widespread production is possible largely because many of the plants are annuals and the length of the growing season may be modified to mature the crop. This may be done by starting the plants under protection and later transplanting them to the open, or the growing may be altered by the use of shades. It is not intended to give a detailed classification of all vegetable crops based upon their temperature requirements and length of growing season; but attention will be directed to the world distribution of a few of the most important. The two big groups comprise the cool-season crops and the warm-season crops.

Cool-season Crops.—There are many kinds of cool-season vegetable crops, but the most important are the potato, cabbage and onion; hence attention is called to the distribution and production of these crops in the principal areas of the world.

Potato.—The Irish potato is probably exceeded only by bread in the number of times per year that it is eaten by the average European and American. It has definitely established itself as a great cool-climate starch food for human consumption. It is probably the plant most universally grown in the vegetable gardens of Europe and America, but its growth as a money crop is quite restricted. Although quick-maturing varieties are grown as fall and spring crops in southern latitudes, its greatest concentration occurs in the most suitable northern latitudes. The most favorable temperature, moisture and soil requirements for the potato are found chiefly in northern United States, southern Canada and northern Europe. The potato is grown in Ireland and on the plains of north Europe, reaching from the northwest point of France, through Holland, Belgium, Germany, Poland (before 1939) and other Baltic states and Russia to the Ural Mountains. Germany with her cool temperature and favorable soil finds the potato one of the best food crops to be grown.

On account of the great bulk and weight of potatoes in proportion to value, and because of their perishable nature, they are much more important in domestic than in international trade.

TABLE 7.—IRISH POTATOES: TOTAL ACREAGE AND PRODUCTION IN LEADING COUNTRIES REPORTING, AVERAGE 1927–1931*

Country	Average area, millions of acres	Average production, billions of pounds	Percentage of total acreage	Percentage of total production
Russia (Europe and Asia)	14	101	30	23
Germany	7	92	15	21
Poland	6	65	14	15
France	3	33	8	8
United States	3	22	7	5
All others	14	119	26	18
Total	47	432	100	100

* "International Yearbook of Agricultural Statistics, 1938," pp. 282–283, Rome, 1938.

Cabbage.—Cabbage grows to perfection in cool damp climates but is also successfully cultivated on the edge of the torrid zone during the wet cooler seasons and is found under cultivation in a broad zone all around the world near the twenty-fifth parallel.

Onion.—For its best development the onion requires cool weather during the early part of its growing period and moderately high temperatures during the latter part. During its early growth it also requires an abundance of moisture. At the present time onions are

TABLE 8.—CABBAGE: TOTAL ACREAGE AND PRODUCTION IN LEADING COUNTRIES REPORTING, AVERAGE 1934–1935*

Country	Average area 1934–1935, thousands of acres	Average production 1934–1935, millions of pounds	Percentage of total acreage	Percentage of total production
United States	193	2,572	24.0	27‡
Yugoslavia	93	565	11.0	6‡
Italy	80	679	10.0	5‡
Germany	78†	1,630†	10.0	17‡
Rumania	68	545	8.0	6‡
Hungary	66	221	7.0	2‡
Czechoslovakia	56	791	7.0	8‡
All others	182	2,561‡	23.0	27‡
Total	816	9,564‡		

* "International Yearbook of Agricultural Statistics 1937," Rome, 1937.

† Includes only white cabbage.

‡ Of the 18 countries reporting acreage Austria, England and Wales, Scotland, Czechoslovakia and Japan did not report production.

grown in all the temperate regions of the world. The crop is of considerable commercial importance in Spain, the United States, Japan and Egypt.

FIG. 15.—Grading onions, Egypt. (*Courtesy W. V. Cruess.*)

Warm-season Crops.—Some of the most important warm-season crops include the tomato, sweet potato, beans, corn, cucurbits, eggplant and peppers.

TABLE 9.—ONIONS: TOTAL ACREAGE AND PRODUCTION IN LEADING COUNTRIES REPORTING, 1934–1935*

Country	Average area, thousands of acres	Average production, millions of pounds	Percentage of total acreage	Percentage of total production
Spain†	94	1,466	16	24‡
United States	92	1,365	15	23‡
Japan	74	961	12	16‡
Turkey†	54	152	9	3‡
Rumania	50	205	8	3‡
Yugoslavia†	49	161	8	3‡
Java and Madeira	43	‡	7	‡
Egypt	39	534	7	9‡
Italy	19	174	3	3‡
Germany	16	‡	3	‡
All others	73	1,029‡	12	16‡
Total	603	6,017‡		

* "International Yearbook of Agricultural Statistics 1937," Rome, 1937.
† Includes garlic.
‡ Of 24 countries reporting acreage Java, Madeira and Germany did not report production. Data are based only on production reported.

Tomato.—The tomato is widely cultivated in Great Britain, Holland, Belgium, the Canary Islands, Italy, the Mediterranean region, Africa, Australia, North and South America and Asia.

Tomatoes are grown extensively under glass in England, Scotland and the Netherlands. Considerable quantities are imported into

TABLE 10.—TOMATOES: TOTAL ACREAGE AND PRODUCTION IN LEADING COUNTRIES REPORTING, 1934–1935*

Country	Average area, thousands of acres	Average production, millions of pounds	Percentage of total acreage	Percentage of total production
United States	609	4,226	69	48†
Italy	109	1,915	12	22†
Spain	63	1,563	7	18†
Mexico	35	114	4	1†
Yugoslavia	18	101	2	1†
All others	49	840†	6	10†
Total	883	8,759†		

* "International Yearbook of Agricultural Statistics 1937," Rome, 1937.
† Of the 14 countries reporting acreage only Germany did not report production. Data are based only on production reported.

England from the Canary Islands and the Netherlands. In the United States and in the Mediterranean regions tomatoes form a most valuable crop for the home markets. Canned tomatoes, tomato juice and ketchup are exported from the United States in large quantities, and canned tomato purée is exported by Italy.

Sweet Potato.—Because of the abundance of starch-producing plants the tropical zone is often said to have great possibilities for the support of human life. One of these plants is the sweet potato, which is a perennial in frostless regions. Although the United States has the largest recorded production, the sweet potato is a universal food crop in the tropical regions, whether it be in the Spanish-speaking settlements of South America, the English-speaking Honduras, the West Indian islands, the coast of Africa or the Malay Peninsula.

Ornamentals

The commercial production of flowers is a part of the agriculture of many countries of the world, but world trade in flowers is relatively

Fig. 16.—Holland is noted for the production of flowering bulbs. Typical tulip field near Haarlem in northern Holland. (*Courtesy Des Moines Register.*)

unimportant. Statistical information on the production of flowers in various parts of the world is quite incomplete; however, 12 countries considered it of sufficient importance to give the extent of acreage in the "International Yearbook of Agricultural Statistics" for the years 1935 and 1936. These countries in order of importance are United States, Italy, England and Wales, Bulgaria, Germany, Netherlands, Australia, French Morocco and Luxemburg.

The most important ornamentals entering world trade consist of various kinds of bulbs; and the most important country exporting these products is the Netherlands, which reported a yearly average export of approximately 97,000 lb. for 1934–1936 valued at approximately 15 million dollars per year. The only country listing export of cut flowers in its total of all exports in the *Report of the U.S. Bureau of Foreign and Domestic Commerce* for 1937 was Italy. It reported an average yearly export of over five million pounds of cut flowers for the period 1934–1936 valued at approximately two million dollars per year.

Nursery Stock and Seeds

One of the important branches of horticulture in all parts of the world is the nursery business. Shipment of nursery stock, however, is relatively unimportant in world trade. Countries reporting acreage designated as nurseries in the "International Yearbook of Agricultural Statistics" in 1935 and 1936 include the United States, Germany, France, Netherlands, England, Australia and Egypt. The principal products in international trade are rose stocks, cuttings of various kinds of plants, seedlings and seeds.

Table 11.—Imports of Trees, Plants, Cuttings and Seedlings for the United States, 1934, 1935, 1936*

Commodity	Amount, thousands		
	1934	1935	1936
Fruit stock—less than	500	6	1
Rose stock and plants	6,371	7,001	6,661
All other	921	765	604

* *U.S. Bureau of Foreign and Domestic Commerce. Statistical Abstract of the United States*, 1937.

Associated with the vegetable industry, the flower industry and the nursery industry is the seed industry. Horticultural seed firms of world-wide reputation exist in many countries. In the United States the principal districts for the production of onion, lettuce, carrot, radish, sweet pea, zinnia and many other seeds are located in California. Similar kinds of seed are produced in some quantities around Saint-Rémy, France, where the climate is comparable to that of the California districts. Denmark produces quantities of cauliflower seed; the Netherlands, culinary peas; and France and Italy, quantities of celery seed. England is noted for the production of cabbage, broccoli and Brussels sprouts seeds.

Labor costs constitute the most important item in seed production, and this factor has been notable in keeping the production of low-priced vegetable seeds, such as spinach and turnips, in those sections of Europe where the climate is especially favorable and the labor costs low. Most of the vegetable seeds planted in the United States, with the exception of spinach, cauliflower and celery, are produced in this country.

TABLE 12.—VEGETABLE SEEDS: IMPORTS, FOR CONSUMPTION INTO THE UNITED STATES, 1932–1934*

Kind of seed	Imports, thousands of pounds		
	1932	1933	1934
Beet, garden	179	572	362
Cabbage	172	174	193
Carrot	34	60	29
Cauliflower	13	11	12
Kale	49	48	84
Kohlrabi	12	15	14
Onion	230	160	210
Parsley	179	44	99
Parsnip	13	24	21
Pepper	1	1	4
Radish	565	445	341
Rutabaga	93	288	125
Spinach	2,386	3,783	3,402
Turnip	482	1,212	1,012

* *U.S. Dept. Agr. Bur. Agr. Econ. Seed Statistics* prepared by the Hay, Feed and Seed Division, Table 99, p. 56, Washington, D.C., 1936.

Review Questions

1. What major phases of life are touched by the horticultural enterprises of the world?

2. What are the two principle environmental factors that determine the geographical distribution of horticultural crops?

3. What are the two principal agencies of climate?

4. What factor has been largely responsible for the decided increase in international trade in fresh fruits since 1900?

5. List in order the five most important fruit crops in the world.

6. Name five important fruits grown in the temperate region.

7. Characterize the climate of, and name some of the important fruits produced in, the subtropical zones.

8. Name the most important fruit of the tropical region, and justify its importance.

9. Name three important cool-season vegetable crops.

10. Name three important warm-season vegetable crops.

11. What vegetable of the tropical region is as important for human food as the Irish potato is in the temperate region?

12. What important tropical vegetable plant is important for forcing in the temperate region?

13. What country is famous for international trade in flowering bulbs?

14. Is the importation of nursery stock into the United States of much importance in our international trade?

15. Does the United States import much of the vegetable seeds that it uses?

Problems

1. Make a bar graph showing the percentage of the total world production of table grapes, wine grapes, apples, oranges and peaches produced by the highest producing country for the year 1935.

2. Make a bar graph showing the percentage of total acreage and the percentage of total production of potatoes, cabbage, onions and tomatoes in the various countries. Which countries show the highest yield per acre for each crop?

3. A number of years ago the federal government found it advisable to prohibit the importation of tulip bulbs. What climatic and soil conditions would one seek in which to produce such bulbs? Where in the United States would one be likely to find such conditions?

4. If you were a large commercial fruit grower, would you advocate a high tariff on fruits? Explain.

5. State and defend your opinion of the federal purchase of fruits and vegetables for distribution to people financially unable to pay the full market price.

Suggested Collateral Readings

1. GOOD, RONALD, "Plants and Human Economics," pp. 67–84, Cambridge University Press, London, 1933.

2. ROBBINS, W. W., "The Botany of Crop Plants," P. Blakiston's Son & Company, Inc., Philadelphia, 1924.

3. SMITH, J. R., "Industrial and Commercial Geography," pp. 226–285, Henry Holt & Company, New York, 1925.

4. WILLIS, O. R., "Practical Flora," American Book Company, New York, 1894.

5. "World Fresh Fruit Production Statistics," U.S. Department of Agriculture, Bureau of Agricultural Economics, Government Printing Office, Washington, August, 1938 (mimeographed).

CHAPTER III

HORTICULTURAL ENTERPRISES OF THE UNITED STATES

Horticulture has participated in the general agricultural and industrial expansion that has taken place in the United States during the last half century. New enterprises have been inaugurated and developed; established ones have been expanded and extended into new territory until practically every known phase of horticulture has been tried and almost every available area has been tested.

Historically it is important to remember that the United States was a pioneer in stock raising and grain farming long before it was possible to test the country out on a large scale for its suitability for fruits and vegetables. The railroads and river barges transported grain and livestock from the earliest days. The former carried fuel into the plain and prairie and made their habitation possible. The same trains took back grain and livestock. It was not until the development of artificial ice and mechanical refrigeration that the railroads, the waterways and the oceans themselves figured largely in the movement of fresh fruits and vegetables to market. In general, the entire United States, except in the vicinity of the large eastern cities, is less than fifty years old in the commercial production of fruits and vegetables. The production of greenhouse products and nursery stock began in those areas adjacent to the large cities; and although these same enterprises are scattered widely over the country today, they still remain concentrated in the areas close to large centers of population.

TRENDS IN HORTICULTURAL ENTERPRISES

During the past thirty years the production of horticultural products has shifted from home and small local enterprises to large commercial enterprises designed to supply not only local but distant markets. The student should note the trends that have taken place in the production of fruits, vegetables and ornamentals since 1909.

Fruit

Although rapid expansion occurred in various fruit enterprises in several sections of the country, it will be noted in Fig. 19 that the aggregate production of fruit for the country as a whole ceased expanding after 1915. It should be remembered, however, that although the fruit production of the country as a whole shows no great expansion since that period, production did increase rapidly in certain areas. For example, in California the acreage in fruit expanded by half from 1909 to 1925; the production of grapes, apples and peaches practically doubled from 1909 to 1936; and the production of oranges more than

Fig. 17.—A fruit enterprise in the Appalachian Mountains. (*Hardie and Co.*)

doubled. The Northwest area also saw a great expansion in fruit during the first half of the period 1909–1936. Apple production alone increased from 10 million bushels per year at the beginning of the period to a level of about 40 million from 1921 to 1936. Other instances might be cited where rapid expansion occurred, such as the citrus area

Fig. 18.—A seed farm in California. (*Ferry-Morse Seed Co.*)

of Texas and the peach area of the Carolinas, but the foregoing examples are sufficient to indicate that expansion has occurred in many specialized areas.

Vegetables

Vegetable production, in contrast to fruit production, showed a rapid expansion during the period 1909–1926, at which point it remained rather constant. This is true for fresh vegetables as well at those processed in various ways. Again rapid expansions occurred

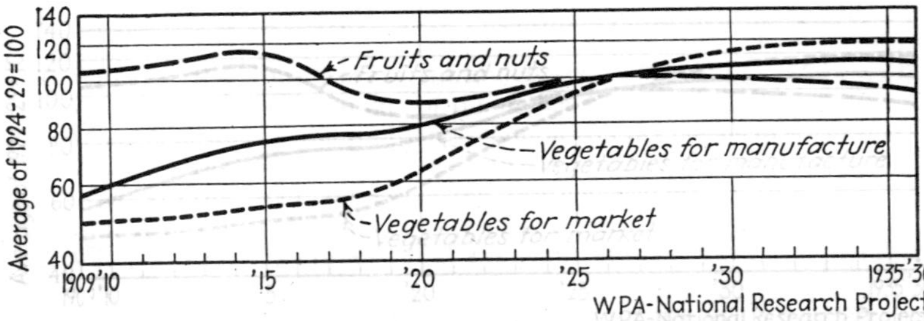

FIG. 19.—Trends in the production of fruits and nuts,[1] vegetables for market,[2] vegetables for manufacture.[3]

FIG. 20.—A floricultural enterprise under glass. (*Metropolitan Co.*)

in certain areas, notably in various areas in the South and in California where vegetable acreage expanded from 38,000 acres in 1909 to 496,000 acres in 1936.

[1] Includes 13 principal fruits and nuts.

[2] Includes 15 principal vegetable crops for market but excludes Irish potatoes and sweet potatoes.

[3] Includes eight principal vegetables for manufacture.

Ornamentals

Official statistics to show trends in the production of ornamentals are available only for census years, and some indication of the expansion in these enterprises is indicated by the increase from approximately 81,000 acres in nurseries in 1909 to approximately 141,000 in 1929—about 56 per cent. The area under glass devoted to both flower and vegetable production increased from about 115 million square feet in 1909 to 173 million square feet in 1929. During this same period the area under glass devoted to vegetables decreased as developments in

Fig. 21.—A nursery enterprise. (*Shenandoah Nurseries.*)

certain parts of the United States accompanied by improved transportation facilities made it possible to produce some vegetables satisfactorily in the open that formerly had been produced for the out-of-season market only under glass. The increased area under glass is now devoted chiefly to flowers.

FACTORS INFLUENCING THE TRENDS

Both the economic and the physical factors underlying the establishment of horticultural enterprises in the United States are so numerous and so complex that detailed discussion would require considerable space. The utilization of the land for horticultural enterprises has always been determined principally by pressure of economic forces against the physical conditions of soil and climate. The student should note a few of the important economic influences that resulted in the expansion of certain commercial horticultural enterprises. These factors include (1) increase in urban population; (2) development of artificial ice, mechanical refrigeration and improved transportation; (3) artificial stimulus of promoters; (4) shifts in food habits; (5) shifts in types of farming and (6) increased interest in amateur gardening.

Economic

The development of commercial horticultural enterprises in the United States first began in the East in the vicinity of large cities. With the increase in population in certain areas which accompanied the centralization of industrial enterprises, the demand for horticultural products rose to the place where they could no longer be produced in sufficient quantities in small gardens. As a result the establishment of commercial horticultural enterprises began on a larger scale.

The stimulation of commercial production of fruit and vegetable products in areas at great distances from consuming centers came about with the more general use of artificial ice, mechanical refrigeration and improved transportation facilities. Although irrigation made it possible to produce many of these perishable horticultural crops in the arid West, there would have been no outlets had it not been for the aforementioned developments.

The establishment of the Floral Telegraph Delivery service made it possible to send flowers hundreds of miles away within a few hours. This service is both national and international in scope. For example, if someone in San Francisco decides to send flowers to someone in New York City, he simply calls his florist, who in turn wires a New York florist who is a member of the F.T.D. service, and the New York florist delivers the order. Similarly, flowers may, for example, be sent from Chicago to London.

The exploitation of land has been to a considerable extent responsible for the complicated and unsatisfactory economic situation that now surrounds the horticultural industry of the country. Land promoters have long used fruit growing as an advertisement for the land. As early as 1887, people were flocking to California to buy tracts of land that had been planted to orange trees. Many groves were planted with no other thought in mind than to attract the eye of the buyer. Oranges were actually left on the trees for many months to add to the attractiveness and to induce the unsuspecting purchaser to buy an orange orchard. No historian has recorded even the outlines of the Georgia Peach Rush, the Georgia Pecan Rush, the Texas Onion Rush, the Northwestern Apple Rush, the Asparagus Rush or any of the many smaller rushes that took place in the United States. Many who rode the crest of the waves made fortunes; others who were caught by the undertow of receding prices met disaster.

The recent readjustments of cotton acreages through soil-conservation and crop-control measures have resulted in a revival of agitation for planting vegetables on some of the acreages released from cotton. This is taking place in some districts wholly unadapted to vegetable production, in spite of the fact that such ventures resulted in failure less than twenty years ago at the time of the boll-weevil infestation. Even if these crops could be produced successfully in the new areas, it would only result in adding to surpluses already produced in areas of established production. It is one thing to produce a commodity successfully and quite another to sell it to advantage. As is usual with exploitation, it was frequently overdone, and as a result the normal growth of legitimate crop production has been diverted frequently and in some cases retarded. By increasing production and by developing marginal territory, promoters have been responsible for at least a part of the present unsatisfactory economic situation that exists in many horticultural enterprises.

Shifts in food habits have undoubtedly been partially responsible for the stimulation of production of such crops as oranges and head lettuce. Earlier in the text it was shown that the per capita consumption of oranges had increased apparently at the expense of apples. Also, the shift to the consumption of green vegetables was apparently at the expense of Irish potatoes. The phenomenal increase in the per capita production of lettuce would not have occurred had it not been for the fact that lettuce was made available every month of the year; yet other products that were available over a long season showed no phenomenal increase. Consequently, the shift to the consumption of oranges, lettuce and other green vegetables at the expense of other fruit and vegetable products was largely responsible for stimulation of production in the aforementioned crops.

Shifts in types of farming usually resulted when it was discovered that a certain crop was more profitable than another. This factor was important in the shift of acreage from cotton to vegetables in certain areas of the South when the ravage of the boll weevil made cotton farming a precarious enterprise. As a result, vegetable production in the South, to supply Northern markets with out-of-season vegetables, increased tremendously.

No discussion of the expansion of American horticulture would be complete without mention of amateur gardening. This interest has developed gradually during the last twenty years. It is important to note that, although it has been chiefly of great social value, it has also been of economic value to the ornamental horticultural enterprises of the country. This is reflected in the fact that the sales reported by nursery companies in 1929 were 184 per cent greater than in 1919 and that the acreage devoted to such horticultural crops increased 174 per cent.

Physical

Generally speaking, in areas with favorable climate and soil, crop specialization yields the most profitable return from the production of horticultural crops. Crop specialization in these areas has characterized the horticultural industry as much as any other single factor during the past twenty-five years. The degree is vividly pictured when it is realized that the state of California produces 40 to 50 per cent of all fruit consumed in the United States, over 90 per cent of the dried-fruit output of the nation, about 70 per cent of the canned-fruit pack and approximately 45 per cent of the fresh fruits. Other examples of specialization are indicated in the production of cabbage, with about 40 per cent of the total commercial acreage located in Wisconsin and New York; and of head lettuce, with 75 per cent of the commercial acreage in Arizona and California. Many of the specialized areas producing horticultural crops are important not because they can produce better crops or larger yields than other regions but because they can produce them during periods of the year when other regions cannot. For example, cabbage, celery, onions, peas and other crops are produced more cheaply in the North than in the South, but the South can produce them at a time when the North cannot. The production of head lettuce on the Pacific coast and in the high altitude of Colorado during the summer months is another example of certain favorable areas' being able to produce a lettuce crop when most areas cannot.

Many commercial areas for the production of horticultural crops were developed by a gradual readjustment of general farms when it was discovered that particular horticultural crops were adapted to the soil and climate of the area. Certain fruit sections of New York, Virginia, Michigan and the Carolinas are largely the result of farmers finding fruit growing more profitable than other agricultural enterprises and gradually extending the area devoted to fruit, until fruit growing became the dominant enterprise of the community. For example, one of the famous apple sections of the country, known as "Apple Pie Ridge," located close to Winchester, Va., had its beginning more than sixty years ago when a farmer planted a large field of apples, which brought ridicule from the neighbors but eventually thousands of dollars to the owner. This started the neighbors to planting apple trees until now this ridge for a distance of 25 miles is almost one continuous apple orchard. In regions such as those just described, fruit farming will vary from the specialized fruit farm, which is devoted nearly exclusively to the production of fruit, through the fruit general farm on which fruit growing is the main but not sole cash-crop enterprise, to the general fruit farm on which fruit growing is of minor importance as a cash crop, being subordinate to dairying, poultry raising, the production of one or more kinds of livestock or the growing of one or more agronomic crops.

The utilization of land for horticultural crops is determined principally by the pressure of economic forces against physical conditions. These physical conditions, such as temperature, moisture and soil underlying the establishment of horticultural areas, are generally of greater significance than with many other agricultural crops. For example, horticultural crops are generally concentrated in smaller areas than such crops as corn, cotton and wheat. Although these physical conditions will be discussed in detail in the chapters on those subjects, it is felt that the student will get more from a study of the maps in the latter part of this chapter if he has a few specific examples where the physical conditions have been most important in the establishment of horticultural areas. These will be discussed under the topics of climate and soil.

Climate.—Temperature is one of the most important of the complex elements of climate that influences the geographical distribution of the commercial production of horticultural crops. Although there are broad regions in the United States in which horticultural crops are grown, there are minor variations that make one part of a region better suited for their production than another. A few examples of horticultural enterprises will be mentioned that are located in rather specialized areas because of favorable local temperatures. Such suitable temperatures are due to proximity to large bodies of water, favorable winds, local altitude and topography. On the eastern shore of Lake Michigan is a large commercial "fruit belt" which varies in width from less than 2 to over 20 miles, whereas on the corresponding western shore in Wisconsin fruit is not grown to any extent. The location of the area in Michigan is largely influenced by temperature that is modified by the prevailing winds that pass over Lake Michigan during most of the winter and spring months. Water and ice change their temperatures much more slowly than do air and the surface of the soil. During the winter the winds may leave Wisconsin at a temperature of 30 or

40° below zero but absorb heat as they pass over the unfrozen water in Lake Michigan and arrive at the fruit belt on the Michigan side at a temperature but slightly less than zero. In the spring the atmosphere over the land and the surface of the soil warms more rapidly than do the ice and cold water in the lake, and the plants would start to grow if the breezes coming across the lake did not absorb cold from the cold water and melting ice and keep the atmospheric temperature of the Michigan fruit belt low enough to retard plant growth until much of the danger due to late frosts is over. The width of this belt is influ-

FIG. 22.—Harvesting cabbage in April at Crystal Springs, Miss. (*U.S. Department of Agriculture.*)

enced greatly by the topography of the land in the vicinity of the lake. As a rule, the ameliorating influence on the temperature is of little significance back of the crest of the slope toward the lake, and this accounts for the variation in the width of the fruit belt. The width and depth of the body of water affects the extent of its ameliorating influence on the temperature of the adjacent land. Lakes Seneca and Canandaigua in New York, which are only about 4 miles wide but very deep, have fruit belts from ¼ to 2 miles in width about their shores. These deep bodies of water act more or less like sponges, absorbing heat when the temperature of the air is higher than that of the surface of the water and liberating it when the temperature of the air is lower than that of the surface of the water. Consequently, the water absorbs heat during all the warm days of summer and liberates it slowly in the fall to the cooler atmosphere. This keeps the temperature of the atmosphere about the lake warmer for a longer period in the fall, with the result that the fruits are matured properly and harvested before danger of injury by frost.

The location of commercial vegetable-producing areas at various places about the Great Lakes is due largely to the fact that such areas, often extending several miles back from the lake, do not receive a fall frost until two weeks later than similar districts whose temperatures are not influenced by bodies of water. The waters of the Gulf of Mexico and the Gulf Stream modify the climate of the Gulf coast and the entire Atlantic coastal region, and this fact is partially responsible for the vegetable areas in these sections. The trucking areas around Providence, R. I., and Boston, Mass., are greatly influenced by the Gulf Stream, thus making it possible to grow vegetables in those regions both early and late in competition with sections much farther south.

Temperature is influenced also by the altitude of the land, and this factor affects the location of certain horticultural areas. For example, head lettuce can be grown to advantage between elevations of 5,000 and 11,000 ft. in certain parts of Colorado and at sea level or even below in the Imperial Valley of California. The high elevation of Colorado provides cool temperature conditions suitable for the production of head lettuce from June to November, whereas the low elevations and location of the Imperial Valley provide suitable temperature conditions during the winter months. Apples, which are relatively a cool-season crop, are grown commercially in the higher elevations of the Appalachian Mountains of northern Georgia and Alabama which otherwise would be too warm for their production.

Rainfall, irrigation and atmospheric moisture are all important in determining the geographical distribution of horticultural enterprises. The distribution of rainfall was an important factor in retarding the development of vegetable-crop production in the East and in the South, but within recent years irrigation has been adopted to supplement the rainfall during the growing season and has been an important factor in the progress of vegetable enterprises in these areas. An inadequate supply of rainfall and difficulty of profitable means of irrigation have resulted in the comparative absence of important fruit enterprises in the Great Plains area. Many horticultural areas of the West, however, were established when economical irrigation made it possible to supply adequate moisture for the growth of the plants. Reports show that 45 per cent of the commercial apple crop of the United States, 40 per cent of the peaches, 65 per cent of the oranges and 60 per cent of the entire commercial fruit crop of the United States is now produced on land that a generation ago was supporting a scant growth of sagebrush, chaparral, and bunch grass. Much of the vegetable acreage of the West is dependent wholly or in part upon irrigation for a satisfactory water supply.

Humidity, especially when considered in relation to temperature, has also played a part in determining the suitability of an area to certain horticultural enterprises. Humidity may work directly through its effect on the water requirement of the plant or indirectly through its effect on insects and fungi which prey upon the plants. The amount of water transpired by a plant is greater in dry than in humid atmosphere. Georgia and Alabama might ship early pears the way California does were it not for the prevalence of a disease known as "fire blight," which is particularly virulent under the combination of heat and high humidity in Georgia and Alabama. The presence of leaf spot and brown rot, two fungous diseases, precludes the commercial production of sour cherries in the Southern states.

Soil.—Soil types have been important in the successful establishment of many horticultural areas. One of the most outstanding examples is that of the apple enterprises in the adjoining corners of the states of Iowa, Nebraska, Kansas and Missouri. It is probably the occurrence of the loess soil in this area that accounts for apple enterprises of considerable size. This soil allows the deep penetration of root systems, which may be a factor in saving trees during several years of successive droughts. The physical characteristics of the sandy loam soils enable them to warm up quickly and to be worked in early spring, and these have been important factors, particularly in the distribution of areas producing early vegetables along the Atlantic Coastal Plain. Peat soils were first used by the Hollanders around Kalamazoo, Mich., for starting early plants in hotbeds and coldframes. From this beginning it was soon discovered that they were adapted to profitable production of vegetable crops. Many vegetable areas in Michigan, Minnesota, Iowa, New York and Florida have been developed on these peat soils.

HORTICULTURAL ENTERPRISES IN THE UNITED STATES

The horticultural industry contains the most complicated and highly intensified specialties in the field of agriculture. It is generally observed that qualifications of successful producers of horticultural crops differ from those of producers of field crops or livestock. Many farmers cannot adapt themselves to the specialized, painstaking and accurate practices required in the successful production of horticultural crops. They prefer the types of farming that give a wider spread of time for harvesting operations and that do not require the degree of skill needed in such operations as necessary spraying and pruning. It is well, then, to review the various types of production that are included in each of the divisions of horticulture. These enterprises will include fruits, vegetables, ornamentals and nursery stock and seeds.

Fruit

Fruit growing concerns itself primarily with the production of fruit crops for sale whether as the fresh or as the processed product. The processing of fruit in drying, canning, wine making and freezing is an important phase of the industry. According to the Fifteenth Census of the United States the total value of fruits was roughly 656 million dollars. This included all small fruits, grapes, orchard fruits, subtropical fruits and nuts. Their relative importance based on value is shown in Table 13.

Table 13.—Percentage of the Total Value of Specified Fruits for the United States, 1929

Commodity	Percentage of Total	Commodity	Percentage of Total
Apples	24.2	Cherries	2.6
Oranges	22.6	Apricots	1.9
Grapes	8.6	Walnuts (English or Persian)	1.9
Peaches	8.4	Raspberries	1.5
Lemons	6.6	Figs	1.0
Strawberries	6.6	Pecans (cultivated and wild)	0.7
Pears	4.6	Almonds	0.3
Grapefruit	3.5	All other	2.2
Plums and prunes	2.8	Total	100

Fruit farming is generally classified in one of the following types:

1. Home fruit gardening. Fruit produced only for home use.
2. Specialized fruit farms. Farms that specialize in the production of one or more kinds of fruit for the commercial market and whose chief source of income is from the production of fruit.
3. Fruit general farms. Farms that produce one or more kinds of fruit for the commercial market. Fruit growing is but one of the major sources of income. It is carried on in conjunction with other major sources of income, such as the raising of cattle, the production of poultry, the operation of a dairy, the growing of grain crops.
4. General fruit farms. Farms where fruit growing is of secondary importance as a source of income to other agricultural enterprises as mentioned above.

VEGETABLES

Vegetables constitute the most valuable group of horticultural crops. The Fifteenth Census Report shows that the value of all vegetables, including potatoes, reported for 1929 was over a billion dollars—

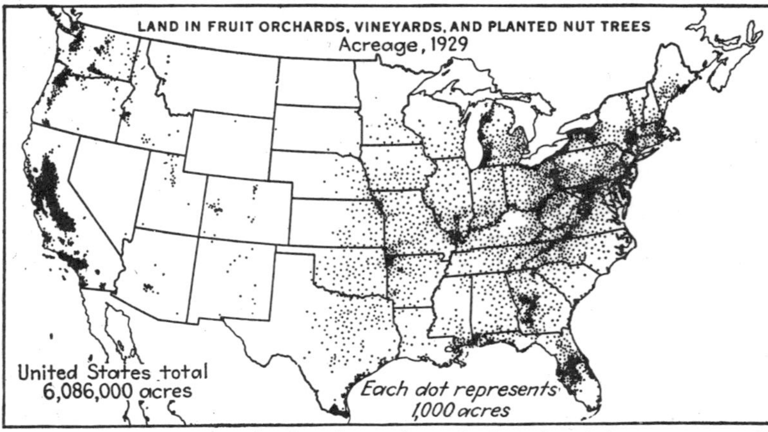

FIG. 23.—Land in fruit orchards, vineyards and planted nut trees, acreage, 1929. (*U.S. Department of Agriculture.*)

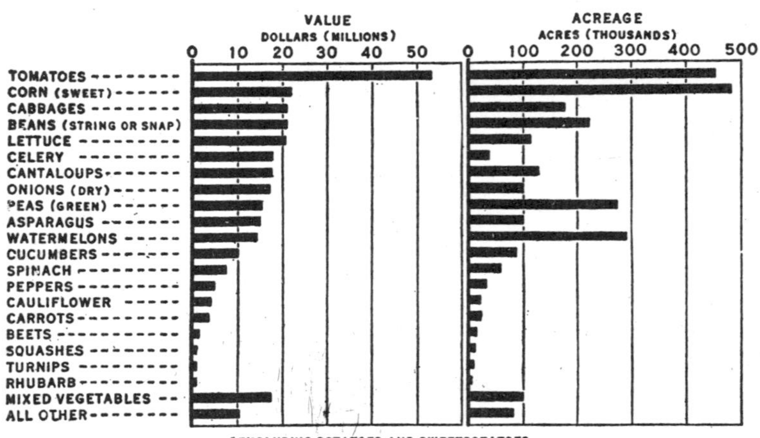

FIG. 24.—Value and acreage of principal vegetable crops grown for sale, United States, 1929. (*U.S. Department of Agriculture.*)

35 per cent greater than that of all fruits and nuts combined. The total value of the vegetable crops was greater than that for the wheat crop. Some idea relative to the importance of the staple vegetable crops may be gained by an examination of Fig. 24.

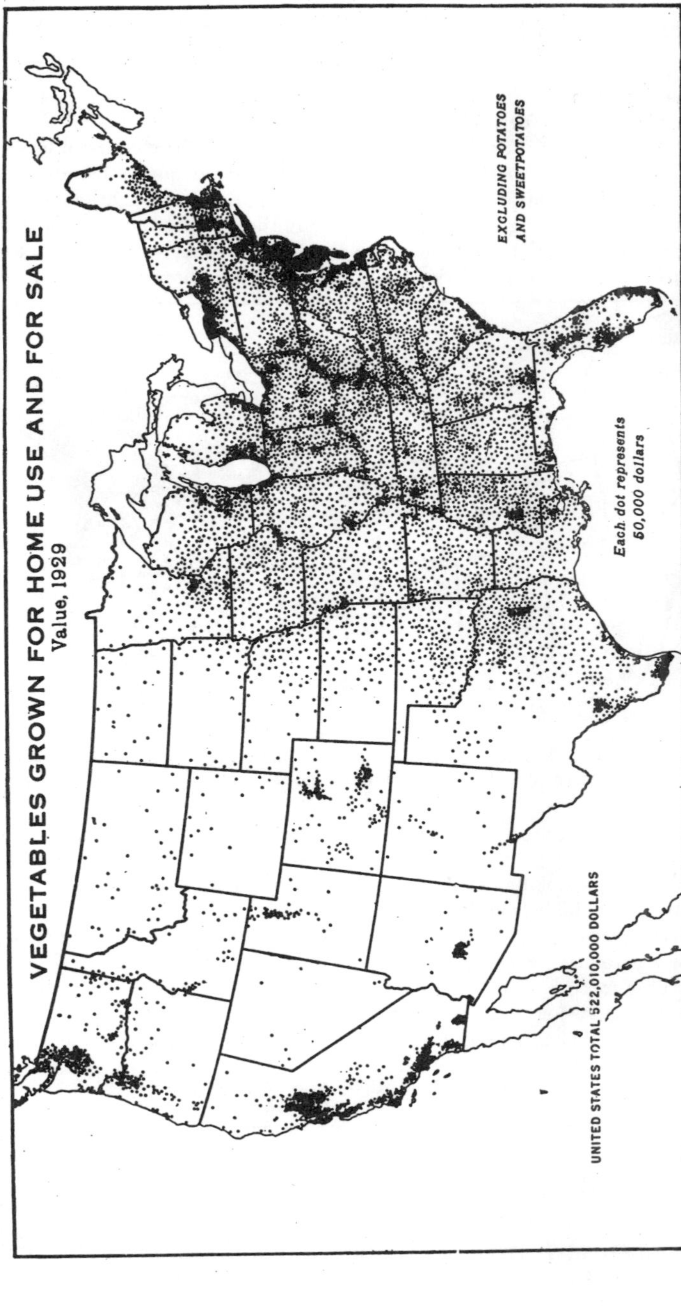

FIG. 25.—Location of vegetable areas, excluding Irish potatoes and sweet potatoes, as indicated by value, 1929. (*U.S. Department of Agriculture.*)

Vegetable growing is generally classified under one of the following types:

1. Home gardening. Small gardens near city, suburban or farm homes that produce vegetables only for that home.

2. Market gardening. Rather large areas (generally in the vicinity of large consuming centers) devoted to the production of commercial quantities of a rather large number of different kinds of vegetables for the local market. Market gardening is still expanding near large cities, but the competition of motor-truck shipments from distant areas with more favorable conditions for the production of specific crops makes this type of vegetable gardening less lucrative than formerly.

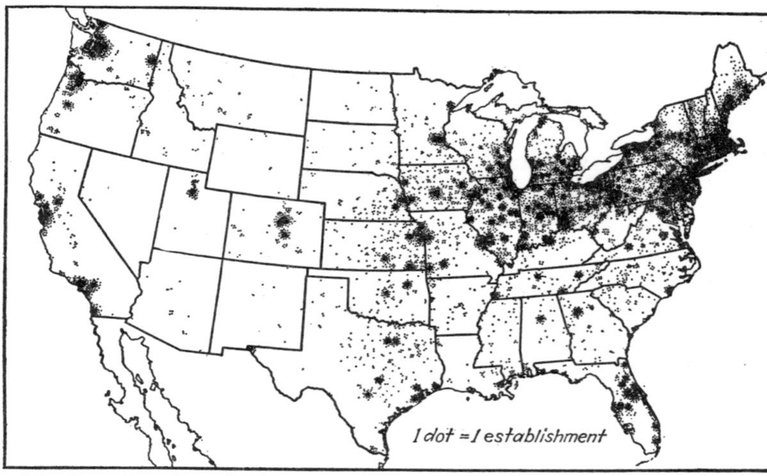

FIG. 26.—Location of establishments producing flowers under glass and in the open, 1929. (*U.S. Census.*)

3. Truck crops. Extensive areas specializing in the production of a very limited assortment of vegetables. These areas are located where the climatic and soil conditions are especially favorable for the production of the specialized crops. They may be many miles from the market.

4. Canning crops. Extensive areas growing specialized vegetable crops under contract for a cannery located in the near vicinity.

5. Vegetable forcing. Growing specialized vegetable crops under glass out of their normal season for that location. Greenhouses are employed largely in Northern sections, but hotbeds and cold frames are satisfactory in Southern regions.

ORNAMENTALS

Few people think of agriculture as concerned with the production of beauty; but according to the 1930 census, American agriculture had invested 290 million dollars in farms and nurseries engaged in growing flowers and ornamental plants outdoors and under glass. There were 13,088 commercial florists producing flowers under glass and in the open. In addition 1,935 farms reported growing flowering bulbs commercially. The number of farms producing flower seeds only was not reported.

Flower growing may generally be grouped under one of the following types:

1. Amateur floriculture. Small privately owned gardens and greenhouses devoted to the production of flowers for personal use and enjoyment. Immense numbers of people are interested in this phase of floriculture.

2. Commercial floriculture. Areas of various sizes devoted to flowers for sale. The crops may be produced under glass or in the open. Some commercial florists could be compared with the market gardeners in that they produce small quantities of a large assortment of flowers for the local trade; others are similar to the growers of truck crops in that they specialize in the production of large quantities of a few kinds of flowers.

3. Conservatory plants. Conservatories in which plants are grown for exhibition purposes only are an integral part of many city, state and national parks and botanical gardens.

NURSERY STOCK AND SEEDS

The nursery enterprise has developed from a small localized business to a national industry in which the sales, according to the 1930 census, amounted to over 58 million dollars. Some idea as to the growth of the nursery industry is indicated in Table 14.

TABLE 14.—NURSERIES—COMPARATIVE SUMMARY FOR THE UNITED STATES, 1899, 1909, 1919, 1929*

	1899	1909	1919	1929
Area, acres	59,492	80,618	51,453	141,133
Increase over preceding census, per cent	......	35.5	−36.2	174.3
Increase in receipts over preceding census, per cent	......	107.9	−2.9	184.7

* Fifteenth Census of the United States, Vol. IV.

The nursery and seed-producing business may usually be grouped under one of the following types:

1. General nurseries. Areas devoted to the production of a large number of different kinds of plants primarily for the local trade. This type of nursery may be compared with the market vegetable garden and the general commercial florist. It often does not grow all its stock but purchases much or all of it from various specialized nurseries. General nurseries sell only to the retail trade.

2. Specialized nurseries. Areas of various sizes devoted to production of large quantities of specialized crops, as fruit plants, ornamental trees and shrubs, evergreens, hardy perennials, grass. These nurseries, like the truck gardens, are located where the climatic and soil conditions are suited to the production of their

special plants. Some of these specialized nurseries, known as "propagating nurseries," propagate only plants and sell them while still too small for the retail trade to other specialized nurseries known as "growing nurseries." Some sell only by wholesale to the general nurseries, but others cater to both the wholesale and the retail trade.

3. Seed nurseries. The production of seeds of vegetable and ornamental plants is usually a specialized enterprise carried on in regions where the climatic conditions are especially favorable for the proper maturing and economical harvesting of the seeds. Tomato seed can be grown successfully under a wide range of conditions and is often produced in those localities where the tomato is grown in large quantities. The same is somewhat true of the cantaloupe, but much of that seed is produced in the Rocky Ford district of Colorado. Other plants, such

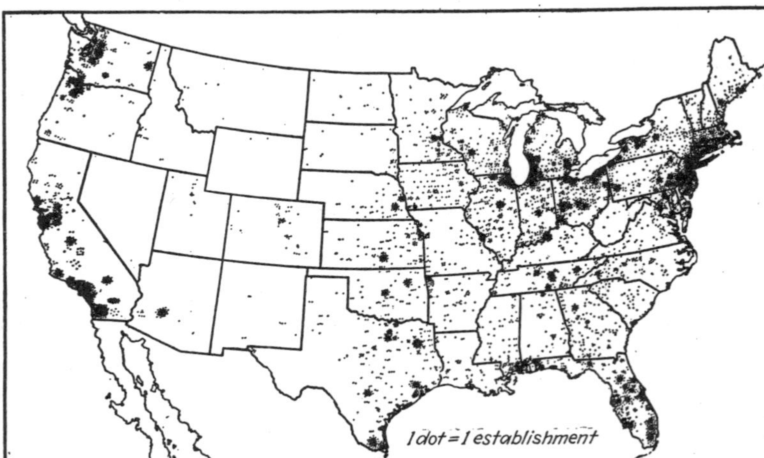

FIG. 27.—Location of establishments producing nursery stock, 1929. Note that the areas are concentrated in each state. 7207 establishments were reported in 1929. (*U.S. Census.*)

as the onion, carrot, parsnip, lettuce, pea and zinnia, are more exacting and require long, dry, warm, windless periods for the satisfactory maturing and harvesting of the seeds. For example, 80 per cent of the onion-seed acreage is located in various sections of California. The seed grower occupies an important position in the horticultural industry because he produces the seeds by which many of the plants are reproduced, and to his lot falls the opportunity to improve the plants by hybridizing, selecting and increasing the seed of the most desirable varieties. Especially during the last thirty years the production of seeds for horticultural crops has been concentrated in those locations which are best adapted for the production of such seeds.

Large quantities of seeds are grown; but because of the fact that the common practice of seed companies is to contract with growers in those areas adapted for seed production, there are few specialized seed farms. Individuals no longer save their own seed stock to any extent. The percentage of the total acreage reported in the 1930 census devoted to the production of vegetable and flower seeds by states is indicated in Table 15.

TABLE 15.—PERCENTAGE OF VEGETABLE- AND FLOWER-SEED ACREAGE OCCURRING IN VARIOUS STATES IN 1929

Commodity	States	Percentage of acreage
Sweet pea	California	Practically all
Nasturtium	California	Practically all
Carrot	California	99
Celery	California	99
Lettuce	California	99
Spinach	Washington	99
Turnip, garden	Washington	99
Beet	Washington and California	97
Muskmelon	Colorado	95
Radish	Michigan and Wisconsin	94
Parsnip	California	90
Aster	California	87
Zinnia	Colorado	87
Onion set	Illinois	83
Onion seed	California	80
Cucumber	Colorado	80
Sweet corn	Ohio, Nebraska, Michigan, Minnesota, Iowa	75
Garden pea	Idaho and Montana	72
Cabbage	Washington	50
Tomato	Indiana	35

A GRAPHIC SUMMARY OF PRINCIPAL HORTICULTURAL CROP AREAS

A graphic summary in the form of maps is now presented to show the location of the principal horticultural crop areas in the United States. No attempt is made to show them all. Table 16 shows that there are many different kinds of fruits and vegetables and that there is a wide seasonal distribution in these products.

TABLE 16.—TOTAL CARLOT UNLOADS (INCLUDES TRUCK RECEIPTS) OF FRUITS AND VEGETABLES ON THE CHICAGO MARKET BY MONTHS, 1937

Commodity	January	February	March	April	May	June	July	August	September	October	November	December	Total
Anise	2	.....	1	.....	.....	.....	.....	.....	.....	.....	1	2	6
Apple	287	318	300	270	192	138	244	197	444	924	565	411	4,290
Apricot	.....	.....	.....	.....	.....	25	54	21	.....	.....	.....	.....	100
Artichoke	1	.....	1	13	10	2	.....	.....	.....	.....	.....	1	28
Asparagus	.....	.....	25	292	94	55	8	.....	.....	8	15	.....	397
Avocado	2	1	.....	1	2	1	.....	.....	.....	.....	.....	.....	7
Banana	355	369	515	539	552	635	546	478	454	452	397	357	5,649
Bean	127	79	53	99	104	157	137	129	100	99	112	91	1,287
Beet	19	19	25	18	18	65	50	36	22	23	17	19	331
Blackberry	.....	.....	.....	.....	.....	.....	.....	.....	.....	.....	.....	.....	0
Broccoli	26	18	24	9	2	.....	.....	.....	6	27	46	42	200
Brussels sprout	16	12	2	.....	.....	.....	.....	.....	2	7	22	20	81
Cabbage	259	237	274	252	266	321	136	113	164	194	161	253	2,630
Cantaloupe	.....	.....	.....	1	13	532	729	199	342	89	.....	.....	1,905
Carrot	177	205	218	190	157	144	89	63	78	100	130	141	1,692
Casaba melon	.....	.....	.....	.....	.....	.....	.....	.....	.....	1	4	.....	5
Cauliflower	94	98	173	143	87	92	45	62	130	108	224	155	1,411
Celery	227	204	240	225	228	139	124	87	71	109	151	162	1,967
Chinese cabbage	4	3	.....	.....	.....	.....	.....	2	4	7	4	13	37
Cherry	.....	.....	.....	.....	22	88	72	5	.....	.....	.....	.....	187
Chickory and endive	18	17	21	10	1	.....	.....	6	13	14	19	12	131
Cranberry	.....	.....	.....	.....	.....	.....	.....	.....	20	40	77	16	153
Corn, green	.....	.....	.....	.....	2	26	81	111	58	.....	.....	.....	278
Cucumber	15	16	39	95	146	244	244	103	60	28	31	15	1,036
Currant	.....	.....	.....	.....	.....	.....	10	3	.....	.....	.....	.....	13
Date	.....	.....	.....	.....	.....	.....	.....	.....	.....	.....	.....	.....	0
Eggplant	1	1	1	3	3	10	14	46	25	12	.....	.....	116
Escarole	10	6	.....	.....	1	1	2	2	9	8	8	11	58
Fig	.....	.....	.....	.....	.....	1	.....	.....	.....	.....	.....	.....	1
Garlic	3	2	2	2	2	2	6	11	16	7	8	3	64
Grapefruit	272	306	332	285	196	43	17	17	67	225	227	265	2,252
Grape	16	3	8	9	13	6	84	198	521	1,085	512	95	2,550
Gooseberry	.....	.....	.....	.....	.....	.....	.....	.....	.....	.....	.....	.....	0

Commodity	January	February	March	April	May	June	July	August	September	October	November	December	Total
Honeyball melon	.....	.....	.....	.....	.....	6	27	11	2	1	.....	.....	47
Honeydew melon	.....	1	1	2	.....	7	106	88	79	67	13	3	367
Horseradish	.....	.....	.....	.....	.....	.....	.....	.....	.....	.....	.....	.....	0
Kohlrabi	.....	.....	.....	.....	.....	32	35	3	3	.....	.....	.....	73
Lemon	69	81	78	84	86	182	153	128	94	50	58	86	1,149
Lime	.....	.....	2	4	6	14	10	6	5	2	2	5	56
Lettuce	326	330	352	529	556	542	478	383	371	380	388	414	5,049
Nectarine	.....	.....	.....	.....	.....	.....	8	31	3	.....	.....	.....	42
Olive	.....	.....	.....	.....	.....	.....	.....	.....	.....	3	1	.....	4
Onion	169	193	99	138	324	329	225	170	491	338	161	147	2,784
Orange	454	485	611	566	527	441	294	280	285	388	569	593	5,493
Parsley	1	2	4	6	9	7	15	22	13	15	16	13	123
Parsnip	13	12	12	7	2	.....	.....	1	4	13	10	23	97
Peach	.....	.....	.....	.....	.....	78	248	752	320	20	1	.....	1,419
Pear	22	17	12	5	3	1	74	230	284	168	84	38	938
Pea, green	21	15	21	55	115	118	93	79	48	31	32	34	662
Peppers	41	36	32	31	17	20	25	67	40	41	41	28	419
Persian melon	.....	.....	.....	.....	.....	.....	.....	5	8	4	.....	.....	17
Persimmon	.....	.....	.....	.....	.....	.....	.....	.....	.....	.....	8	1	9
Pineapple	19	9	18	66	229	99	5	8	5	1	5	6	470
Plum and fresh prune	.....	.....	.....	.....	.....	37	112	77	165	35	.....	.....	426
Pomegranate	.....	.....	.....	.....	.....	.....	.....	.....	.....	16	3	.....	19
Potatoes, Irish	1,039	1,027	1,135	1,173	1,529	1,920	1,482	1,238	1,592	1,697	1,350	1,070	16,252
Quince	.....	.....	.....	.....	.....	.....	.....	.....	.....	5	.....	.....	5
Radish	18	14	32	92	110	95	53	43	27	25	33	19	561
Raspberry	.....	.....	.....	.....	.....	17	3	.....	.....	.....	.....	.....	20
Rhubarb	31	38	37	76	61	71	30	7	3	3	6	3	366
Rutabaga	93	75	72	33	20	32	31	21	58	95	84	73	687
Shallot	18	23	31	34	57	66	58	42	21	11	6	7	374
Spinach	101	113	129	153	156	93	57	80	46	120	119	82	1,249
Squash	8	1	.....	2	.....	4	10	21	33	18	29	20	156
Strawberry	65	94	62	182	467	569	45	2	.....	.....	.....	.....	1,486
Sweet potato	129	147	103	86	44	21	15	53	132	198	236	183	1,347
Tangerine	90	71	12	.....	.....	.....	.....	.....	.....	.....	.....	2	113
Tomato	119	174	189	208	317	542	671	457	507	323	191	142	3,940
Watermelon	.....	.....	.....	1	19	239	1,462	733	162	3	.....	.....	2,619

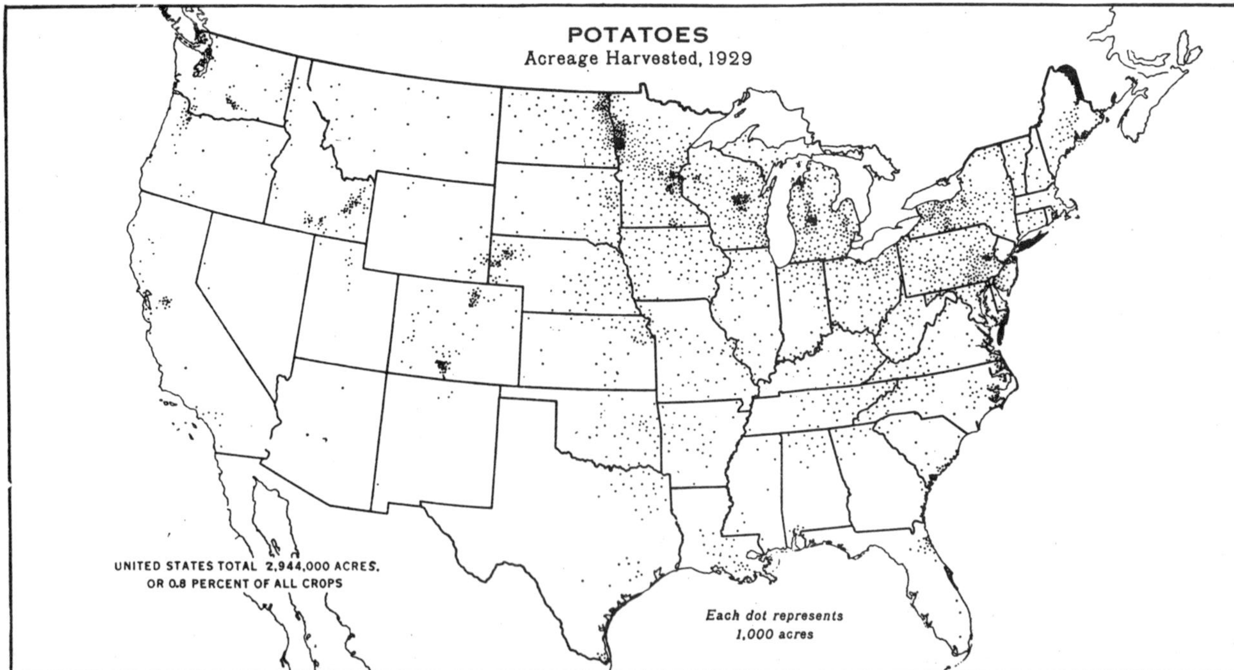

FIG. 28.—The districts of heaviest production of potatoes lie in latitudes north of the corn belt. This is partly because the quality and yield of potatoes are better in cool climates and partly because corn requires labor at the same time, is very productive and on the average has given a better return. Many of the potato-producing centers are in districts of sandy or loamy soils. Aroostook County, Maine; Long Island, New York; New Jersey; eastern Virginia; western Michigan; central Wisconsin and eastern Minnesota are leading areas of production. In recent years potato production has become more important in the South and there is now a fairly continuous gradation of regions which dovetail into each other in shipping season as early, intermediate and late. (*U.S. Department of Agriculture.*)

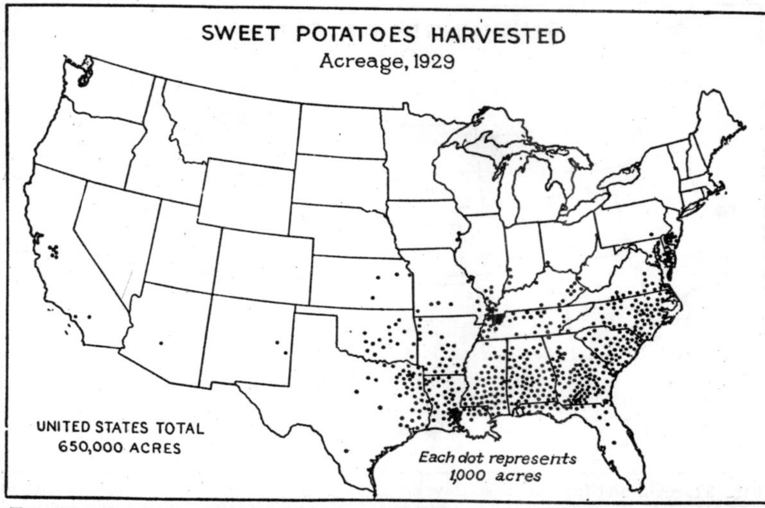

FIG. 29.—Sweet potatoes are a warm climate crop but are grown in sandy soils as far north as Muscatine County, Iowa, and southern New Jersey. In the eastern and central cotton belt sweet potatoes largely replace Irish potatoes as a staple food of the people. It will be noted that the four intensive areas of sweet potato production are western Tennessee, southern Louisiana, and the eastern shore area of Virginia, Maryland and Delaware. (*U.S. Department of Agriculture.*)

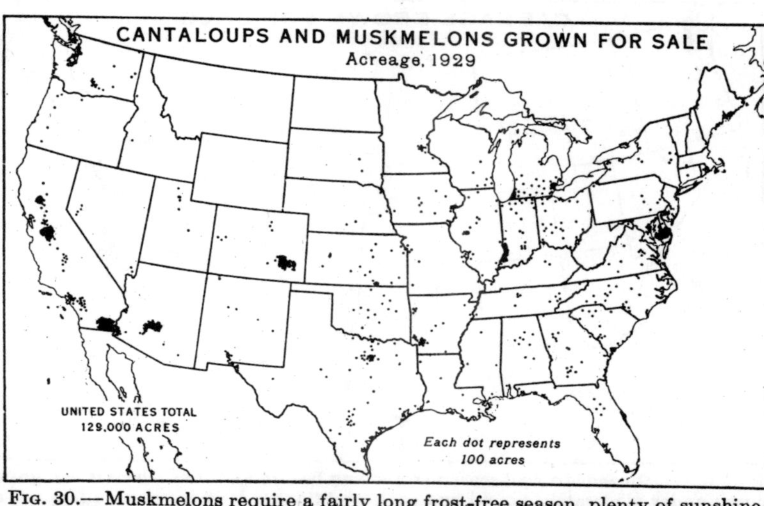

FIG. 30.—Muskmelons require a fairly long frost-free season, plenty of sunshine and heat, dry atmosphere and sufficient soil moisture. California is the leading state in acreage followed by Arizona and Colorado. There are 26 states of importance in the production of cantaloupes and muskmelons. (*U.S. Department of Agriculture.*)

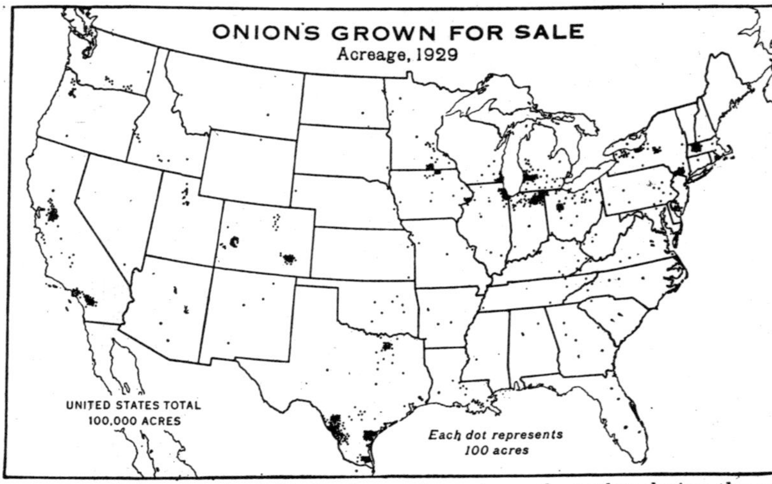

FIG. 31.—Onions for the best development require cool weather during the early part of the growing season and moderately high temperature during the latter part. Onions are grown mainly on sandy loams, silty loams and muck. An abundance of moisture is required in the early growth of the plant, and the soil should be kept moist so that the plant will continue to grow, for when growth is checked and then stimulated again it often causes splits and doubles. Although onions are produced in several rather concentrated districts, Texas, Michigan and New York ship half the nation's commercial crop. (*U.S. Department of Agriculture.*)

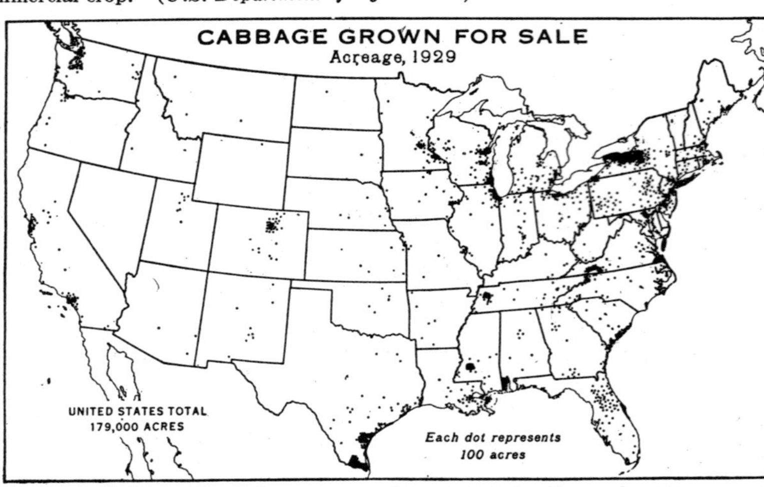

FIG. 32.—Cabbage is a cool-season crop doing best in localities where it can grow to maturity under moderately humid and cool conditions. It is grown as a late fall and winter crop in the South, as a summer crop in the North and as a spring and fall crop in the intermediate states. The largest late cabbage districts are in western New York and eastern Wisconsin. Intermediate areas are Long Island, New York, and southern New Jersey; southwestern Virginia; and Muscatine County, Iowa. Early cabbage is raised mostly in Texas, California, the Gulf states, the Carolinas and the Tidewater area of Virginia. (*U.S. Department of Agriculture.*)

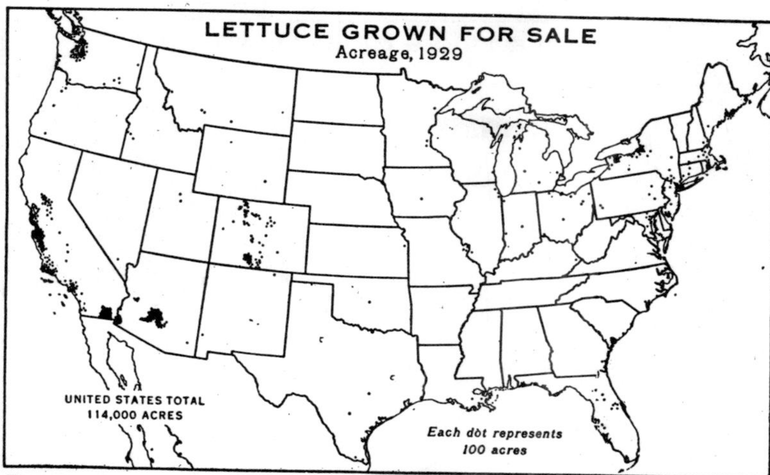

FIG. 33.—Lettuce requires a relatively cool temperature. It is principally an early spring, fall and winter crop in the South and a summer crop in the high altitudes of the West and near the coast in California, Oregon and Washington. California is the leading state in production, followed by Arizona, Colorado and New York. (*U.S. Department of Agriculture.*)

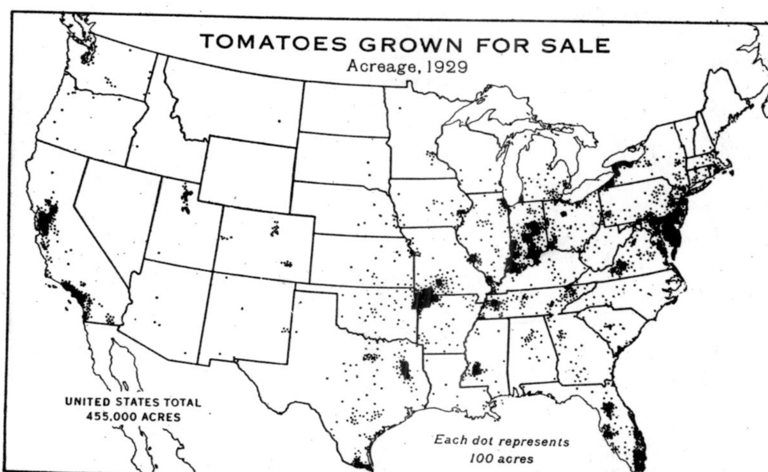

FIG. 34.—The tomato is a warm long-season crop. Good crops are produced on a wide range of soils but sandy soils are usually selected for early crops, heavy soils for late market and canning crops. Muck and peat soils which are high in nitrogen usually give an excessive growth of vines. Some of the principal states producing tomatoes include Maryland, Delaware, New Jersey, New York, Virginia, Indiana and Arkansas. Florida, southern Texas and California produce the winter crop. Copiah County, Mississippi, and Cherokee County, Texas, are important early districts. (*U.S. Department of Agriculture.*)

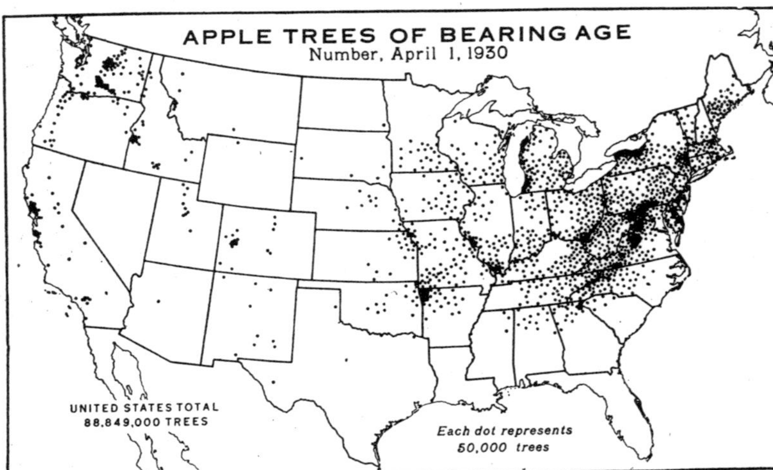

FIG. 35.—The apple requires a comparatively cool, long growing season, but can be grown on a wide variety of soils. The southern boundary of successful commercial production is limited by a mean summer temperature of 79° F., which extends only a little beyond the northern limit of cotton, and the northern boundary is limited by a mean winter temperature of 13° F. Most of the apple trees in the East are located in the Appalachian Mountains and the Piedmont Region, and around the shores of the Great Lakes where spring frosts are less injurious than in the interior. (*U.S. Department of Agriculture.*)

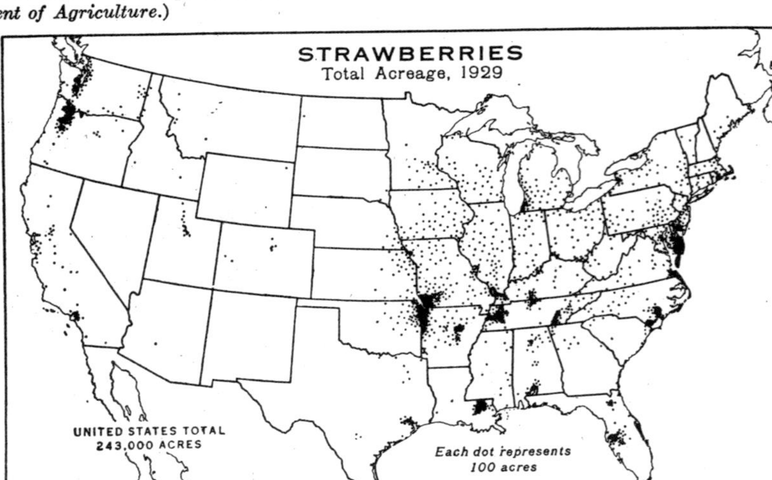

FIG. 36.—The strawberry is a short cool-season crop and this factor plus adaptibility of specific varieties to definite areas makes it possible to grow it in every state in the United States. The early areas are in Florida, Alabama, and Louisiana. As the season advances the districts in eastern North Carolina, Tennessee, and in White County, Arkansas, begin shipping; still later the large Ozark region, southern Illinois and the important Norfolk-Eastern Shore district ship; and later southern New Jersey places her crop on the market. Production areas of the late crop are widely scattered, only western Michigan and the north Pacific coast showing notable concentration. (*U.S. Department of Agriculture.*)

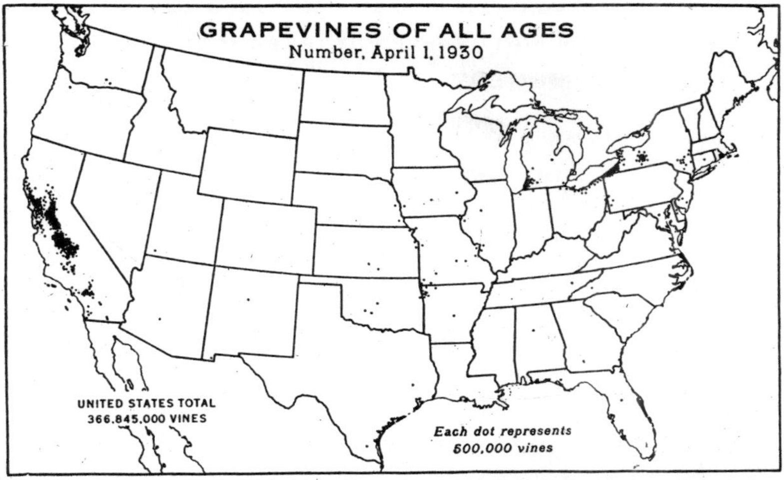

FIG. 37.—The grape produced in the eastern United States is a cool-season crop, but the grape (*V. vinifera*) produced in California requires a warm season for its best development. Nearly nine-tenths of the nation's grape production is in California. Roughly 22 per cent of the tonnage is of wine varieties, 18 per cent table varieties, and 60 per cent raisin varieties of which only about three-fourths are dried. The eastern grapes, produced principally in New York, Pennsylvania, Ohio and Michigan are mostly consumed fresh or made into grape juice. (*U.S. Department of Agriculture.*)

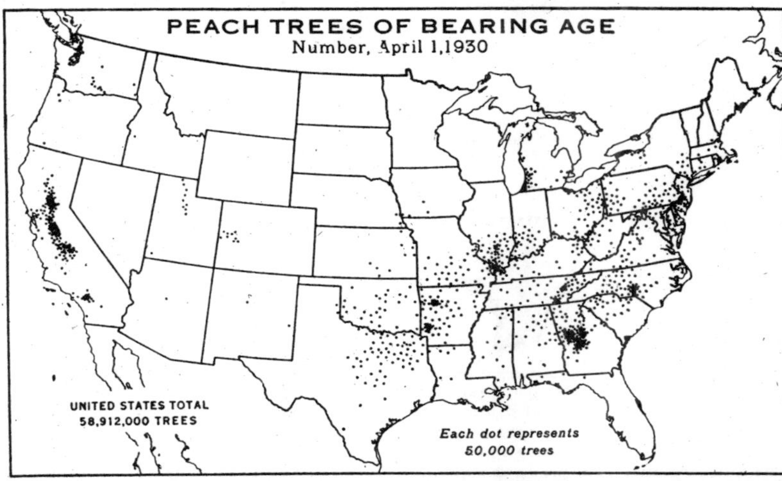

FIG. 38.—The peach is a warm-season crop adapted to lighter soils than the apple. Cold dry winters prevent peach production northwest of a line drawn from Chicago to Omaha and thence to Amarillo, Texas. Note the location of peach districts on the leeward shore of the Great Lakes, where winter temperatures are moderate and growth in spring is retarded until danger of frost is past. California is the largest producer of peaches, producing nearly all of the canned peaches. Georgia generally ranks second in production. Other important centers of production are southern New Jersey, the Appalachian Mountains and the Piedmont sections, the Lake Ontario shore of New York, the Lake Michigan shore of Michigan, southern Illinois and Indiana, upland Arkansas and eastern Texas. (*U.S. Department of Agriculture.*)

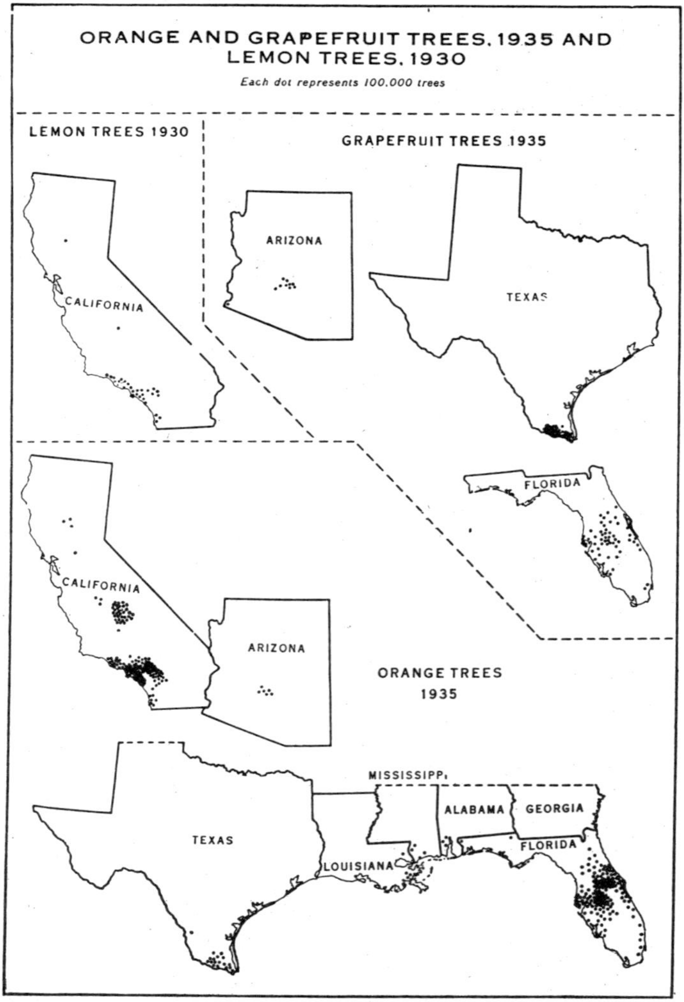

FIG. 39.—Citrus fruits are adapted to a subtropical climate. Practically all of the citrus fruits of the United States are grown in California, Arizona, Florida and the Gulf coast. California produces two-thirds of the oranges and nearly all the lemons. Florida produces nearly one-third of the oranges and two-thirds of the grapefruit. Texas and Arizona both produce considerable quantities of grapefruit. Production of citrus fruit has developed from a few small areas, until now it equals that of the commercial production of apples. (*U.S. Department of Agriculture.*)

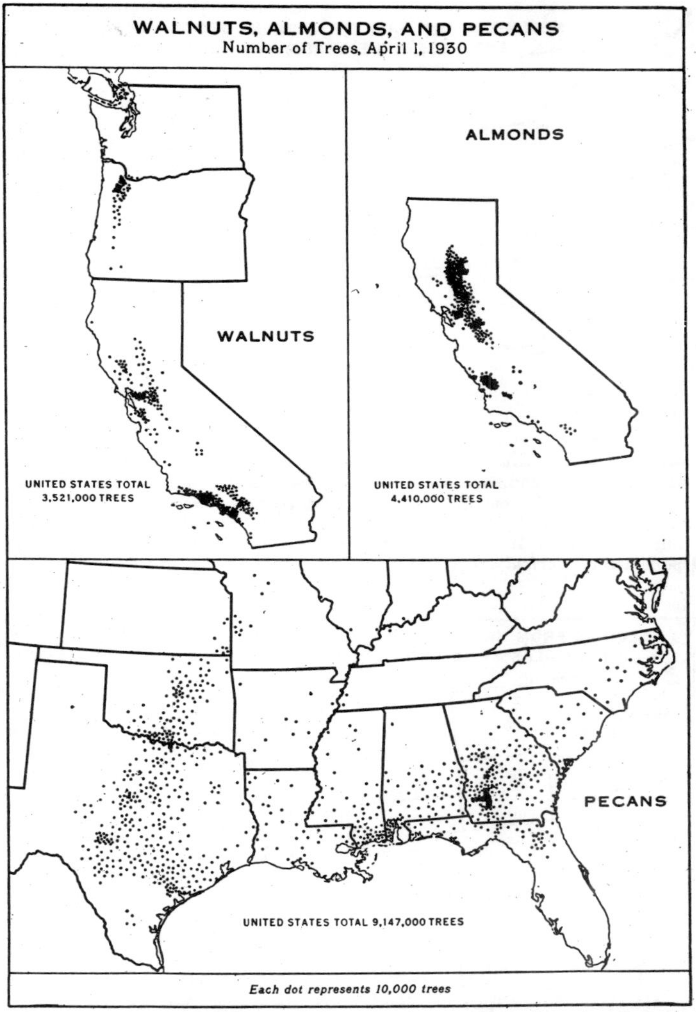

FIG. 40.—Nuts are principally warm-season crops. Almonds and walnuts are grown almost entirely in the Pacific states and principally in California. Pecans, on the other hand, are a product principally in the southern states from Texas to the Carolinas but also extend northward to southern Indiana, Illinois and Missouri. Much of the pecan crop in Texas and Oklahoma is from wild trees. Walnuts (English, French or Persian), are produced in southern California, the valleys of central California and in the Willamette Valley of western Oregon. Almonds are grown mostly on the foothills of the Great Valley and San Luis Obispo County, California. (*U.S. Department of Agriculture.*)

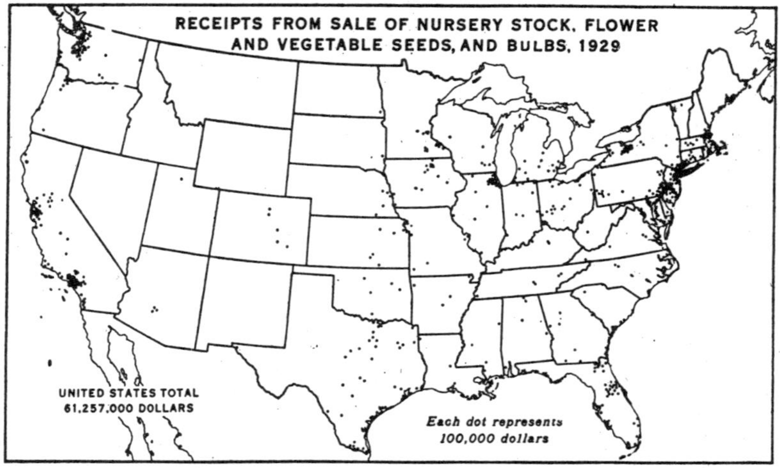

FIG. 41.—The Bay counties and Los Angeles, Calif., supply most of the nursery stock for that state and most of the flower and vegetable seeds of the United States. The lower Rio Grande Valley in Texas and central Florida also supply much subtropical nursery stock. In general, the principal nursery areas are near large cities. (*U.S. Department of Agriculture.*)

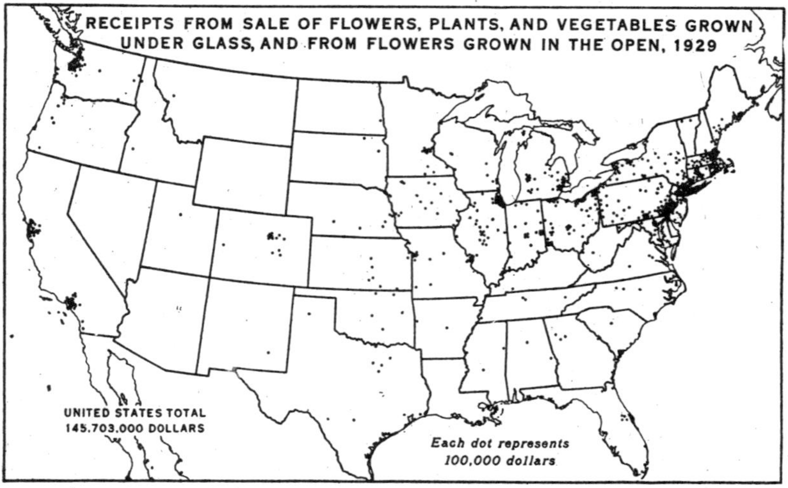

FIG. 42.—Most of the greenhouse products and cut flowers are produced in the northeastern quarter of the country adjacent to large cities. The major belt of production includes the Atlantic coast from Baltimore to Boston. Next in importance is the Great Lakes Belt from Rochester to Milwaukee and Chicago. Then come the river cities of Cincinnati, St. Louis, Kansas City, Omaha, and the Twin Cities, St. Paul and Minneapolis. More important, however, are the Pacific coast cities, Los Angeles, San Francisco, Portland and Seattle. The three inland cities of Denver, Indianapolis and Columbus deserve notice. (*U.S. Department of Agriculture.*)

Review Questions

1. Did the horticultural industry in the United States develop before or after the commercial production of grains and livestock?

2. What has been the trend in the production of fruit crops in the United States since 1915?

3. What has been the trend in the production of vegetable crops in the United States since 1910?

4. What was the trend in the production of ornamental horticultural plants in the United States from 1909 to 1929?

5. What two major factors have influenced the location of horticultural enterprises in the United States?

6. What six important economic factors influenced the expansion of horticultural enterprises in the United States?

7. What two physical factors are largely responsible for the production of horticultural crops in localized areas?

8. Name and define the various types of fruit farms.

9. Name and define the various types of vegetable farms.

10. Name and define the various types of floricultural enterprises.

11. Name and define the various types of nurseries.

Problems

1. How can one justify the commercial production of Irish potatoes in Florida and cabbage in Mississippi when these crops can be produced so much more economically in New York?

2. Mr. A is considering the purchase of a farm for the production of peaches. He has the choice of two farms equally suitable from the standpoint of soil, market and price. One is located near Kenosha in southeastern Wisconsin, and the other directly across Lake Michigan near Benton Harbor, Mich. Explain to Mr. A which farm you would advise him to buy.

3. Outline a tour for a foreign delegation through the principal commercial horticultural areas of the United States.

4. Mr. A lives in Alabama where cotton is the major crop. He is advised to shift to the production of vegetables, as they are more profitable. Advise him what to do.

5. Using Table 16, make a bar graph showing the monthly carlot arrivals of apples, bananas, oranges and strawberries on the Chicago market in 1937.

Suggested Collateral Readings

1. CORBETT, L. C., *et al.*, "U.S. Department of Agriculture Yearbook," pp. 125–131, 151–452, 719–727, 1925.

2. GARDNER, V. R., *et al.*, "Fundamentals of Fruit Production," pp. 701–771, McGraw-Hill Book Company, Inc., New York, 1939.

3. LAURIE, ALEX, and L. C. CHADWICK. "The Modern Nursery," pp. 1–32, The Macmillan Company, New York, 1931.

4. WARE, G. W., *et al.*, "Southern Vegetable Crops," pp. 1–10, American Book Company, New York, 1937.

5. WHITE, E. A., "The Florist Business," pp. 1–13, The Macmillan Company, New York, 1933.

CHAPTER IV

HORTICULTURAL ENTERPRISES OF THE HOME

The horticultural features of a home are integral parts of its development. The basic problem in planning the home grounds is to create an effect in which the various features, including buildings, plants, walks and drives, are developed in an orderly, logical manner with regard for both convenience and appearance. It is beyond the scope of this book to discuss the many problems associated with the arrangement of buildings, walks and drives and with the principles of design, but the student should know some of the basic and elemental principles involved in planning the home grounds. Consequently the problem is approached from the viewpoint of the home grounds as a whole; and planning the fruit garden, the vegetable garden and the ornamental plantings, respectively, is a part of one unified plan for the horticultural plantings about the home.

It should be emphasized that no two home developments will be exactly alike. For example, the number, sizes and uses of farm buildings will vary with conditions found on farms of different types, such as dairying, cattle raising, general farming and fruit growing. Homes vary in style of architecture; in size of house and in size, shape and topography of the surrounding land. Finally, the tastes and desires of individuals vary sufficiently to warrant different types of developments. Regardless of the section of the country, the size of the property or the type of property or individual tastes, the few basic principles remain the same.

THE PLAN

To avoid errors in arrangement and costly alterations a well-prepared plan drawn to scale is a prerequisite for the satisfactory development of the home grounds. This should show first the locations of the permanent features, which will include the buildings, roads and walks. One may then proceed to locate those features which call for the use of plant materials, such as the lawn, shade trees, shrubbery, flowers, fruits and vegetables.

Basic Requirements of the Plan

The primary consideration in planning the home grounds, whether it is a farmstead or a city lot, is the combining of usefulness with attractiveness. Since the grounds should be both useful and beautiful, it is well to consider some of those factors which are related to the attaining of these characteristics.

Utility.—The factors contributing to utility in the plan are (1) arrangement of buildings in relation to exposure and to the routing of routine work, (2) arrangement of plant units for convenience and (3) maintenance of building and plant units.

If a new home is being planned, the first consideration is the location of the house. This, in the case of a farmstead, should be built 100 ft. or more from the highway upon a spot having good natural drainage and providing pleasant views. The location should allow ample space for the barns and feed lots to leeward of the prevailing summer winds. In the Middle West, as a rule, the prevailing breezes in spring and summer are from the south, southwest and west; and in order to avoid the objectionable odors from the barns and feed lots, these areas should not be placed to the windward of the house.

Any house should be planned to provide the sunniest exposure and best views for those rooms most frequently occupied by the family. Facing the house toward the road may not produce these conditions.

In the Middle West the long axis of the barn should be north and south in order to give each side of the barn the benefit of sun for one-half day during the winter and to allow for cooling summer breezes through the barn during the summer. The hog houses and poultry houses should face south and run east and west, thereby giving the animals benefit of the south sun and some protection from prevailing winds during late winter. The corncribs should run north and south for the best circulation of air and maximum sunlight.

The machine shed, shop, dairy house, garage and scales should be arranged for functional convenience. The machine shed should be located close to the lane from the fields; the shop, close to the machine shed and garage. The water trough, the well house and the dairy house should be close together and along the path to the barn, and the scales should be so located as to be easily accessible. The garage should be close to the house.

The fruit and especially the vegetable garden should be placed conveniently close to the house. The outdoor living area in which there may be flowers and other ornamental plants should also be in a convenient place and connected in some easily accessible way with the living room. A windbreak is used on many farmsteads in the colder regions and should be located along the north and west of the farmstead and about 200 ft. from the buildings as a protection from cold northwest winds during the winter.

If the grounds about the home have been well planned and planted, good maintenance will insure the permanency of the work that has been done. On every farm there will gradually be an accumulation of wire fencing and posts; parts of machinery too valuable to discard and other bulky, little-used equipment that does not justify shed storage. This unattractive collection of useful material can be suitably protected and effectively concealed by the proper planting of trees and shrubs.

Foresight will save an individual much expense in the maintenance of garden areas. The type of development that looks best with the minimum of expense for maintenance consists mainly of lawn with trees and shrubbery grouped informally. The cost increases with formal hedges which require clipping, with pergolas and arbors which must be painted and with ornamental vines which require training. Flower borders add to both the beauty and the cost of upkeep. Formal gardens are almost prohibitive to the individual of average means and limited time.

Unity.—The home grounds should possess unity. Unity requires the complete orderly arrangement of the buildings and entire grounds and presents a pleasant picture from whatever position the homestead is seen. It demands that from every viewpoint there should be a central feature with details properly subordinated to the main feature and to one another. For example, the lawn should not be spotted with showy flower beds or specimen shrubs which claim more attention than the central feature, the house. The buildings should have a unity in design, and the plantings should be used to tie the various buildings and other permanent features together in one harmonious unit.

A few localities in the United States have developed rather typical styles of farmstead architecture. Examples are found in New England, with the house and barn connected by a woodshed; in central New York, with a story-and-a-half house and a moderate-sized barn; in southeastern Pennsylvania, with the bank barn of stone, stable high with an "overshoot" on the south and a moderate-sized dwelling; in the South, with large houses and other farm buildings subordinated; and in the Middle West, with its large barn and attached silos. Regardless of the locality or the type of development or whether it be a farmstead or a city lot, the house is always the center of interest of the home grounds. If a barn or any other farm building or group of plantings attracts more attention than this central feature, the principle of unity is violated.

The masses of plants used on the home grounds should possess unity in relation to the grounds as a whole and in relation to each individual group of plants. For example, a flower bed might possess unity in itself yet when planted in the middle of the front lawn be entirely out of place in relation to the grounds as a whole. A similar flower bed might be used in a formal area as a point of interest and not violate the principle of unity. It is important to keep in mind that the general effect of the mass of plants is of more importance than individual specimens. For example, tall slim plants such as the Lom-

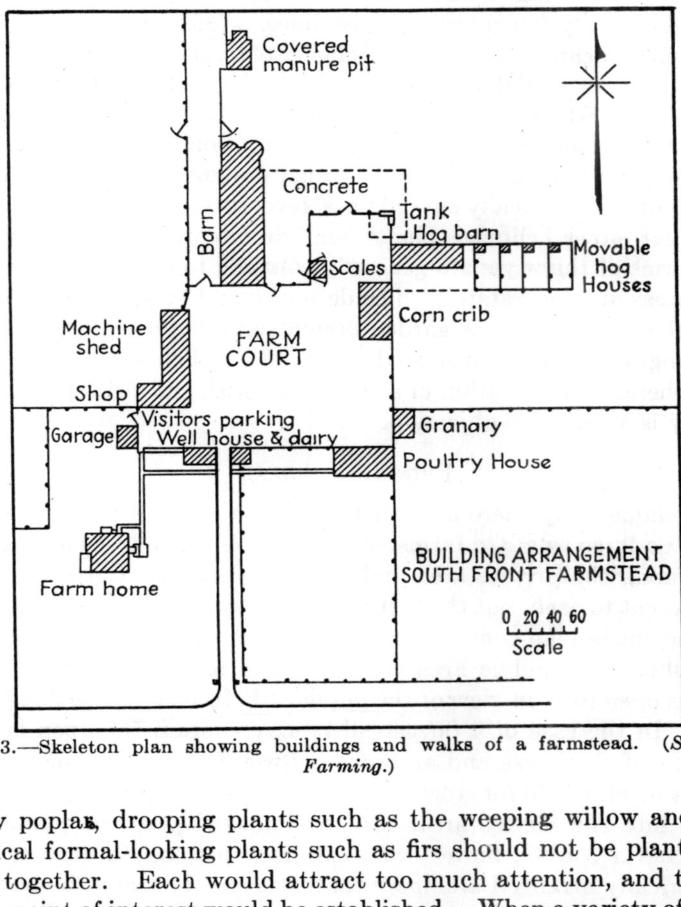

FIG. 43.—Skeleton plan showing buildings and walks of a farmstead. (*Successful Farming.*)

bardy poplar, drooping plants such as the weeping willow and symmetrical formal-looking plants such as firs should not be planted too close together. Each would attract too much attention, and thus no single point of interest would be established. When a variety of plants of different textures and widely differing shapes are used, they may be united by placing plants of intermediate characters between them. The lines of the different parts of the mass should flow into one another without too great contrasts. Transitions in color and texture should also be gradual.

Type of Plan

The first step in making a plan for the home grounds, whether for a new place or for one that is to be remodeled, is to determine the type or style of the design. Most garden designs are considered as formal or informal, although elements of both types are often found in the same plan. The informal design is often termed the "natural style" and characterized by irregularly curved lines. This type of design conforms most nearly to the arrangement of plants as they are found in nature. As stated previously, this style is the more economical both in the first cost and in maintenance. It is well adapted to the needs of the farm home and to the small city home, although the latter may use a more formal design. Formal designs are characterized by straight or symmetrically curved lines, level surfaces, geometric balance of similar areas, clipped plants and architectural embellishments. Pure forms of this style are generally confined to city homes, parks or small areas or large estates. The design should be appropriate for the particular conditions. A garden design suitable for a city home may be incongruous when transferred to the informal charm of the country atmosphere. The creation of a farmstead with the atmosphere of the country is a desirable objective.

Principal Areas

Fundamentally there are but three basic areas in any home plan, be it for a large estate, a farmstead or a small city lot. These are the public area, the private area and the service area. Certain features are inherent to each, but the three should be so arranged and combined that the entire plan constitutes a serviceable and attractive unit.

Public.—The public area comprises the foreground of the house which is open to or in view of the public. It furnishes a setting for the home. In the case of a farmstead it may contain the open lawn to the front of the house and an orchard toward the front and side or a show or display field for stock at one side. In this case the public area merges into the service area. In the informal plan the public area should never contain flower beds, specimen plants, statuary or other geometrically arranged architectural features. It should be bordered by trees and shrubs which serve to give privacy to the other areas. The walks and roadway that may be in this area are utilitarian in purpose and should attain their objects in the most direct way. A curved walk or drive without reason is unjustifiable.

Private.—This area is an outdoor living room reserved for the private enjoyment of the family and friends. It is enclosed by trees, shrubs and other features that will provide privacy from the casual passer-by. It should be easily accessible from the house, be attractive and possess features of convenience and comfort, depending upon the type of development and the personal wishes of the family. Some may wish only a good turf, an attractive shrubbery border and shade trees

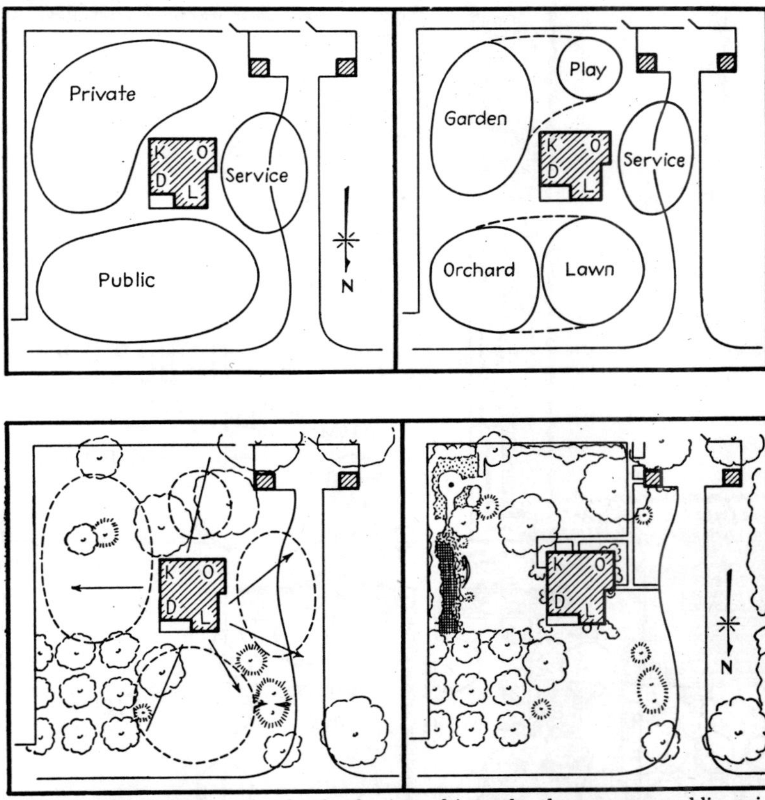

Fig. 44.—Division of the grounds of a farmstead into the three areas—public, private and service. (*Adapted from Iowa Ext. Serv.*)

and a play area; others will desire flower borders or gardens, a pool, garden furniture, an outdoor fireplace or other features that would add to their enjoyment.

Service.—The service area is strictly utilitarian. On the city lot it will include the drive, the garage, the laundry yard and the vegetable garden. On the farm it will be the largest of the three. Other than the house, all the buildings, the feed lots, the kitchen garden and the fruit garden will be in this area. It should be attractive and be a part of the entire plan although screened somewhat by suitable plantings. The basic principle dominating this part of the general plan is convenience in arrangement and suitability of construction to facilitate

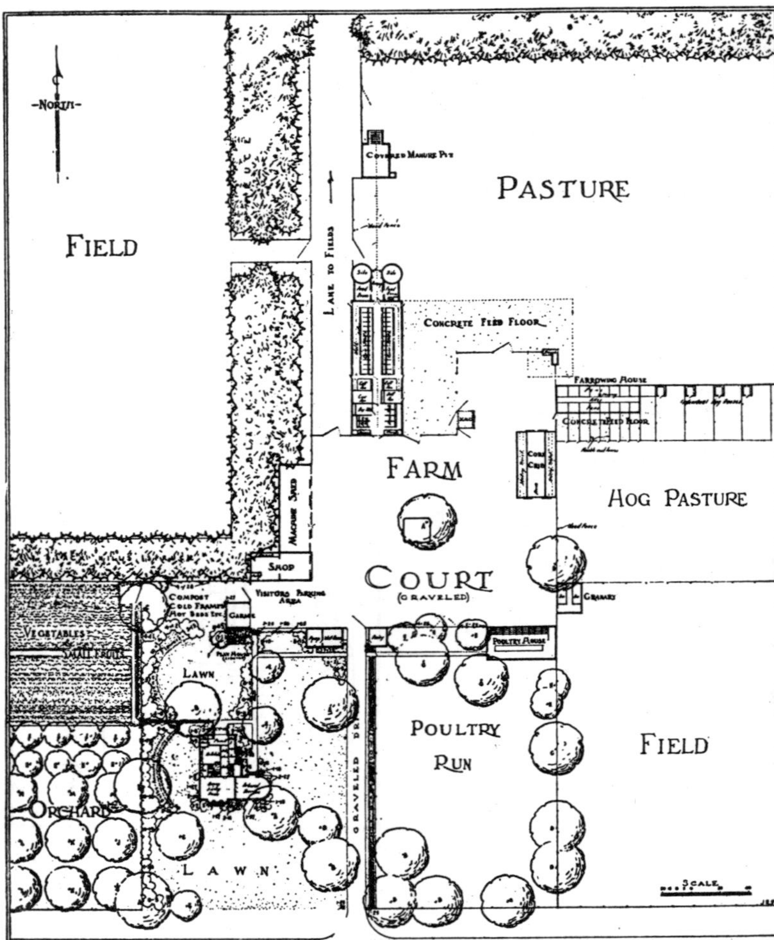

FIG. 45.—A suggested plan for a farmstead. (*Successful Farming.*)

the efficiency of the work to be done. It should be emphasized that careful planning of this area on the farm home in relation to the routine of the work to be done on that particular farm will be found decidedly economical of both time and labor.

PREPARING THE PLAN

After deciding upon the style of design, the next step is to place the plan on paper according to scale. A simple method is first to locate those features which are permanent—the house, the garage, drive,

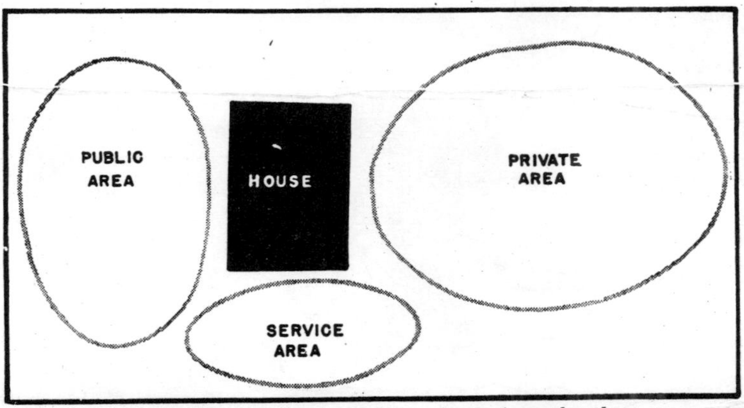

FIG. 46.—Division of the grounds of an urban home into the three areas—public, private and service. (*Suggested from Better Homes and Gardens.*)

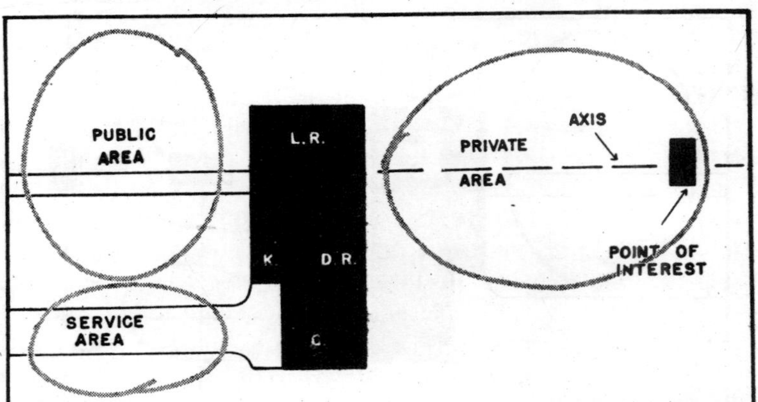

FIG. 47.—Principal areas of home grounds showing main axis. (*Suggested from Better Homes and Gardens.*)

walks, existing features that are to remain and possibly some trees. As it takes many years to grow a tree, all large trees that are suitable should be retained. After the permanent features are located, the grounds may be divided into the three areas by a series of three ovals, each indicating one area: public, private or service. Theoretically, all the space indicated within the ovals should be planted to grass;

all the space outside the ovals should be planted with trees, shrubs and flowers. This idea cannot be followed too literally, because many spots will be left for paths and open places where a planting might be too wide or might screen a beautiful view of a distant wood, a field or possibly an orchard. Each of the three areas can be subdivided by

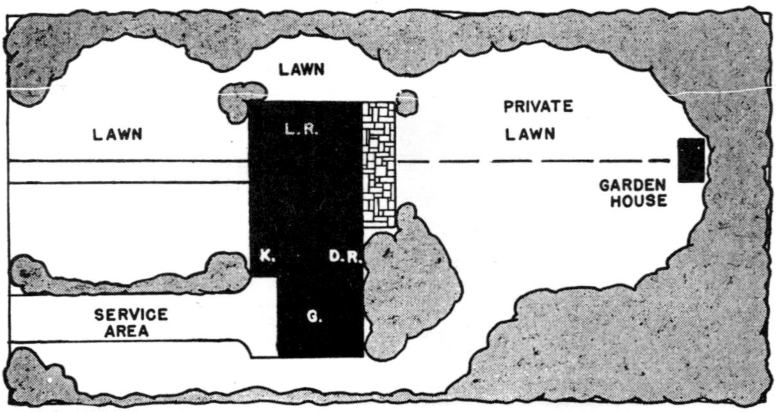

Fig. 48.—Completed plan showing plant areas. (*Suggested from Better Homes and Gardens.*)

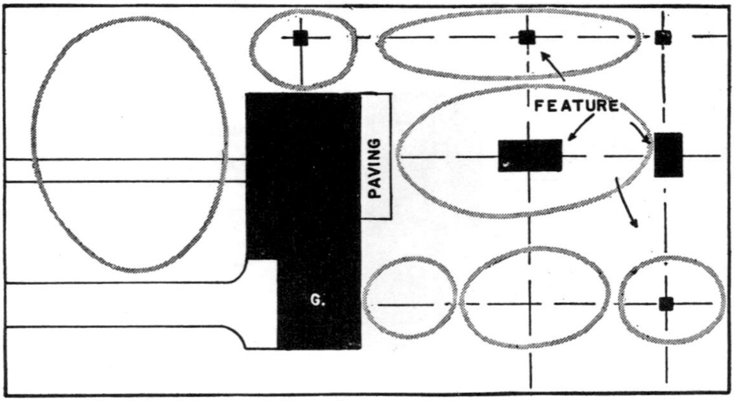

Fig. 49.—A more elaborate plan than that shown in Fig. 47. Note the axis and subdivisions of the private area. (*Suggested from Better Homes and Gardens.*)

other ovals to indicate special features. For example, the public area of the farm grounds may be divided into an open area and a display lot. The private area of either farm grounds or city grounds may be subdivided to indicate the location of a flower garden, a pool or other features. The service area may be subdivided to indicate a laundry yard or possibly a vegetable garden.

After establishing the three major areas with their chief subdivisions, the next problem is to unite them into a harmonious unit. Then the features within each of them must be combined in a similar fashion to produce unity within the separate areas. The procedure here entails the establishment of areas or lines of view which bind the

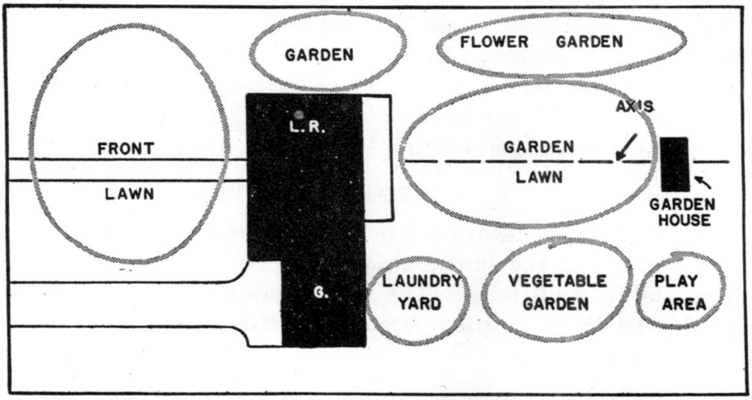

FIG. 50.—Principal areas of home grounds showing possible subdivisions. (*Suggested from Better Homes and Gardens.*)

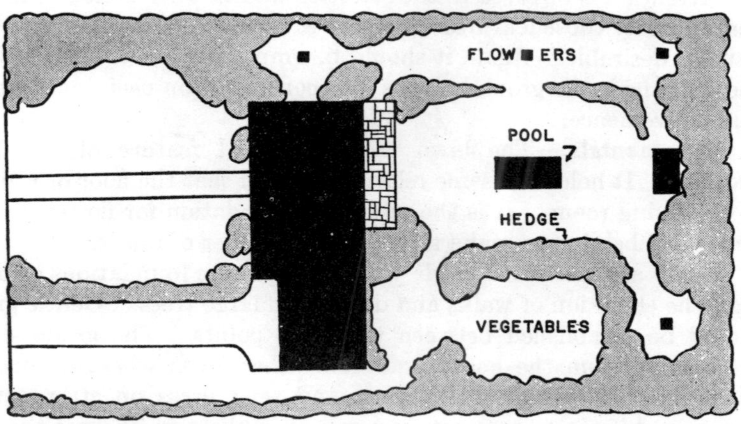

FIG. 51.—Completed plan showing plant areas. (*Suggested from Better Homes and Gardens.*)

various units together and about which other features revolve. Subordinate areas are established for each of the three major areas and often for minor divisions within them. The method of handling the private area will serve as an example.

This procedure entails the drawing of an axis, indicated by a line, from a door or window of the house that has a view of the private area to a point of interest within the area. Generally the most logical place to start it is from the living-room window. This axis is an imaginary line which indicates a line of view and about which the garden centers. If one looks out of an important window or door and obtains this view, then the garden and house are tied together and make a unit. If the view is screened, unity is incomplete. The axis connects two things: the start of the view and the end of it. Some point of interest should be at the end—a garden house, a rose trellis, a group of trees or a distant view of the landscape. Whatever is used at the end of the axis must have mass enough to hold the interest.

THE PLANTS

The plan of the farmstead in Fig. 45 shows the following horticultural features: the lawn; ornamental trees, shrubs and flowers; and vegetables and fruits. The lawn, trees and shrubs are essential features; the flower garden, vegetable garden and fruit garden are desirable ones.

Selecting Plants

Attention is directed to the selection and arrangement of the plant materials for those features which are essential and for those which are usually desirable. Again it should be emphasized that good arrangement of the home grounds takes into consideration usefulness, beauty and convenience.

Ornamentals.—The lawn is an essential feature of the home grounds. It holds the same relation to them that the floor or rug does to the living room. It is the setting or foundation for flowers, shrubs and trees, helping to make the house the center of interest.

Every area has fixed grade points such as the foundations of buildings, the elevation of walks and drives and large trees. Gentle grades should be established between the fixed points. The grade should slope away from the house and from other areas where drainage is necessary. It is difficult to grow and tend grass on steep terrace slopes; and if the terrace area is small, it is likely to be drier than the surrounding area. Proper preparation of the soil is essential for a satisfactory lawn. The subsoil should be retentive of moisture but must provide adequate drainage. The surface soil should be a fertile friable loam at least 3 to 5 in. thick. Of over 30,000 species of grass, only 30 are very satisfactory for turf purposes. The grasses differ in their climatic and soil requirements, and one specific kind of grass or a mixture of several kinds should be selected for a given locality. The bent grasses do well in the moist, slightly acid soils and cool climate of the northeastern United States and in a narrow strip along the western coast north from San Francisco. Bermuda grass is limited to the warmer southern part of the country. The fescues are quite variable in their requirements and are often used as special-purpose grasses. The bluegrasses are the most generally adaptable turf grasses and can be grown successfully over most of the northern part of the United States where the soil is neutral and fairly fertile.

The various lawn grasses may be grouped conveniently into three classes: basic grasses, those which will live for many years and will eventually predominate; nurse grasses, those short-lived grasses which

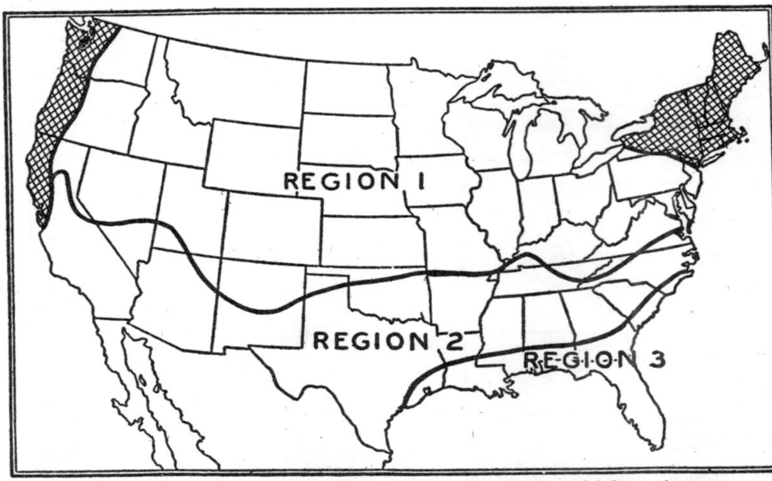

FIG. 52.—Map showing regions of the United States to which various grasses are adapted; region 1, bluegrass—crosshatched areas represent those in which bent grasses are most likely to succeed; region 2, Bermuda grass; region 3, Bermuda grass and carpet grass. (*U.S. Department of Agriculture.*)

furnish favorable conditions for the development of the basic grasses; and special-purpose grasses, those especially adapted to grow satisfactorily in shady, wet, dry or sandy locations. Because of variations in soils, exposures, etc., mixtures of grasses are often planted rather than single species or varieties. Mixtures may contain two basic grasses but never more than 40 per cent of one or more nurse grasses and should not contain any special-purpose grasses. A suitable seeding on new lawns will be obtained by using 3 to 4 lb. per 1,000 sq. ft. Only the basic or special grasses for particular areas, should be used when reseeding an established lawn. The seed can be planted either in early spring or in late fall, but best results are obtained when there is a cool moist period for six or more weeks after planting. The following mixtures have proved successful when planted in favorable localities. Mixtures are given on the basis of weight.

No. 1
Bluegrass (Fertile Neutral Soil)

	Per Cent
Kentucky bluegrass	60
Redtop	25
Chewing's fescue	15

No. 2
Bluegrass

	Per Cent
Kentucky bluegrass	80
Redtop	20

No. 3
Bent Grass (Fertile Acid Soil)

	Per Cent
Colonial bent	80
Redtop	20

Numerous other mixtures are used with success in the same and different regions. In regions where bluegrass is the basic grass, 5 to

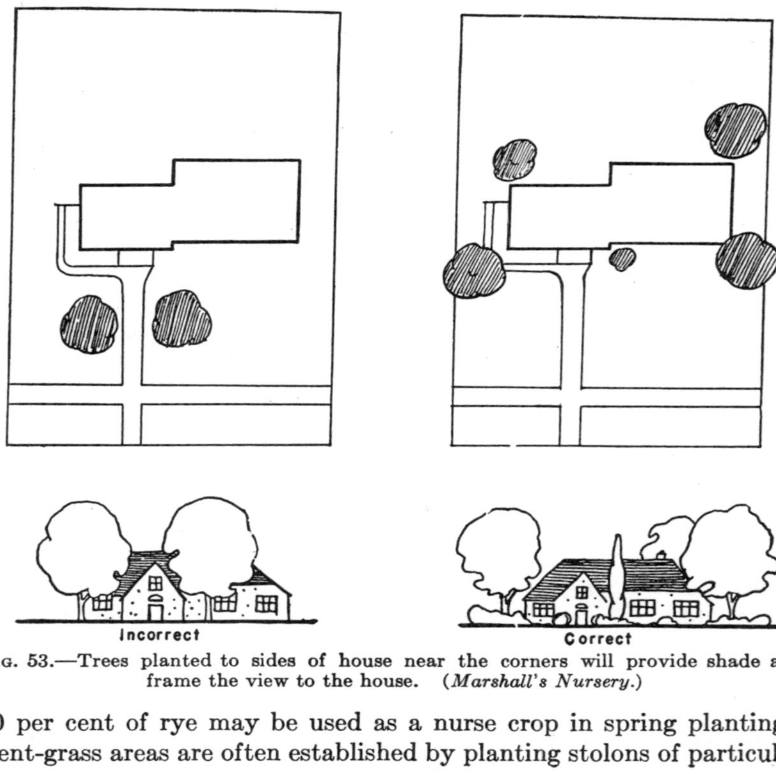

FIG. 53.—Trees planted to sides of house near the corners will provide shade and frame the view to the house. (*Marshall's Nursery.*)

10 per cent of rye may be used as a nurse crop in spring plantings. Bent-grass areas are often established by planting stolons of particular strains of bent grass.

Trees are an essential permanent feature of the home grounds. They are used for framing views, as points of interest, as a background for the house or other feature, for shade in summer, as protection from cold winds in winter, to screen unsightly objects, as accent points and as specimen plants. Two or more of these objects may be served by the same tree or group of trees. The selection of the kind of trees to plant depends upon the purposes for which they are used and their adaptability to the environment. Trees planted to the side of a house, near the corners and at the back serve to frame a view to the house, provide a background for it and supply shade. On farms it is often advisable to protect the farmstead from cold winter winds. For this purpose, windbreaks or shelter belts of trees are planted to the west and north of the buildings. Evergreens, such as Austrian pines, are

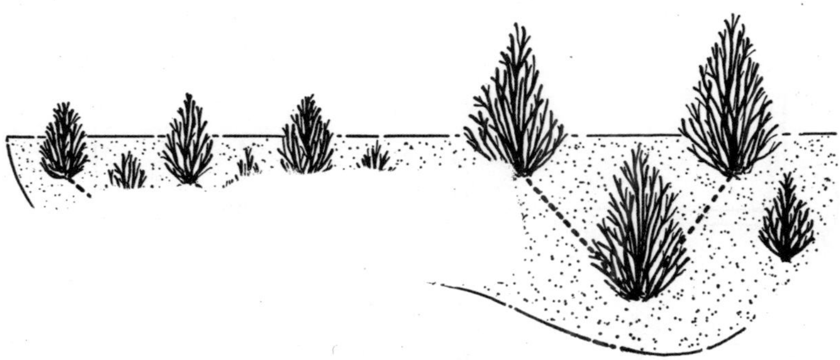

FIG. 54.—Stagger or zigzag shrubs in planting beds. The taller shrubs should be in rear with smaller ones in front.

good trees for this purpose, but many windbreaks are composed of a mixture of deciduous and evergreen trees. The effect of the windbreak extends horizontally ten to twenty times its height.

Trees are often used alone or in conjunction with shrubs to screen unattractive objects and areas such as buildings, feed lots and farm machinery. Columnar trees, as the Lombardy and Boleana poplar, may be planted singly or in groups of three or five as accent points in the skyline. Single trees may be planted as specimen plants and furnish the special feature at the end of a garden axis. Many of the evergreens, as the Colorado blue spruce, or a flowering tree, as a species of crab apple, make good specimen trees.

Shrubs are used to separate areas, to screen unattractive objects and areas, as foundation plantings, as specimen plants and as backgrounds.

The shrub border usually designates the boundaries of the home grounds. It is usually arranged informally with higher growing plants at the back and lower growing sorts at the front. The back row is generally planted in a straight line, but the plants in front of this row should be staggered in order to break the monotony of a straight line and form the bays and promontories that lend so much interest to the informal style.

The informal shrub border should have a depth of 10 ft. at the narrowest portion and 15 to 20 ft. at the wider portions if the space is available and the materials employed require that much space for suitable growth. Such a planting should form long flowing curves, never short jagged straight lines. In some instances, however, the design may call for a long straight line. The selection of materials for a shrub border is important. In the same planting it is better to

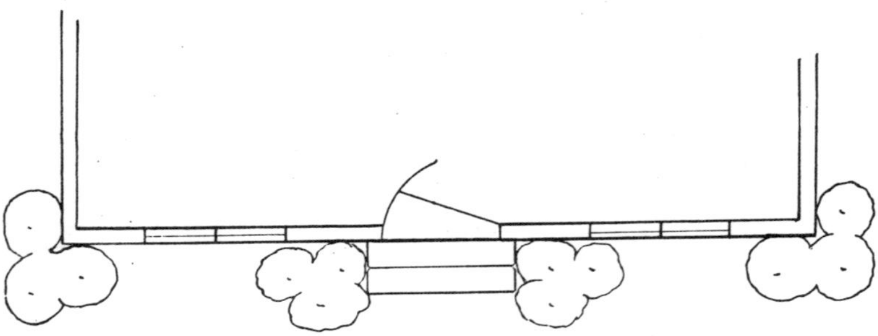

Fig. 55.—Foundation planting—ground plan.

select materials from a few groups like lilacs, Philadelphus, viburnums and spiraeas than to choose a wide variety of sorts. The problem of harmonizing various colors of foliage, shapes of plants and textures of twigs and leaves is much simpler with large masses of a few different kinds of plants than it is with small masses of a large assortment.

Shrubs are used in both the trimmed and the untrimmed hedge. The trimmed hedge is in keeping with the formal garden and requires more care than the untrimmed hedge which is better suited to the informal garden. The hedge may be used to separate the various areas of the home grounds, as a screen for the service yard and in place of a fence for marking the boundaries of the grounds and furnishing privacy. It is really a form of border planting. Shrubs are sometimes used singly or in groups to screen unsightly objects or areas. They are used very commonly for close-to-the-house foundation planting, the function of which is to harmonize the house with the rest of the grounds. It should knit the house to the lawn. The plants used around the foundation of the house should be in harmony with those in the other areas of the home grounds and in harmony with the style of the house.

When the house is high, low-growing shrubs should not be used, as they will accentuate its height. A low, rambling house calls for low-growing plants. The style of the planting may be formal or informal depending upon the type of house. Formal evergreens do not go well with an informal rambling type. A great variety of plant material is available for use in the foundation planting. The foundation, however, should not be hidden completely; neither should it be embellished with too many different kinds of plants. Complete hiding of the foundation makes the house appear as though supported by a brush pile. Generally three kinds of plants are sufficient for the average-sized house; for a larger house, five kinds should be sufficient.

Set the plants in groups around the house. Place the taller growing sorts at the corners and in the angles. This blends the perpendicu-

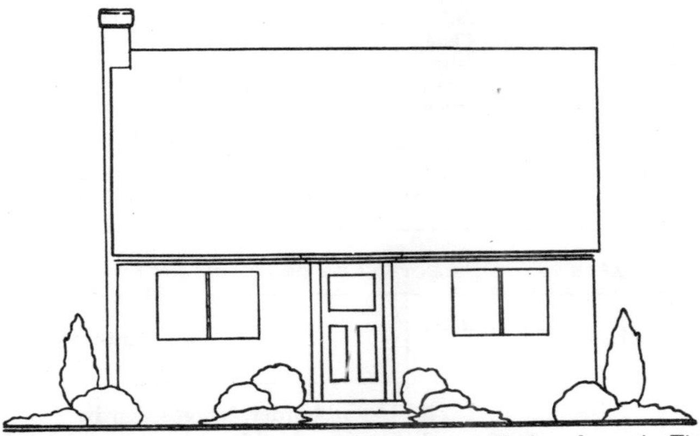

FIG. 56.—Foundation planting, perspective of ground plan shown in Fig. 55.

lar line of the house with the horizontal line of the ground. Low-growing sorts may be planted under the windows and about the taller types to tie them to the ground more smoothly. The type of plant material used may consist of evergreens which are particularly desirable for winter effect, of deciduous material or of a combination of the two. If only evergreens are used in the foundation planting and only deciduous material is located on other parts of the grounds, the transition may be too great, disrupting the harmony of the planting as a whole. A combination of deciduous plants such as lilacs with the mugho pine or Pfitzer juniper will work out well. Such plants as spruces, pines and firs should not be used in a foundation planting, as they eventually become too large. The average homeowner should be careful in the selection of plants that require trimming in order to keep them in proportion with the house, because neglect may produce plants that are not in harmony with the house. Shrubs may be used as a background for the flower border and also for flower beds and as specimen plants in a manner similar to that suggested for trees.

Many variable factors, as type of soil, amount of moisture, purpose for which the plant is grown, enter into the question of planting distances for shrubs. The final spread of the shrub for a particular locality generally determines the planting distance. A rule quite often

TABLE 17.—A SELECTED LIST OF HERBACEOUS BIENNIALS AND PERENNIALS

Name of plant	Height, inches	Color	Name of plant	Height, inches	Color
Aster, hardy	8 to 48	Various	Flax	18	Light blue
Bleeding heart	18 to 36	Deep rose	Forget-me-not	8 to 12	Light blue
Campanula	24 to 36	Various	Gaillardia	15 to 24	Various
Candytuft, hardy	6 to 10	White	Gypsophila	24 to 36	White, pink
Chrysanthemum	18 to 40	Various	Hollyhock	60 to 80	Various
Columbine	18 to 36	Various	Lupine	15 to 48	Blue, various
Coreopsis	24 to 36	Yellow	Oriental poppy	24 to 36	Various
Cynoglossum	18	Blue	Phlox	30 to 40	Various
Delphinium	36 to 60	Blue, various	Primrose, hardy	8 to 15	Various
Dianthus (pinks)	8 to 18	Pink, various	Veronica	8 to 18	Blue, white
Evening primrose	18 to 24	Primrose	Viola	6 to 8	Blue, various

TABLE 18.—A SELECTED LIST OF DECIDUOUS SHRUBS

Name of plant	Height, feet	Comments
Barberry	5	Use in border, hedges; hardy
Buddleia*	6 to 8	Best in shrubbery border; not hardy in cold regions
Deutzia*	5	White flowers; specimen or shrub border
Flowering quince	6 to 8	Pink to red flowers
Forsythia	10	Early; yellow; hardy
Honeysuckle	10	Many forms; hardy
Kerria	8	Yellow or white flowers; early summer to fall
Kolkwitzia	8	Pale pink blossoms
Lilac	20	Many fine hybrids and original species
Philadelphus	10	Several forms; shrub border; white flowers
Privet	5 to 7	Many forms; clipped hedges
Spiraea	4 to 6	Many forms for many places
Sweet shrub*	6 to 8	Chocolate-brown flowers
Viburnum	5 to 15	Many forms; shrub borders
Weigela*	5 to 7	White, pink or red flowers

* Subject to winter injury in regions of long cold winters like the Middle West.

followed is to set the plants at a distance from one another that equals two-thirds of their final height. If smaller plants are to be set close to taller growing plants, allow for the spread of the taller growing plant

TABLE 19.—A SELECTED LIST OF EVERGREEN SHRUBS

Name of plant	Height, feet	Comments
Abelia	2 to 4	Pinkish-white flowers
Azalea*	18 to 48	Must have acid soil
Boxwood*	12	Great variety of uses
Common juniper	4	Wide-spreading; several forms
Cotoneaster*	1 to 5	Several practically evergreen species
Daphne*	1	Fragrant pink flowers
Mountain laurel*	5 to 15	Acid soil
Pfitzer juniper	5	Spreading
Rhododendron*	4 to 12	Many species and hybrids; acid soil
Sargent juniper	2 to 3	Spreading

* Subject to winterkilling in regions of long cold winters like the Middle West.

TABLE 20.—A SELECTED LIST OF ANNUAL HERBACEOUS FLOWERING PLANTS

Name of plant	Height, inches	Color	Name of plant	Height, inches	Color
African Daisy	12 to 24	Various	Gaillardia	24 to 30	Maroon, bronze
Ageratum	6 to 18	Blue	Larkspur	24 to 40	Blue, various
Alyssum	4 to 10	White, lilac, yellow	Lobelia	4 to 10	Blue, various
			Marigold	12 to 40	Yellow, gold
Aster, china	18 to 30	Various	Nasturtium	15 to 72	Various
Babies's breath	12 to 15	White			
			Petunia	15 to 24	Various
Balsam	16 to 30	Various	Portulaca	6 to 10	Various
Browallia	15 to 30	Blue	Salpiglossis	20 to 30	Various
Calendula	12 to 18	Yellow, orange	Salvia	24 to 30	Blue, red
California Poppy	10 to 12	Various	Scabiosa	24 to 30	Various
Calliopsis	18 to 36	Yellow, red			
			Snapdragon	8 to 30	Various
Centaurea	18 to 36	Various	Stock	12 to 20	Various
Candytuft	10 to 18	Various	Strawflower	12 to 24	Various
Clarkia	24 to 36	Rose, various	Verbena	8 to 10	Various
Cosmos	36 to 72	Pink, various	Zinnia	15 to 36	Various
Drummond's phlox	6 to 15	Various			

in estimating the distance between the two. Many mistakes are made in planting too closely. It is time well spent to study the growth and spread of the particular plants in your own locality before determining the planting distance.

TABLE 21.—A SELECTED LIST OF ORNAMENTAL DECIDUOUS TREES

Name of tree	Height, feet	Comment
Beech	25 to 80	Purple and European most popular
Black walnut	90	High branching; provides light shade
Dogwood	30	White or pink flowers.
Elm, American	90	One of best for all landscape purposes
Ginkgo	60	Specimen and street use
Hawthorn	15 to 25	Specimen and hedge use
Hickory, shagbark	75	Picturesque; best in background
Honey locust	90	Rapid grower; best in groves and background
Horse chestnut	50	Pyramidal; makes dense shade
Linden	80	Symmetrical form; provides dense shade
Maple	50 to 100	Several species; mostly tall
Oak	60 to 100	Several species; shade and specimens
Plane tree	90	Often called "sycamore"; picturesque
Sweet gum	40	Native; symmetrical; spreading
Tulip tree	100	Mastlike trunk; specimen and shade

TABLE 22.—A SELECTED LIST OF ORNAMENTAL EVERGREEN TREES

Name of tree	Height, feet	Comments
American arborvitae	50	Flat, fernlike foliage; specimen hedge; screen
American holly	30	Red berries; specimen
Austrian pine	50 to 75	Dark glossy green; various uses
Canada hemlock	60	Symmetrical; pyramidal
Carolina hemlock	50	More handsome than foregoing; specimen
Cedar	30 to 50	Numerous varieties; slender tree; various uses
Douglas fir	60 to 75	Dark bluish-green foliage; specimen
Hinoki cypress	2 to 20	Dense dark green; many forms
Koster blue spruce	40 to 50	Silver blue-green; background; distant accent
Norway spruce	60	Dark green; rapid grower; screen or windbreak; tall hedge
Oriental arborvitae	50	Dark green foliage; uses same as American
Scotch pine	30 to 40	Picturesque; specimen; background
White fir	70 to 90	Concolor a variety; silvery-green foliage; specimen; background; windbreak

TABLE 23.—A SELECTED LIST OF VINES

Name of plant	Height, feet	Comment
American bittersweet	15	Deciduous; woody; perennial
Boston ivy	50	Evergreen; woody; perennial
Clematis	12	Herbaceous; annual
Englemann creeper	50	Deciduous; woody; perennial; adhesive disks
English ivy	50	Evergreen; woody; perennial
Morning-glory	10	Herbaceous; annual
Virginia creeper	50	Deciduous; woody; perennial; requires support
Wisteria	40	Deciduous; woody; perennial

TABLE 24.—A SELECTED LIST OF HERBACEOUS PERENNIALS PROPAGATED WITH BULBS, ROOTS, TUBERS

Name of plant	Height, inches	Color	Name of plant	Height, inches	Color
Canna	36 to 60	Various	Oxalis	4 to 6	Various
Crocus	4 to 6	Various	Peony	24 to 48	Various
Dahlia	18 to 72	Various	Royal lily	48 to 60	White, tinted
Gladiolus	24 to 48	Various	Scilla	6 to 10	Various
Grape hyacinth	4 to 8	Blue, white	Snowdrop	4 to 6	White
Iris	24 to 48	Various	Tiger lily	48	Orange
Madonna lily	36 to 48	White	Trillium	4 to 18	Various
Narcissus	4 to 18	Various	Tulip	6 to 30	Various

Perennial and annual herbaceous flowering plants offer a large amount of plant material for color and bloom on the home grounds. Due to the fact that most home grounds have an informal design, flower borders meet the requirement better than flower beds. The flower border should have a background. A mixed shrubbery planting, a hedge or a vine-covered fence may serve this purpose. The flower border may be from 3 to 12 ft. wide, but an average width of 6 to 8 ft. facilitates cultivation. The length is governed by the extent of the home grounds and the desires of the owner. The sides of the border may be straight and parallel or curved and variable in width.

The use of the flower border is similar to decorating the wall of a room of a house with a picture, the background, such as the hedge, taking the place of the wall, and the border becoming the picture. The arrangement of the flowers should be simple in design. Use tall plants in the corners and in the middle of the back of the border, followed by medium-height plants and low ones in front.

Other factors are involved in the planning of a flower border than height of plant. One should note the length of life of the plants (annual, biennial or perennial); growing habit (spreading, prostrate,

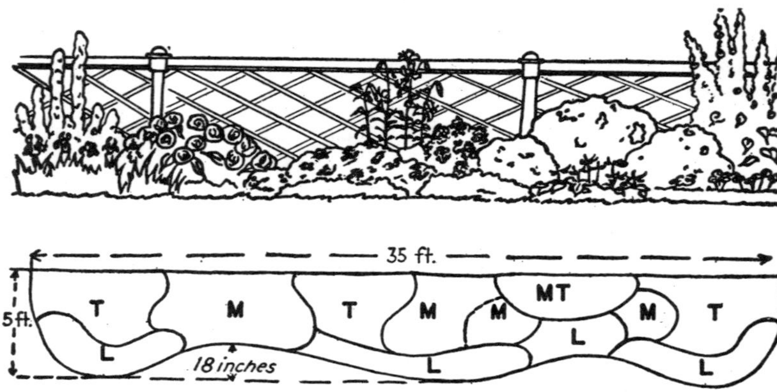

FIG. 57.—Arrangement of plants in a flower border. (*Suggested from Marshall's Nursery.*)

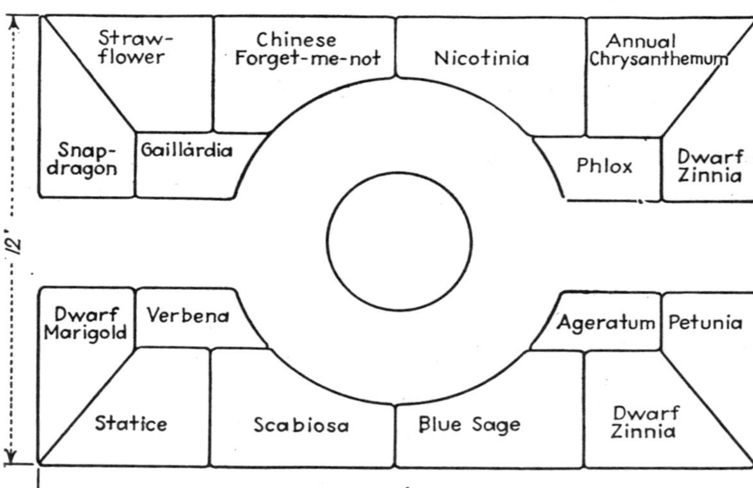

FIG. 58.—Arrangement of plants in the formal flower garden. The diagram shows one half of two suggested groupings.

erect or climbing); time and length of flowering period; color and texture of flowers and foliage; hardiness and method of propagation.

It is important to use enough plants in a mass or group to give a decided effect. The height of flowers used in a border planting depends somewhat on the width of the border. In general, tall plants should be used in wide borders. Some flowers should be in bloom in the garden during the entire season, but not necessarily all the same kind in one place. Certain color combinations are more effective than others. For example, good results are obtained when any of the following two colors are used together: scarlet and blue, orange and violet, orange and blue and yellow and purple. White flowers may be used to separate clashing colors.

Flower beds are suited to formal grounds and for areas from which flowers are to be cut. Beds are usually more elaborate than borders. Special kinds of gardens such as rock gardens, water gardens or pools and wild-flower gardens are of interest to many, however, one should be sure that the area is suited for such special gardens before constructing any of them.

The distance at which flower plants are spaced varies according to conditions previously discussed relative to spacing of shrubs. Again a good general rule to follow is to keep the plants at distances of two-thirds of their final heights.

Vines are not used so much in America as in Europe, but they do give individuality to the home grounds and should be used more frequently in this country. They are used as ground covers and to cover trellises, walls, fences and pergolas. There are many different kinds of vines. These include woody and herbaceous kinds, evergreen and deciduous kinds and also annual and perennial types.

The selection of a vine for a particular purpose often depends upon its method of climbing. Some vines, such as the English ivy, climb with aerial roots which attach themselves to rough surfaces; others, such as the Englemann creeper and the Boston ivy, climb by means of adhesive disks which can cling to wood, brick or stone. Both these types are good for covering walls. Vines, such as some species of the grape, attach themselves to a support by means of twining tendrils. Those like the clematis climb with the aid of leaf petioles which twine around any suitable support. Those like the bittersweet twine their stems around a support. The vines that twine their stems are adapted for a variety of purposes but seem to do best on vertical supports such as strings and slender poles.

Fruits.—The owner of a farmstead as well as the owner of a town or city lot often desires to grow fruit for home use. Some grow fruit trees not particularly for the fruit but because of the pleasure of having fruit trees in blossom around the place during the spring. One must keep in mind that the growing of fruit requires a certain amount of intelligent care and that the old method of planting the trees and then harvesting the crop is now past. Pests are more prevalent at present than in the past, and the necessary means of controlling such pests should be considered seriously before planting a home orchard or fruit garden.

The fruit garden for the average suburban or city lot usually consists only of the small fruits, and the suggestions for the farmstead will also apply here. The fruit trees on the small area may be placed at the back or to the side of the lot. Also, some may prefer to use fruit trees in places where others use ornamental trees. If one desires a succession of fruits of different varieties, such as apples, he might graft or bud several varieties of apples on the same tree.

Perhaps the principal reason for a fruit garden or orchard on a farmstead is that the farm family will consume more fruit if a sufficient quantity is produced at home than if it has to be purchased. The orchard itself becomes a permanent point of interest and is associated with country life. The fruit garden on the farmstead should be carefully planned. Generally the tendency is to overplant, which may result in the whole planting's being neglected. It is better to plant less and take care of it properly than to overplant and neglect proper care.

The planting should be located near the farm home on land reasonably fertile and well drained. The advantage of placing the fruit garden between the windbreak and the farm buildings, especially in regions subject to severe cold, is that it lessens wind damage and injury due to extremely low winter temperature.

The home fruit garden should include most of the kinds of fruit adapted to the locality. Apples, pears, cherries, plums, peaches, grapes, strawberries, raspberries, blackberries, currants, gooseberries and nuts are the principal kinds from which to choose in temperate-zone regions. In the warmer parts of the temperate zone and in the subtropical areas, figs, citrus fruits, etc., would be a part of the fruit garden.

The choice of varieties should be based upon adaptability to the location, high quality and personal preference. With some fruits, such as the apple, attention should be given to succession in ripening of varieties, because fresh fruit can then be provided for the table over a greater period of time. A succession of apples, for example, may be provided by a proper selection of summer, fall and winter varieties. Because the home orchard should be kept small, only a few varieties of each kind of fruit should be chosen. Varieties adapted to various states and even to parts of the same state differ; hence it is suggested that one obtain the recommendations of the state experiment station before making a final selection.

Close planting is a common mistake in fruit growing. It results in dwarfing the plant; makes spraying, pruning and harvesting difficult;

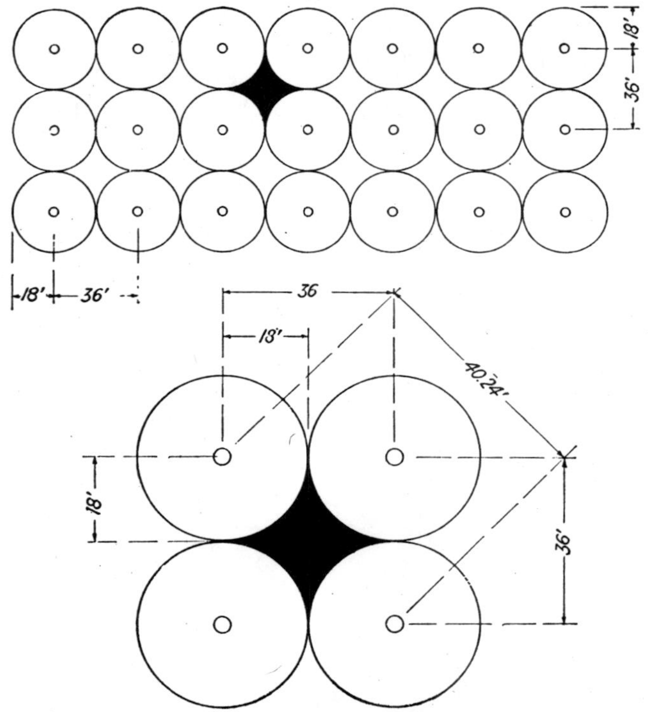

Fig. 59.—Square system for planting an orchard. Black area shows theoretically waste or unoccupied land.

and causes poorly colored fruit on lower halves of the trees. The planting distances for different parts of the United States will vary

Table 25.—Planting Distances of Fruit Plants

Name of Plant	Distance Apart, Feet
Apple	30 to 40
Pear	20 to 30
Cherry, sour	16 to 20
Plum, European	30 to 35
Plum, other	16 to 25
Peach	16 to 22
Grape	8 by 10
Red raspberry	8 by 3
Black raspberry	8 by 3
Blackberry	8 by 3
Currant and gooseberry	8 by 4
Strawberry	3¼ to 4 by 1½ to 2

according to soil, climatic conditions and kind and variety of fruit. The same variety will require more room on fertile than on less fertile soils.

Choice of the planting plan rests with the grower. Orchard trees are commonly set in squares, rectangles or equilateral triangles. The first is the simplest and commonest method in use in home planting.

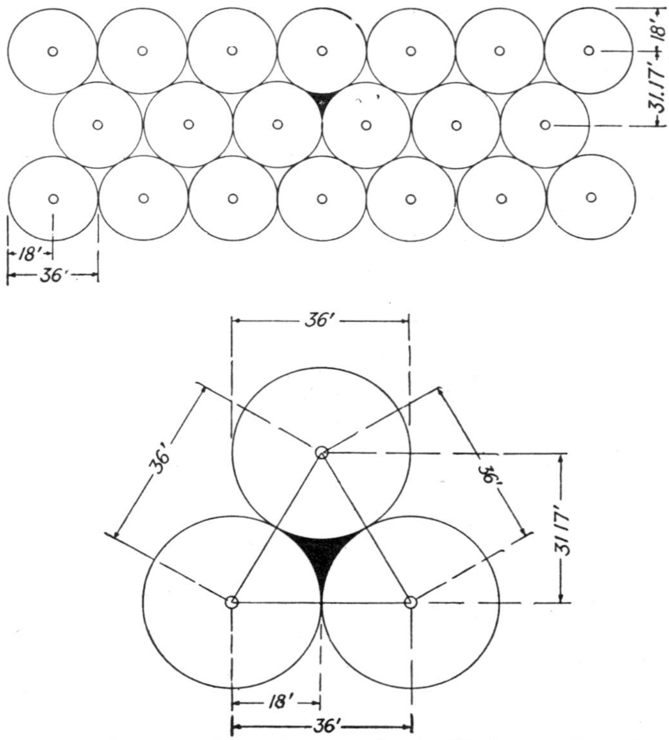

FIG. 60.—Hexagonal system of planting an orchard. Black area shows theoretically waste or unoccupied land.

In the hexagonal (triangular) method, trees are planted in triangles and are the same distance from each other. This system allows for about 15 per cent more trees per acre than does the square system.

Dwarf fruit trees have been used in Europe for years and are now gaining some attention in the United States. Where the space is limited, as in backyards of city lots, and environmental conditions permit, dwarf apple and dwarf pear trees may be preferable to standard trees.

For several reasons, dwarf trees are not used to any great extent at the present time: (1) They cost much more; (2) a great many more trees are required to provide sufficient fruit than with standard trees; (3) a limited number of varieties are available; (4) in regions of severe winters dwarf trees are subject to damage from low temperature; (5) one cannot be sure of getting true dwarf trees, as the stocks have not been standardized.

In small gardens and in instances of intensive culture, strawberries are grown in hills by removing all runner plants. The most usual practice, however, is to grow the plants in matted rows, a system in which many or all of the runners are allowed to set, but the plants are kept in rows with cultivated strips between rows of strawberry plants. In this system the matted rows of strawberry plants occupy about half the land and the cultivated strips the other half. The spaced runner system is a modification of the matted row in which some of the runner plants are removed while young and those remaining are spaced about 6 in. apart in the matted row.

Red raspberries and blackberries are usually grown in hedgerows because of their habit of sending up new plants from the roots. The hedgerows of plants are 18 to 24 in. wide (at the base) and are separated

TABLE 26.—APPROXIMATE AGE AT INITIAL BEARING AND APPROXIMATE YIELD OF FRUITS IN THE HOME FRUIT GARDEN

Kind of fruit	Approximate age at initial bearing	Approximate annual production at full bearing age, with good care
Tree fruits:		
Apple	5 to 8 years	7 bu. per tree
Pear	5 to 8 years	5 bu. per tree
Peach	3 to 4 years	3 bu. per tree
Plum	4 to 7 years	2 bu. per tree
Sour cherry	4 to 5 years	60 qt. per tree
Sweet cherry	5 to 7 years	70 qt. per tree
Small fruits:		
Grape	3 years	15 lb. per vine
Strawberry	1 year	1 pt. per plant set
Black raspberry	2 years	1 qt. per plant set
Red raspberry	2 years	1 qt. per plant set
Blackberry	2 years	1 qt. per plant set
Gooseberry	3 years	5 qt. per plant set
Currant	3 years	5 qt. per plant set

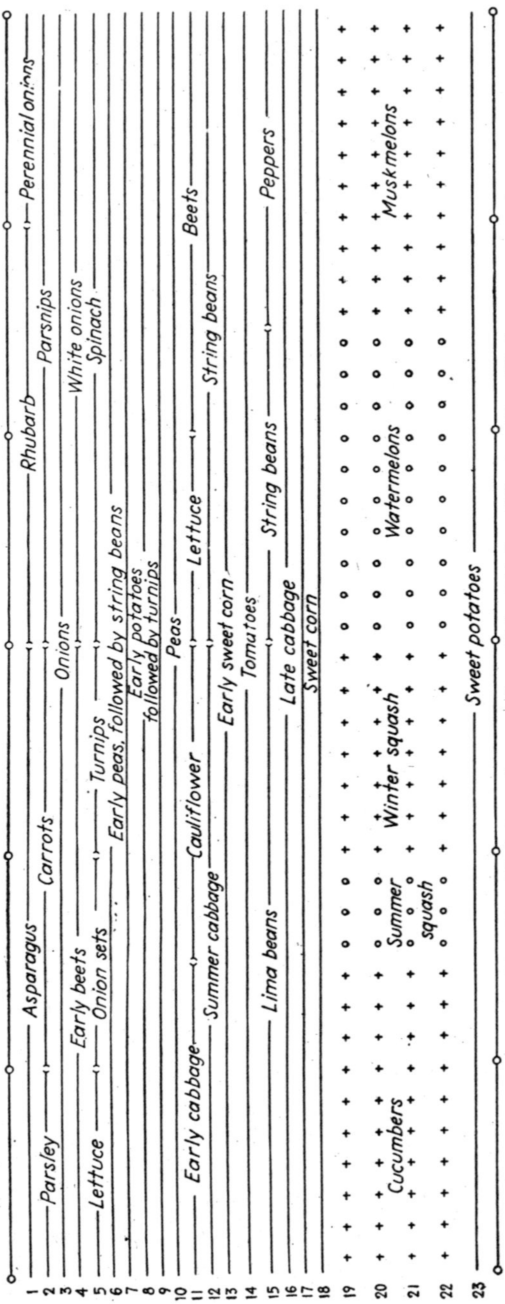

FIG. 61.—Plan for a farm garden, 90 by 240 ft. (Ill. Agr. Exp. Sta.)

by cultivated strips 4 to 6 ft. in width. Since black raspberries and the purple cane varieties do not sucker from the roots but reproduce by tip layerage, these plants are grown by what is termed the "linear system." They are kept in rows with the cultivated strips between the rows, and each year three to five canes are permitted to develop from the crowns of each of the original plants in the row.

Vegetables.—The primary purpose of the home vegetable garden is to supply fresh, canned and dried vegetables for the family table. Whether on a farm or in the city, the home garden may become a place of interest, pleasure and profit. The care of a vegetable garden is easier if a well-organized plan is made which designates the crops to be grown and their location and the distance between rows and plants

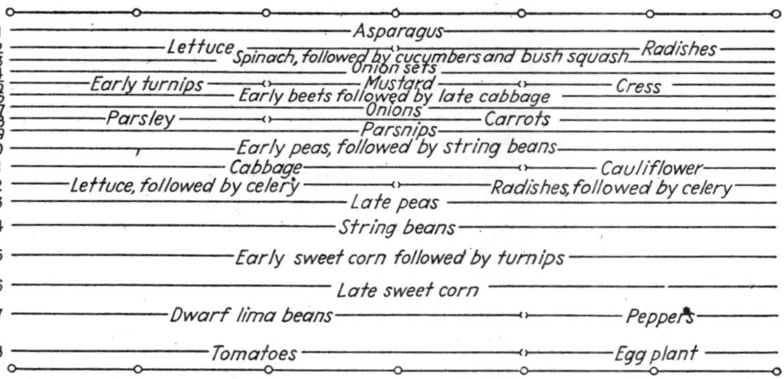

FIG. 62.—Plan for a small home garden, 30 by 60 ft. (*Ill. Agr. Exp. Sta.*)

in the row. The size of the garden will depend upon the number of people who are to be fed and the amount of land available. A small garden well cared for is better than a large one neglected. Regardless of size, however, the garden should be planned to be economical of labor. In the small garden the wheel hoe can be used, and in the large one the rows should be far enough apart to permit horse cultivation; particularly is this true in the large farm garden. In general, the rows of the vegetable garden should run north and south for optimum light.

The perennial vegetables such as rhubarb, asparagus and globe artichokes, which remain in one place for several years, should be located on one side of the garden so that they will not interfere with the cultivation and rotation of the land where the annual crops are grown.

A planting calendar is an aid in planning the home garden. Such a calendar generally gives planting dates, spacing of plants, length of time to mature the crop and other useful information. It is generally adapted to limited areas because of variations in climate and soil.

Satisfactory planting calendars are usually available at the different state agricultural experiment stations, and suggestive ones are often given in seed catalogues.

Succession cropping in the vegetable garden refers to the growing of two or more crops in succession on the same land in a single season. This method would include a greater variety of crops in regions where the growing season is longer than where the growing season is comparatively short. For example, in some districts in the South, cool-season vegetables, such as lettuce, peas, carrots, spinach and beets, are grown during the winter and are followed in the summer by warm-season

TABLE 27.—THE HOME GARDEN AND THE YEAR'S FOOD SUPPLY*

Foods (vegetables)	Servings per person			Quantity per person per year	Amounts to plant per person (with fair yields under favorable conditions)	
	Weekly	Yearly			Spring	Fall
Tomatoes	6	300		30 qt.	15 plants	
Cabbage	.....	...	180	50 lb.	18 plants	18 plants
Lettuce	6	310	90	.....	9 ft.	6 ft.
Spinach	.....	...	40	6 pt.	15 ft.	15 ft.
Carrots	.....	...	120	30 lb.	15 ft.	15 ft.
Turnips	.....	...	60	15 lb.	10 ft.	15 ft.
Beets	6	315	60	10 lb.	10 ft.	10 ft.
Parsnips	.....	...	15	10 lb.	6 ft.	
Onions	.....	...	60	12 lb.	20 ft.	
String beans	.....	...	15	8 qt.	60 ft.	
Asparagus	.....	...	10	.....	8 crowns	
Broccoli	1 to 2	80	10	.....	4 plants	4 plants
Celery	.....	...	10	.....	8 plants	
Peas	.....	...	30	4 pt.	45 ft.	
Beans, dried	3	120	40	7 lb.	105 ft.	
Corn, can	.....	...	50	10 pt.	100 ft	
Corn, dry	.....	...	50	1 lb.	100 ft.	
Potatoes, white	.....	...	685	3 bu.	400 ft.	
Potatoes, sweet	.....	...	50	½ bu.	50 ft.	
Rhubarb	15	815	50	4 qt.	1 plant	3 plants
Squash	.....	...	30	10 lb.		

* Ross, R. C., *et al.*, Shall We Move to the Country? *Univ. Ill. Cir.* 479, p. 13, 1937.

crops such as tomatoes, melons and eggplants. In Iowa such crops as bush beans, beets, early cabbage, radishes, carrots and cauliflower occupy the ground for only a part of the growing season and can be followed by late crops.

Companion cropping, which consists of growing two crops of vegetables on the land at the same time, may include the growing of quick-maturing crops such as spinach, lettuce, green onions and radishes between rows of slow-growing crops like peppers and tomatoes. The city gardener often uses both successive cropping and companion cropping because of a limited amount of land.

Obtaining Plants

The chief problems in obtaining plants are: (1) the quantities of the various plants to obtain, (2) the source of the materials, (3) the quality of the various plants and (4) the season of the year for purchase and delivery.

Quantity.—Attention has been called to several factors that govern the quantity of the various kinds of horticultural plant materials to be obtained. Information was given for spacing distances of fruits, flowers, vegetables and ornamentals and also for the amount of fruits and vegetables required for each member of the family. Although this information is a good guide for the quantity of material to purchase, quite often the expense becomes a limiting factor. Many farms have sufficient space in the vegetable garden to set ornamental plants known as "lining-out" stock. Lining-out stock is small plants usually sold in quantities of not less than 100 to be grown under cultivation for one or more years until they are sufficiently large to be planted in their permanent locations. One may purchase considerable quantities of plants cheaply and grow them in a home-garden nursery plot.

Source.—Seeds are used to start annual flowers and most vegetables. Plants are used for other horticultural materials as trees, shrubs, general herbaceous perennials, special vegetables as tomatoes and cabbages, and those annual herbaceous flowers for which an early start is desired.

The sources from which the plants are obtained depend upon the kind of plant; but in general, plants will be purchased from nurseries or from greenhouses, dug from native plantings or received as gifts from neighbors.

The success of a nursery depends upon its ability to continue to produce and sell plants at a profit; therefore, the nursery is the most reliable source from which to obtain trees, shrubs and general peren-

TABLE 28.—PLANTING TABLE FOR VEGETABLES IN IOWA*

Crop	Seed per 100 ft. of row		Planting distance between rows, inches		Depth to plant, inches	Distance in row, inches	Plant seed		Transplant to field	Days required to mature from seed
	Seed	Plants	Horse cultivation	Hand cultivation			In open field	In hotbed		
Asparagus roots	......	66	36 to 48	36 to 48	8 to 10	18 to 24	....	....	....	3 years
Beans, bush	1 qt.	....	30 to 36	18 to 24	1 to 1½	2 to 3	May 1	....	....	45 to 65
Beans, bush lima	1 qt.	....	30 to 36	18 to 30	1 to 2	4 to 6	May 1 to 15	....	....	50 to 70
Beans, pole	¾ pt.	....	36 to 48	36	1 to 2	36 to 48	May 1	....	....	45 to 65
Beans, pole lima	¾ pt.	....	36 to 48	36	1 to 2	36 to 48	May 1	....	....	65 to 80
Beets	2 oz.	....	24 to 36	12 to 18	1	2 to 3	Apr. 1	....	....	60 to 110
Brussels sprouts	¼ oz.	66	36	20 to 28	¼ to ½	18	....	Feb. 15	Apr. 1 to 15	90 to 110
Cabbage, early	¼ oz.	66	24 to 36	20 to 28	¼ to ½	18	....	Feb. 1 to 15	Apr. 1 to 15	90 to 120
Cabbage, late	¼ oz.	50 to 65	36 to 42	24 to 32	½	24 to 30	June 1	....	July 1 to 15	100 to 135
Cabbage, Chinese	⅓ oz.	100	24 to 28	18 to 24	½	12 to 15	July 1 to 15	Feb. 15	Apr. 1 to 15	80 to 100
Carrots	½ oz.	....	24 to 28	12 to 18	½	2 to 3	Apr. 1†	....	....	60 to 120
Cauliflower	¼ oz.	66	36 to 42	24 to 30	½	18	....	Feb. 15	Apr. 1 to 15	100 to 120
Celeriac	¼ oz.	200	36	24	⅛	6	....	Feb. 15	Apr. 1 to 15	125
Celery	¼ oz.	200	36 to 48	20 to 24	⅛	4 to 8	....	Feb. 1 to Apr. 1	Apr. 15 to June 15	120 to 150
Chard, Swiss	1 oz.	....	24 to 30	15 to 18	1	6 to 8	Apr. 1 to 15	....	....	50 to 120
Corn, sweet	½ pt.	35 to 65	36 to 42	30 to 36	1 to 2	18 to 24	May 1 to 15	....	....	75 to 90
Cucumber	½ oz.	20 to 25	48 to 60	48 to 60	1	48 to 72	May 15	....	....	90 to 130
Eggplant	¼ oz.	50	30 to 36	24	½	24	Mar. 1	....	May 15	150 to 160
Endive	½ oz.	100	24 to 28	14 to 18	½ to 1	8 to 12	Apr. 1	....	....	60 to 90
Horseradish roots	......	70	30 to 36	24 to 30	3 to 4	14 to 20	....	....	Apr. 1 to 15	120 to 140

Kale	¼ oz.	100	24 to 30	15 to 24	½ to 1	12 to 24	Apr. 1†	....	....	90 to 1[illegible]0
Kohlrabi	¼ oz.	150	24 to 28	15 to 24	½	4 to 8	Apr. 1	Feb. 15	Apr. 1 to 15	
Leek	½ oz.	....	24 to 28	12 to 18	1	4 to 6	Apr. 1	....	....	60 to 90
Lettuce	½ oz.	200	24 to 28	12 to 15	½	4 to 8	Apr. 1†	Feb. 15	Apr. 1	140 to 180
Muskmelon	½ oz.	20 to 25	48 to 60	48 to 60	1	60 to 72	May 15	Apr. 15	May 15	90 to 120
Mustard	¼ oz.	....	24 to 28	12 to 15	¼	2 to 3	Apr. 1†	....	....	60 to 90
Okra	1 oz.	50	30 to 36	30 to 36	1	24	May 15	....	....	90 to 140
Onion, seed	1 oz.	....	24 to 28	12 to 15	½ to 1	2 to 3	Apr. 1	Feb. 1	Apr. 15	140 to 160
Onion, set	2 qt.	....	24 to 28	12 to 15	1	2 to 3	....	....	Apr. 1	45 to 75
Onion, Bermuda	1 oz.	400	28 to 36	12 to 15	1	2 to 3	....	Feb. 1	Apr. 1 to 15	130 to 150
Parsley	¼ oz.	100 to 200	28 to 36	12 to 18	⅛	3 to 6	Apr. 15	....	....	90 to 100
Parsnips	½ oz.	....	28 to 36	18 to 24	½ to 1	3 to 5	Apr. 1 to 15	....	....	140 to 160
Peas	1 to 2 pt.	....	30 to 36	24	1 to 2	1 to 2	Apr. 1	....	....	75 to 100
Peppers	¼ oz.	66	30 to 36	24 to 28	¼	15 to 20	....	Mar. 1	May 15	140 to 160
Potato, Irish	5 to 8 lb.	....	28 to 36	12 to 18	3 to 4	12 to 18	Apr. 1†	....	....	120 to 140
Potato, sweet	......	65 to 75	36 to 48	36 to 48	3 to 4	14 to 18	....	Apr. 1	May 15 to 30	140 to 150
Pumpkin	½ oz.	10 to 15	8 to 12	8 to 12	1	84 to 108	May 15	....	....	90 to 120
Radish	1 oz.	....	24 to 36	12 to 18	½ to 1	1½ to 2	Apr. 1†	....	....	30 to 65
Rhubarb plants	......	33	36 to 60	36 to 60	2 to 3	36 to 60	....	....	Apr. 1	365
Salsify	1 oz.	....	30 to 36	18 to 24	½ to 1	2 to 4	Apr. 1	....	....	140 to 160
Spinach	1 oz.	....	30 to 36	12 to 18	1	2 to 4	Mar. 15 to 30	....	....	60 to 80
Spinach, New Zealand	1 oz.	....	30 to 36	30 to 36	1	12 to 18	Apr. 15 to 30	....	....	60 to 150
Squash, summer	½ oz.	20 to 40	36 to 48	36 to 48	1	36 to 48	May 15	....	....	60 to 65
Squash, winter	½ oz.	10 to 35	84 to 108	84 to 108	1	84	May 15 to 30	....	....	125 to 140
Tomatoes	¼ oz.	25 to 35	36 to 60	36 to 48	¼	36 to 48	....	Mar. 15 to 30	May 15 to 30	150 to 170
Turnips	½ oz.	....	24 to 36	18 to 24	¼ to ½	......	Aug. 1‡	....	....	60 to 80
Watermelon	1 oz.	33	96 to 144	96 to 144	1	84 to 108	May 15 to 30	Apr. 1 to 15	May 15	100 to 130

* Haber, E. S., The Home Vegetable Garden, *Iowa Agr. Expt. Sta. Cir.* 115, pp. 12, 13, 1929.

† Succession plantings may be made.

‡ Planting date for fall crop.

nials. A nursery may deal in a large assortment of plants or specialize in the production of a specific kind and be well known for its varieties of roses, peonies, iris or dahlias. If one desires an extensive variety of one kind or the latest improved varieties, it is desirable to buy them from established specialized growers.

Generally there is native material growing wild in every community that is suitable for planting. These trees and shrubs are usually more difficult to transplant successfully than the same sorts purchased from a nursery. This is because in the wild state the roots have never been pruned and extend long distances without branching, with the result that, when the plant is dug, a much larger proportion of the root system is cut off and left in the soil than when the plant is taken from a nursery where it has been root pruned frequently. Root pruning results in a compact root system.

Trees, shrubs and herbaceous materials may often be obtained from a neighbor, who when he decides to thin out or remove some plants is pleased to have someone make use of them. Such plants may often be divided into a number of smaller ones.

Seeds should be obtained from a first-rate firm, or they may be ordered from catalogues of reliable seed houses. It is usually most economical to buy good ones because such seeds are true to their name, free of weed seeds and of high viability. Some save the seeds from home-grown plants; although these may not always hold true to variety and may not be so viable as purchased seed, they do offer a cheap source of materials for planting the next season's garden.

Quality.—Since the original cost of the plant is very little when compared with its ultimate value, it is economical to buy only first-grade nursery stock. Such stock is well grown, is free from insects and diseases and possesses an adequate root system in relation to the top, which enables it to withstand the shock of transplanting much better than the poorer grades that do not have such satisfactory root systems. With suitable plants the expense of replacing becomes negligible, provided the necessary care is used in planting.

Ornamental deciduous trees up to 10 ft. are usually priced according to height, and larger trees according to the diameter of the trunk near the base. Ordinarily, first-grade deciduous shade trees 2 in. and less in diameter have good root systems in relation to the top and make a large tree more quickly than the smaller sizes.

In the selection of shrubs, the final height and spread to which they will grow is an important consideration. Ornamental deciduous shrubs are sold according to height. For shrubs whose final height is low (2 to 4 ft.), purchase the 1½- to 2-ft. size; for those of medium height (4 to 8 ft.), purchase the 2- to 3-ft. size; and for the higher growing (8 to 15 ft.), purchase the 3- to 4-ft. size. Shrubs should have bushy tops and bushy root systems. The heights specified indicate first-grade stock which is generally from two to four years of age and

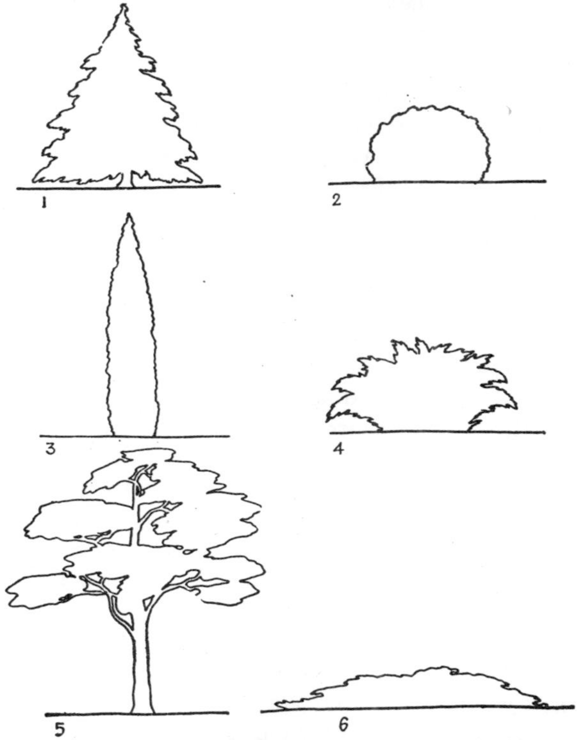

FIG. 63.—Evergreens grouped according to shape.

Group 1. Adapted for specimens on lawn; massing in groups; screen and windbreak planting; includes Douglas fir, white fir, Colorado blue spruce, Austrian pine

Group 2. Dwarf growers; more or less globular; includes mugho pine, globe arborvitae

Group 3. Upright, columnar; used for accent; formal effects; includes Colorado junipers, Chinese junipers, pyramidal arborvitae

Group 4. Dwarf growers; used for foundation plantings; includes Pfitzer juniper, Sabin juniper

Group 5. Picturesque; specimen, background; Scotch pine

Group 6. Creeping varieties of juniper; used on slopes and wherever a low carpet of evergreen foliage is desirable; includes all forms of creeping junipers

has been transplanted or root pruned and top pruned at the proper stages of development. Such stock easily withstands the shock of transplanting.

Many different kinds of evergreen are produced by nurseries. To many people all evergreens are spruces or pines; to others all evergreens are similar in shape and final size. There are two big divisions of evergreen: (1) the broadleaf kinds, such as rhododendron, mountain laurel and boxwood; and (2) the narrow-leaf kinds, such as spruces, pines, junipers and firs. The former group is restricted to rather specialized areas, as they require acid soils and are not hardy to extremely low temperatures. The latter contains many plants adapted to extensive areas.

The best grades of evergreens have been transplanted frequently in the nursery row and have been root pruned and when necessary the tops staked and trained. All narrow-leaf evergreens should be balled and burlapped. This means that the plant is dug with a ball of earth surrounding the roots and the ball is wrapped in burlap for shipping. Perhaps the most important factor in deciding the size to purchase is price. Evergreens are quoted by height or by spread, depending on whether they are erect growing or spreading. The expense of transportation becomes quite great when large plants with their necessarily large balls of earth are shipped a few hundred miles.

Fruit-tree grades are based on size. Size is expressed as height in feet or trunk caliper in sixteenths of an inch shortly above the ground. Age refers only to the top growth of the tree. The fact that the small tree did not prosper in the nursery in competition with the others may indicate that it is inferior. Nurserymen sometimes cut these smaller trees back and allow them to remain in the nursery another year and then sell them as one-year "cut-backs." Fruit trees are usually sold when one or two years old. Generally the one-year-old tree is the one to buy. It is cheaper; easier to transplant because it is smaller; more likely to live because a larger percentage of the root system was retained in digging than would have been the case with larger and older trees; and fruits as early as the two-year-old tree.

Nursery stock often possesses specifications other than grade that confuse the prospective purchaser. Among these are (1) budded or grafted plants; (2) double-worked trees; (3) sports, strains and pedigreed plants. Choice between budded and grafted plants occasionally presents a problem with apple trees and rose plants. In regions where root killing is important because of low winter temperatures, root-grafted apple trees are preferable. These are generally propagated from a short root and a long scion, and when first planted most of the scion is below ground. This places the point of union between stock and scion several inches deeper in the soil in the permanent location than it would be for budded trees. The underground portion of the scion may send out roots, and because of their favorable position they may become the chief roots of the tree. If the variety selected is hardier than the seedling root used in its propagation, and if the scion eventually develops a root system, then it follows that the tree will have a root system better able to resist cold. Apple trees grown from whole seedling roots show no consistent advantage over those from piece roots. Available data indicate no distinct advantage of grafted over budded roses.

Double-worked trees are of particular value with certain fruits. The term "double-worked" signifies twice grafted or budded. The method used in propagating double-worked trees is discussed in the chapter on Propagation of Horticultural Plants.

Bud-selected, or "pedigreed," stock means that the nurseryman has selected either scion wood or buds from so-called "high yielders." Jonathan apple trees may be offered for sale that were propagated from wood selected from a high-yielding tree. The latter might have been planted in soil that was particularly fertile or well drained, or it might have been on a seedling stock that was particularly congenial. Actual evidence indicates no permanent yield variations arising from a single branch of a tree or from an entire tree that would be inherent in the scion. With citrus fruits, however, there is often considerable variation in different parts of the same tree; and with these fruits the selection of buds from desirable wood is a universal and highly desirable practice.

Recent studies have rendered possible the identification of varieties of the same kinds of plants in the nursery row by leaf, shoot, twig and stem characteristics, thus permitting the elimination of plants not true to variety. Some nurseries have had their stock plants inspected and are offering plants propagated from these trees as "certified" plants. This means that the plants are certified to be true to name.

Certified "seed" in the case of Irish potatoes means that the tubers have been grown under specified conditions and are free of diseases such as mosaic and spindle tuber. These potatoes are inspected several times during the growing season and again after digging. If they have been properly grown and meet the legal requirements of freedom from disease, they are certified.

Time.—The time of the year for ordering seeds and plants varies according to the section of the country and with the particular plant material desired. Generally the order for plants should be sent to the nursery several months before the planting season. A nursery often depletes its stock of certain varieties and of certain kinds of plants long before it fills all its orders.

Planting

The death of plants at the time of transplanting is due to drying out of either roots or stems before opportunity is given the plants to become re-established in their new location. This drying may be due infrequently to improper handling before the plants are received. The loss of plants that are in good condition when they are received from the nursery, however, is generally due to improper handling by the purchaser.

Where several days' delay may occur before trees and shrubs received from the nursery can be set in their permanent places, the crate or bundle should be opened, and the plants "heeled in" until they can be planted properly. They should be separated, and the roots placed in a trench with the tops inclined to shade the trunks.

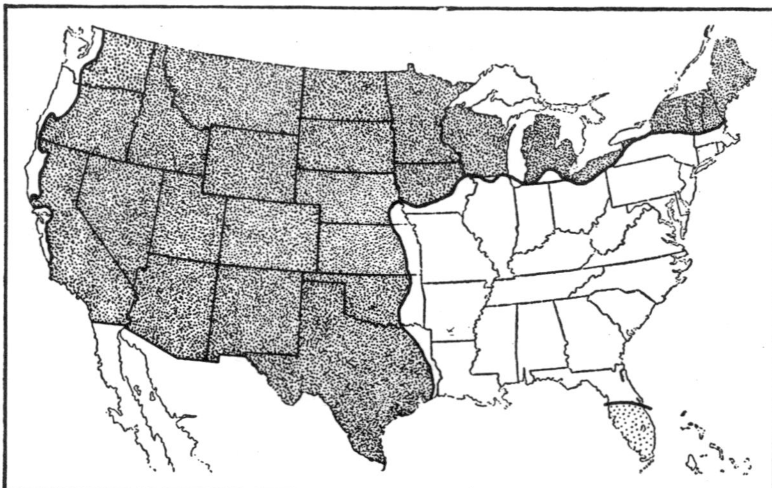

FIG. 64.—Transplanting map of the United States. The white areas indicate where fall is at least as favorable a time as spring for transplanting most deciduous trees and shrubs. In the darkly stippled areas transplanting should be done only in spring unless care is exercised. In the lightly stippled areas transplanting may be done whenever moisture conditions are favorable. (*U.S. Department of Agriculture.*)

The roots should be covered with moist soil packed around them to keep them from drying out.

Season of Year.—The season of the year for planting will vary according to the kind of plant and the section of the country. The U.S. Department of Agriculture divides the country into 32 climatic regions each of which differs in climate, soil and altitude. For that reason it is difficult to give the exact season of the year for planting. For every 100 ft. above sea level and for each 15 miles north, spring arrives one day later, and fall arrives one day earlier. Seeds may be sown on the coast of South Carolina in March, but in the mountains of the same state similar seed should not be sown until May.

The time for sowing seed also varies according to whether the seed is sown directly in the field or started in a structure such as a hotbed, coldframe or a container indoors. The time of sowing seeds for those plants which are started early and transplanted outdoors later is determined largely by the time when it is desired to set the plants in the field. In milder sections of the South, where winter production of vegetables is important, many plants are started in summer or early fall, whereas in sections farther north similar seeds are started under protection in early spring. Many people start forcing plants too early, and this results in stunted plants or tall and "leggy" ones which are difficult to transplant. Some plants like petunias and cosmos reseed themselves in the fall, and the seeds lie dormant over winter and germinate the following spring.

Spring-flowering herbaceous perennial plants, including the tulip, crocus, daffodil, narcissus, peony and iris, are generally planted in late summer or fall, because if the planting is delayed until spring there would be insufficient time for suitable growth to give early blooming. Woody deciduous plants are planted in the fall, spring or winter depending on the section of the United States.

All living plants are giving off moisture, whether the plant is in leaf and growing or dormant. Growing plants and dormant plants under normal conditions are able to replace the loss of moisture by absorption through the roots. A consideration of this principle is the determining factor in deciding the season of the year when plants should be set. Particularly is this true with trees and shrubs. In those climates where winter moisture is insufficient, planting should be done in the spring. When the plant is set in the fall, it must develop enough new roots to absorb sufficient water to keep the plant from drying out. If freezing weather occurs before the root has had enough heat and moisture to develop sufficiently, the plant will dry out. Where winter winds are very drying and the soil moisture is limited, evaporation from the stem of the plant is likely to be in excess of the moisture supplied by the root, and the plant will die.

The month or six-week period just preceding freezing weather is the best time for transplanting deciduous trees and shrubs in large parts of eastern and southern United States and some areas on the Pacific coast. The other most desirable time for planting is in the spring as soon as the ground is dry enough for suitable mechanical operations. Planting should be done as early as possible in order to give the root system an opportunity to develop somewhat before warm weather forces the top into growth. Where cold temperature rather than lack of moisture is the limiting factor in time of planting, a heavy mulch will often permit fall planting. The mulch delays the penetration of frost, keeping the soil temperature higher later in the season and also retarding the evaporation of moisture from the soil. Small evergreens, such as those used for windbreaks, are generally planted with bare roots, but larger ones are generally balled and burlapped. Balled and burlapped evergreens are sometimes planted in the fall in the colder regions of the United States after their active growth has ceased for the season. When this practice is followed, it is always advisable to mulch, for the reasons mentioned previously and also to prevent alternate freezing and thawing of the soil, which causes heaving and results in damage to the root system. Deciduous trees may be protected from drying by wind by wrapping the trunks and large branches with burlap. Recently various waxy coverings have been applied to both deciduous and evergreen plants to retard evaporation.

Evergreens which are usually transplanted with a ball of earth attached to the roots and deciduous trees which are sometimes moved under similar conditions may be transplanted at any season of the year. It is generally advisable, however, to transplant them at that season when loss of moisture is at a minimum.

A tree of sufficient size to furnish shade can often be obtained quickly by transplanting a large tree. In such cases it is advisable to move the tree with a considerable amount of the root system preserved in a ball of earth. The character of the soil, the rooting habit of the tree and the amount of preparatory treatment are all important in determining the size of ball that is necessary to take. A suitable tree 4 to 5 in. in diameter is a satisfactory size, although larger trees can be moved successfully. In late fall a trench is dug around the tree deep enough to cut off the side roots. The volume of soil inside the trench, commonly referred to as the "ball," should be 4 to 5 ft. in diameter for 5-in. trees. After digging this trench, pack it with leaves. When the block of dirt is frozen, the rest of the roots are cut, and the tree and the ball of dirt are pulled over at a sharp angle. The hole is partly filled under the ball of dirt, and the roots and the tree swung back and forth a few times in this way raising the ball of roots out of the hole. The tree can then be rolled on a sled for transportation to the new hole which should have been dug in the fall before the ground was frozen. Defective crotches should be eliminated, and the amount of leaf area reduced by thinning some of the branches. This can be done before the tree is replanted.

Manner of Planting.—Death of plants due to carelessness on the part of the planter in transplanting is quite common. The method of planting will vary according to the type of plant and texture of the soil.

Seeds may be sown in hills, rows or broadcast, depending upon the kind of plant. The operation of seeding may be done by hand or with a seeder. The average layman will most likely seed by hand or with a mechanical hand seeder. Regardless of the method, seeds should be placed at the proper depth, and the soil should be left compacted about them. The depth of planting seeds depends on the manner of germination, size of seed, texture of soil, availability of moisture and length of time required for germination. Under ideal conditions seedlings will grow best when planted at a depth of about four times their diameter. In a sandy soil which dries out readily and does not form a crust after rains, seed should be planted approximately twice as deeply as in a soil with a clay surface. When soil moisture is deficient,

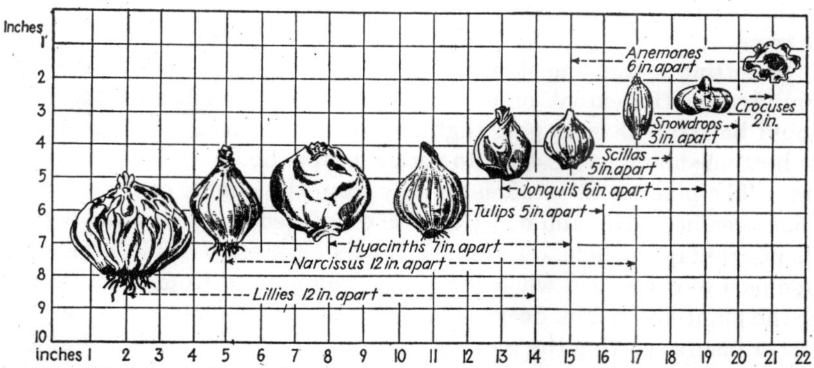

Fig. 65.—Diagrammatic sketches showing suggested depths and distances for planting various bulbs.

seeds should be planted twice as deep as when ample moisture is available. Some seeds germinate slowly and under dry conditions should be planted more deeply than those which germinate quickly.

The average layman will use plants, tubers, bulbs, rhizomes or roots in planting some of the herbaceous material that is used in the vegetable and flower gardens. The depth of planting is an important consideration in planting tubers, bulbs and rhizomes. Observations extending over many years have established certain practices as being best. Rhizomes of the iris are generally set singly in shallow holes just deep enough to permit covering. When planted deeper they may suffer from rhizome rot. Roots of the peony are set in holes large enough for the root system but with the eyes or buds 2 in. below the top of the soil. Deeper planting may keep them from flowering. Gladiolus corms are planted from 3 to 6 in. deep in heavy clay soil, 6 to 8 in. in light sandy soil. The size of bulb and texture of soil determine the depth of planting most of the plants propagated from bulbs. In general, crocuses are planted at a depth of 3 in., tulips 4 to 6 in., narcissi 5 to 6 in. and lilies 6 in.

In transplanting petunias, cosmos, marigold, tomatoes, cabbage and similar material that has soft succulent stems, care should be taken to press the soil about the root after the plant is placed in the hole. Pressure against the stems might injure or kill the plant. Water may be applied after transplanting to settle the soil about the roots. Shading the plants for a day after transplanting is also an advantage.

All the steps involved in setting plants are performed in order to facilitate the development of roots and the absorption of moisture and thereby prevent the dessication and death of the plants. In order to attain the foregoing objectives the following procedure is suggested for trees.

In setting a tree one should dig a hole that is 3 to 6 in. wider all around than the extent of the root system. The sides of the hole should be straight or sloping slightly outward at the top. The soil in the bottom should be loosened to permit proper bedding of the root system. The plant is then taken directly from the bundle or from where it has been heeled in, and badly broken or injured roots removed before it is inserted in the hole. Retain all the root system possible, because a large part of it was lost when the plant was removed from the nursery. Set the plant about 2 in. deeper than it stood in the nursery row, as this further insures getting the root system well covered with soil. Fill the hole about two-thirds full of fertile or surface soil while packing firmly about the roots. If desirable the hole can then be filled with water, after which sufficient soil should be added to fill it. Soil should not be mounded about tree.

Deciduous trees and shrubs require pruning back of the tops at the time of planting, but training and pruning of horticultural plants will be discussed in later chapters of the text.

Review Questions

1. What are the two basic requirements of the plan for the home grounds?
2. What three basic factors are required in obtaining utility in the plan for the home grounds?
3. What are the requirements for unity in the plan for the home grounds?
4. What are the two principal types of plans for the home grounds?
5. What are the three principal areas of the plan for the home grounds?
6. What are the characteristic features of each area?
7. What procedure is followed to combine the three areas into a unit?
8. What procedure is followed to combine the major feature of a single area into a unit?

9. What kinds of horticultural plant material are used for the home grounds?

10. What basic grass should you recommend for your own locality?

11. What are seven uses of ornamental trees on the home grounds?

12. What are six uses of ornamental shrubs on the home grounds?

13. How should the shrubs be arranged in an informal shrub border?

14. What determines the distance at which plants are spaced?

15. Name three kinds of each of the following: (*a*) deciduous ornamental trees, (*b*) evergreen ornamental trees, (*c*) deciduous ornamental shrubs and (*d*) evergreen ornamental shrubs.

16. Name three kinds of each of the following plants: (*a*) annual herbaceous flowers, (*b*) perennial herbaceous flowers propagated from seed or by division and (*c*) perennial herbaceous flowers propagated from a tuber or bulb.

17. What chief factors should be considered in planning the home fruit and vegetable gardens?

18. What determines the quantity of the horticultural plant material to be used in the home plan?

19. From what sources can such material be obtained?

20. At what seasons of the year should deciduous plants be set?

Problems

1. Make a diagrammatic sketch, to approximate scale, showing the three principal areas on your own or another known home grounds.

2. Make a diagrammatic sketch showing the kinds and locations of deciduous shrubs used in making an informal border 100 ft. long used as a screen between the private and the service areas in your home plan.

3. Draw a plan for a farm orchard in your locality showing the system of planting and the number, kinds and varieties of trees.

4. Place an order for the plants used in the foregoing plans.

5. You have a piece of land 378 by 350 ft. that you wish to plant to apple trees. You wish to put the largest number of permanent trees possible on the area, but all trees must be 36 ft. apart, and no tree must be closer than 18 ft. from the boundary. Draw a sketch showing the arrangement of planting. Give the number of trees planted. Show all figures used in the calculations.

6. You are intending to buy 1,000 Delicious apple trees. The nursery has two-year-old trees which appear exactly alike, but one group was propagated from a Delicious tree that has yielded an average of 10 bu. of good apples for the past ten years, whereas the other was propagated from another Delicious tree growing just 36 ft. from the first Delicious tree; but the second tree has produced an average of but 1 bu. of good fruit per year for the same ten years. You can buy for $1 each the young trees propagated from the high-producing tree, and the trees propagated from the low-producing tree for 50 cts. each. State which you would buy, and justify your choice.

Suggested Collateral Readings

1. Beach, George A., Improving Colorado Home Grounds, *Colo. Agr. Expt. Sta. Bul.* 445: 1–49, 1938.

2. Bottomley, M. E., "The Design of Small Properties," The Macmillan Company, New York, 1926.

3. Chadwick, L. C., Ornamental Evergreens, *Ohio State Univ. Agr. Ext. Serv. Bul.* 113: 1–56, 1936.

4. FREE, MONTAGUE, "Gardening," Harcourt, Brace & Company, New York, 1937.

5. HALLIGAN, C. P., Landscaping the Home Grounds, *Mich. State Col. Ext. Bul.* 199: 1–60, 1939.

6. JONES, H. A., and S. L. EMSWELLER. "The Vegetable Industry," pp. 34–40, McGraw-Hill Book Company, Inc., New York, 1931.

7. MULFORD, FURMAN LLOYD, Transplanting Trees and Shrubs, *U.S. Dept. Agr. Farmers' Bul.* 1591: 1–34, 1929.

8. SILCOX, F. A., *et al.*, Possibilities of Shelterbelt Planting in the Plains Region, pp. 33–38; 59–76, The Lake States Forest Experiment Station, U.S. Forest Service, Washington, 1935.

9. VOLZ, E. C., "Home Floriculture," The Macmillan Company, New York, 1928.

10. WESTOVER, H. L., and C. R. ENLOW, Planting and Care of Lawns, *U.S. Dept. Agr. Farmers' Bul.* 1677: 1–18, 1935.

11. WYMAN, DONALD, "Hedges, Screens and Windbreaks," McGraw-Hill Book Company, Inc., New York, 1939.

CHAPTER V

STRUCTURES AND FUNCTIONS OF HORTICULTURAL PLANTS

A clear understanding of the principles involved in various horticultural practices requires a knowledge of the principles of plant growth. Such an understanding of the activities of the plant requires an elementary knowledge of plant structure.

Most horticultural plants conform to one of three general forms: trees, shrubs or herbs. Great variations exist, however, in the appearance of these plants, because in some the stems are below ground; in others they run horizontally on the ground; and in others they stand erect. Stems may be woody or may be herbaceous and tender with

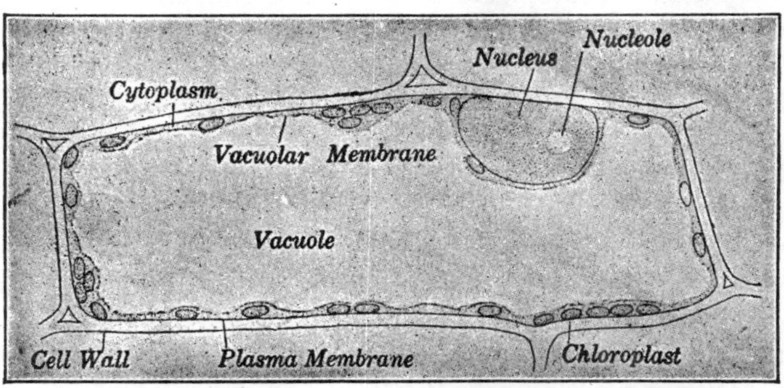

FIG. 66.—Diagram of a cell. (*Gardner, Bradford and Hooker.*)

the woody tissue much reduced or absent. With all this variation in form and structure, however, the many economically important horticultural plants exhibit one fundamental body plan. Always there is the general organization of the plant into the stem and root, each with its characteristic appendages, and into those structures associated with sexual reproduction, the flower and fruit. All plants are composed of organs that are made up of tissues, and the tissues in turn are composed of individual cells.

THE CELL

The cell is the structural unit of the plant. In the succeeding discussion, however, the term "cell" will be used to include not only true cells and their surrounding walls but also those elements which are made up of similar kinds of cell of which, in many cases, only the dead walls remain.

Cells vary in size, structure, composition, arrangement and function. They may be large or small, may have thick walls or thin walls, may vary widely in their chemical composition, may be arranged close together like bricks in a house or loosely with air spaces between; and they may function as supporting cells, storage cells or conducting cells. Cells of similar type combine to form a tissue which performs a special function. There are tissues for growth, for protection, for support, and others for the conduction of water, mineral elements and synthesized food.

Parts of the Cell

A cell is generally defined as a protoplast which consists of a nucleus, cytoplasm and various inclusions. Some authorities consider the enclosing cell wall as a part of the cell.

Protoplasm.—Protoplasm is the living substance of the cell. It appears to be a complex, ever changing colloidal protein. The portion of the protoplasm outside the nucleus is known as the "cytoplasm," and the protoplasm within the nucleus is termed "nucleoplasm." The cytoplasm constitutes the main mass of the protoplast and is surrounded by a membrane known as the "plasma membrane." In very young cells the cytoplasm is dense and contains numerous minute vacuoles; but in older, larger cells these minute vacuoles have united into one or more vacuoles which occupy most of the space within the cell. The vacuoles are filled with cell sap.

The Cell Wall.—The cell wall is formed by the protoplasm. In young, actively growing cells it is thin and is composed of cellulose, a carbohydrate material which allows the wall to stretch as the protoplasm expands. In older plants, especially woody plants, lignin, a complex chemical substance, is deposited in the wall, and this gives rigidity and strength. Fatlike substances are often deposited in the cell walls of protective structures such as the bark of trees and the skin of a potato.

Inclusions.—The nucleus and plastids are the principal inclusions of the cell.

The nucleus is a compact mass of protoplasm enclosed within a membrane known as the "nuclear membrane." The nucleus contains water, dissolved substances and chromatin. When cell division is about to take place, the chromatin forms distinct bodies called "chromosones." The chromosones, which undergo complex maneuvers during cell division, are believed to carry and transmit the heritable characters of the plant.

In the cytoplasm occur a number of small bodies known as "plastids." Plastids containing chlorophyll are known as "chloroplasts"; the colorless plastids are known as "leucoplasts." The yellow and a few of the red pigments of many flowers and many vegetative structures may be localized in the plastids; other pigments responsible for color in plants are in solution in the cell sap.

Growth of Cells

The number of cells in a single plant, as an apple tree, is incomprehensible. An idea can be obtained from the following illustration. If a human being at birth should begin a journey toward one of the nearest fixed stars at the rate of 1 mile per minute, he would have covered half the distance by the time he was ninety years old. If this passenger had taken an average-sized apple leaf with him on this hypothetical journey and removed a cell from the leaf each minute of his journey, he would discard the last cell of the single leaf on his ninetieth birthday. It has been reported that an average-sized apple leaf contains 50 million cells. When one recalls that an apple tree contains thousands of leaves in addition to roots and stems, it is difficult to estimate the total number of cells in a plant. The increase in size of an organism and the modification of tissues that take place in various organs are the sum of the activities of the cells. In general, the life cycle of a cell involves cell division, cell enlargement and cell maturation.

Cell Division.—Certain cells of the plant remain alive and retain the ability to divide and form new cells. This multiplication accounts for the principal increase in the size of the plant. Such live, dividing cells constitute but a small percentage of the total cells in the plant. These cells generally have quite definite locations in the plant and are known as "meristematic cells" or, if grouped into tissues, "meristematic tissues."

Cell Enlargement.—Cells located in various immature parts of the plant have the ability to increase in size by stretching the cell walls. Many cells just back of the growing tips of roots and stems and in the region of the cambium are in a zone of cell enlargement. The increase in size of many fruits, after a certain period, is due to an increase in the size of the parenchymatous cells and intercellular spaces rather than to an increase in the number of cells. The increase in the size of a tomato fruit after it is about ⅓ in. in diameter is due largely to the increase in the size of the cells rather than to an increase in the number of cells.

Cell division in the fruit of the peach continues for about 30 days after it is first formed, and further increase in size is due to the enlargement of those cells. Cell division in the apple fruit mostly ceases about six to eight weeks after the blossoming period, but during the remainder of the growing season the small cells increase from seven to ten times their size. This increase in cell size which is accompanied by considerable enlargement in the intercellular spaces accounts for the great increase in size of the developing fruit.

Cell Maturation.—Cell maturation is associated with both physical and chemical changes that occur in them after they reach their final size. The thickening and lignification of cell walls and the storage of food are examples of maturation activities. The proper maturing of cells is an important factor in lessening damage by low temperatures. Climatic conditions and cultivation practices in the late summer may stimulate the formation of young tender cells which will be more subject to damage from low temperature than mature cells will be.

TISSUES

It is important to remember that cells work in groups. A group of cells of common origin having essentially the same structure and function is known as a simple tissue. Aggregations of simple tissues having structural or functional unity are generally regarded as complex tissue. The cells constituting the principal tissues vary; hence attention is called to the characteristics of those constituting the principal tissues of the multicellular plant. The principal tissues may be grouped according to function into meristematic tissues, parenchymatous tissues, conductive tissues, mechanical tissues, protective tissues.

The horticulturist is concerned with the characteristics and the location of cells that constitute the tissues of a plant, because often in the culture of the plant these tissues may be modified to increase the size and number of organs that are used as food; or the tissue may be modified in a particular organ to improve the quality of the product. A few horticultural problems associated with modification of tissues, or with their locations and functions, will help to show the importance of knowing something about tissues.

The fleshy part of pears has unique cellular structures which are called "stone cells," or "grit cells." The occurrence of these cells in large quantities in a particular variety of pear known as the "Kieffer" lowers its quality to the extent that few people like to eat it. If this variety is harvested at the usual time and stored at a temperature of 60°F. until softened to 4 or 5 lb. pressure test, it attains a good flavor and much improved texture because the insoluble pectin substances forming the grit cells become soluble.

Stringiness in celery is objectionable to the consumer. It is associated with a particular tissue which is composed of cells known as "collenchyma" which have thickened walls at the angles. This condition is more pronounced in some varieties than in others; and although stringless celery does not exist at the present time, it may be obtained in the future by proper breeding and selecting from the present varieties.

The increase in size of fruits and vegetables after certain periods is due largely to the development of parenchymatous tissue.

To be performed intelligently, the horticultural practices of ringing, pruning and grafting and budding require a knowledge of various plant tissues.

Meristematic Tissues.—At the tips of roots and stems of plants are groups of cells that have the ability to divide and grow. The increase in length of both stems and roots is due to the division and elongation of these cells. Just outside the cylinder of wood is a layer of meristematic cells, known as the "cambium," which is responsible for the increase in the circumference of the stem or root. Another cambium, sometimes known as the "cork cambium," arises from parenchymatous cells within the cortex and gives rise to the characteristic bark formation. Meristematic cells are usually cubical in form and are characterized by having dense cytoplasm, small or no vacuoles and thin walls.

Parenchyma Tissues.—The parenchymatous tissues of the plant are composed of cells that are closely related to meristematic cells. Parenchymatous differ from meristematic cells in that they generally have thicker walls, less dense cytoplasm and an apparently smaller nucleus. Parenchyma cells occur scattered throughout various plant organs, and they vary in shape, size and arrangement according to the tissue of which they are part. Parenchyma occurs in the pith, cortex and rays of the stem; in the cortical and pith areas of the root; in the tender edible parts of fruits and vegetables and in the leaf. Parenchyma in the leaf containing chlorophyll is referred to as "chlorenchyma."

The parenchyma tissues function in various ways. Being a potential meristem, its cells may divide to form other tissues. Roots, and often adventitious shoots, arise from a parenchyma tissue known as "pericycle." Chlorenchyma in the leaf is associated with the manufacture of carbohydrates. Parenchyma in the pith and cortex functions in the storage of food. The large proportion of parenchymatous tissue in the tubers of Irish potatoes is one of the principal characteristics that renders this plant of great economic value as a food plant.

Mechanical Tissues.—Strengthening tissues of a plant are composed of both live and dead cells; and although the cell types vary, they generally are elongated and have much thickened walls. The principal strengthening tissues are collenchyma, fibers and stone cells.

Collenchyma tissue is composed of cells that have many characteristics of parenchyma, the principal difference between the two cell types being thickening of cell walls chiefly in the angles of the collenchyma. Collenchyma cells retain their protoplasts at maturity and are capable of division. The function of collenchyma is to afford mechanical support, especially evident in the petioles of herbaceous vegetables, such as celery and rhubarb.

Fibers after reaching maturity are composed of elongated dead cells with much thickened walls. They occur in various parts of the plant and are named according to their locations. In general, they are elastic and can be stretched without losing their power to return to their original length. Stone cells are similar to fibers but are not elongated.

Conductive Tissues.—The longitudinal conductive tissues of the plant are of two kinds, xylem and phloem, which occurring together form a vascular bundle. Both xylem and phloem are complex tissues.

The xylem tissue may include tracheids, vessels, fibers and parenchyma. The chief function of the xylem is the conduction of water and nutrients from the roots upward in the plant. This conduction takes place in tissues (tracheids and vessels) composed of dead cells. These conductive cells of the xylem were originally alive; but as they matured, the walls became impregnated with lignin, and the protoplasm disintegrated. These cells form tubes or vessels in the plant body and extend from the roots through the stems and into the leaves.

The phloem tissue may contain sieve tubes, companion cells or parenchyma cells. The main conductive tissue is the sieve tube, and the principal function of the sieve tube is the conduction of synthesized food. The sieve tube cells are elongated and pipelike but do not have the thick walls exhibited by the conductive elements of the xylem. The end walls between successive cells are not completely dissolved away but contain small openings through which food passes. The sieve tubes also contain a small amount of protoplasm. Apparently the upward movement of water and soil nutrients through the conductive element of the xylem is more rapid than the downward movement of food through the phloem. The phloem, like the xylem, extends to all parts of the plant; and wherever in root, stem or leaf there is a bundle of xylem cells, next to it is a bundle of phloem cells.

Protective Tissues.—The cells of the protective tissues usually have thickened outer walls. The principal protective tissues are known as "epidermis," "cortex" and "bark." The most common of these is the epidermis. It may consist of one or more layers of cells. Most plants first form an epidermis for protection. Later this is replaced in the older portions of roots and stems by cork tissue, which is usually thicker than the epidermis. On young tissues the epidermal cells excrete a waxy substance, known as "cuticle," which forms a continuous protective covering over all exposed parts of the plant.

THE PLANT STEM

The stem of the plant is chiefly above ground and is one of the two principal parts of the plant. The primary functions of the stem are the support of leaves and the conduction of food, water and mineral nutrients to and from the leaves. Secondary functions are food manufacture and storage.

Attention was called earlier in this chapter to the fact that stems vary in appearance. Those of horticultural plants that occur underground are referred to as "modified stems." Their tissues are similar to those of aerial stems but have become modified to meet certain functions. Underground stems of horticultural plants modified for food storage include iris, tulip, asparagus, Irish potato, dahlia, peony and banana.

Dicotyledonous Stems

When based on structure there are generally two types of dicotyledonous stem: the woody and the herbaceous.

The Woody Stem.—Woody stems exist in most trees and shrubs. An examination of the cross section of the trunk of an apple tree will give one a concept of a woody stem. The outer layer of a one-year stem is called the "epidermis," and this is replaced by corky bark as the stem grows older. This outer tissue serves as a protective coat against evaporation and the entrance of destructive organisms. Under the outer bark is the cortex, which is composed of parenchyma cells which serve as a place for food storage. Lying next to the cortex is the phloem which conducts the food upward and downward. Beneath the phloem is the cambium. The true cambium is only one cell in thickness, and it is responsible for the increase in the circumference of the stem. New phloem cells are formed on the outside of the cambium, and new xylem cells on the inside. Xylem cells are formed more frequently than the phloem cells. The cambium varies in activity at different seasons, being more active during the period of growth in spring and early summer when more moisture is usually available Most types of ringing and budding are done when the cambium is active and the bark can be easily separated from the wood. On the inside of the cambium is the xylem, which conducts water and nutrients upward. The portion of the stem on the inside of the cambium is commonly called wood and is often divided into sapwood and heartwood. The sapwood, in which are located the xylem vessels, is next to the cambium. The heartwood, which is composed chiefly of non-functioning xylem, is older and is generally distinguished by its darker color, which is due to the deposition of resins and gums. The heartwood is composed of dead cells and functions only in giving strength.

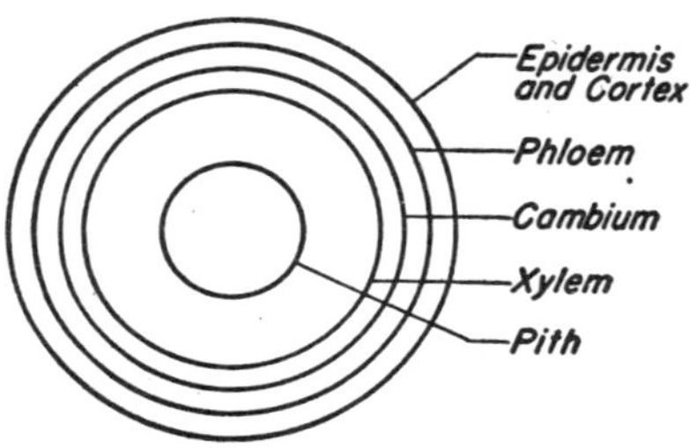

FIG. 67.—Diagrammatic sketch of the cross section of a woody stem.

Ray cells, which originate from the cambium and radiate much as do the spokes of a wheel from a hub, serve as passageways for the horizontal transfer of food and water in the stem. They are also important in food storage. These parenchyma-ray cells extending into the xylem are termed "xylem rays," and those extending into the phloem are known as "phloem rays."

The Herbaceous Stem.—The aboveground part of the stems of most of the annual and perennial vegetables and flowers are herbaceous and die after the growing season. These stems do not develop the thick-walled wood cells so common in woody plants. In the strictly herbaceous stem the tissues are similar to those in the very young stems of woody plants. The cortex makes a wider zone; in general, the vascular bundles are smaller; a concentric ring of cambium is lacking; and the pith area is wider. The older stems of herbaceous plants approach the structure of woody stems. The difference between the two types of stems is, in the main, a difference in quantity and degree of development of various tissues rather than in kind of tissue.

Monocotyledonous Stems

Horticultural plants like the banana, date palm, sweet corn, asparagus and lily belong to the group of flowering plants known as "monocotyledons." The characteristic stem of this type of plant has no vascular cylinder and no cambium. The vascular bundles consisting of phloem and xylem are scattered throughout the cross section of the stem. The diameter of such stems shows little increase after the early stages of growth. Growth in thickness is due mainly to enlargement of cells derived from the apical meristems.

Classes of Stems in a Woody Tree

An examination of a deciduous tree, such as an apple tree, after the leaves have fallen, will reveal a number of different parts above the ground. If one begins at the ground line, one will notice first a part called the "trunk," which is the oldest section of the tree. Arising from this trunk are sections similar to the trunk but younger, called "primary branches," because they are the first to grow from the trunk. These primary branches in turn rebranch, and these branches are generally referred to as "secondary branches." The secondary branches, in turn, have branches that are called "tertiary branches"; and so on until the divisions of branches reaches a great number in old trees. The growth made during the past season at the very end of a branch is referred to as a "twig." Twigs that occur at unexpected places, on wood older than two years of age, are termed "watersprouts." The term "shoot" refers to the twig while it still possesses leaves. Closer examination of the branches will reveal short growths scattered here and there. These slow-growing short growths are called "spurs." They are particularly characteristic of apple and pear trees. A spur is generally defined as a stem that made 2 in. or less growth during any season. Its total length may be 20 in., but it might have taken ten years or longer to grow to that length. A shoot becomes a twig after the season's growth and is known as one-year-old wood. The next year the same twig is referred to as two-year-old wood, the next year as three-year-old wood and so on until it eventually becomes difficult to tell its age. All the aforementioned parts are referred to as stems; thus it is evident that the tree contains a number of kinds of different stems, namely, the trunk; primary branches, secondary branches, etc.; shoots; twigs; watersprouts; and spurs.

An encircling zone of scars on the young twig ordinarily marks the end of one season's growth and the beginning of the next. These scars are rather long and narrow and extend at right angles to the axis of the stem. As the bud starts growth in the spring, the scales surrounding the bud are pushed off, leaving a narrow zone of scars.

Modified Stems

Buds, leaves and flowers are special forms of stems.

Buds.—A bud is an undeveloped shoot. It contains a growing point of meristematic tissue, surrounded by embryonic leaves or blossoms. These, in turn, are encased by a protective envelope composed of bud scales.

When terminal growth stops for the season, all the common woody plants such as the apple, peach and cherry form a terminal bud. Buds are also formed on the sides of shoots and are known as "lateral buds." Two or more buds are often produced collaterally on shoots of plants such as the peach.

Some buds known as "shoot buds" produce only vegetative growths. Others producing only flowers are called "fruit," or "flower," buds. Some contain both shoots and flowers, and these are called "mixed" buds." The peach-flower bud ordinarily produces only one flower, and there is no provision for shoot growth from that bud. A shoot must be produced from a separate and distinct shoot bud for further vegetative growth. The apple-flower bud differs from the peach in that it contains both shoots and flowers. The vegetative growing point allows for further terminal growth even though a fruit is produced.

All the buds on a plant do not show the same degree of activity during the growing season. Some of them that were formed during the previous growing season remain quiescent, showing no visible signs of growth for one or more seasons. These buds are known as "latent buds" and act somewhat as reserve buds, because they may grow years later when some stimulus occurs and often produce what is known as "watersprouts." Some buds never function, because they die.

Buds on plants in the temperate zone go through two kinds of comparative inactivity annually. Dormant ones show no visible activity during that part of the year when the environmental conditions are unfavorable for growth. Resting buds, on the other hand, show no visible signs of activity even though the environmental conditions are favorable. For example, the buds formed in the axils of the leaves on new shoots during the growing season normally do not grow until the following season, even though the environmental conditions are favorable for growth. Such buds are in a state of rest. If a twig from this same tree is brought into a warm greenhouse in

November and placed in water, the buds will still fail to grow. In both cases conditions for vegetative growth were satisfactory, but the buds did not grow because they were in a state of rest.

Leaves.—Every horticulturist knows that a large healthy leaf surface is needed for the satisfactory development and maintenance of a good plant.

The leaf is a modified stem, and its primary function is the manufacture of food. Although many variations exist in form, the essential tissues are much the same. The epidermis is a protective layer usually one cell in thickness covered with a waxy, waterproofing substance known as the "cuticle." In the interior of the leaf blade are a large number of cells, many of which contain small green bodies known as

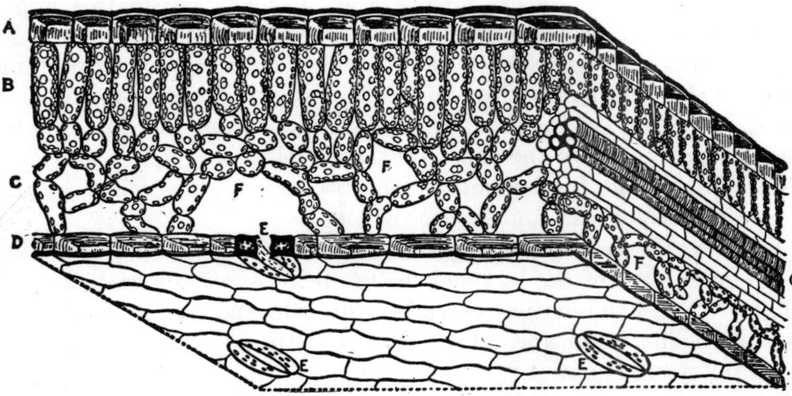

FIG. 68.—Diagram of a portion of the leaf. *A*, upper epidermis covered by cuticle in black; *B*, palisade layer; *C*, spongy parenchyma; *D*, lower epidermis; *E*, stomates; *F*, air space; *G*, vein. (*Gardner Bradford and Hooker.*)

"chloroplasts," essential for the manufacture of carbohydrates. The palisade cells are comparatively large, elongated and compactly arranged with their long axes at right angles to the epidermis. This layer, which may be one or more layers of cells in thickness, is directly below the epidermis but is often lacking on the underside of the leaf. The loosely arranged cells located between layers of palisade cells or the upper palisade and the lower epidermis are known as the "spongy parenchyma." The epidermis is pierced by an immense number of small openings called "stomates." Stomates may occur in both the upper and the lower epidermis, but in most leaves they are much more abundant in the lower side of the leaf. Most fruit plants have stomates only on the under surface, whereas many vegetable plants have them on both surfaces but with the greater number on the undersurface. Each stomate is surrounded by two guard cells, and the activity of these guard cells varies in horticultural plants. Guard cells of the leaves of plants like the apple, peach, pear, cantaloupe, turnip and radish close the stomatal opening before the plant wilts; but guard cells of the leaves of Irish potatoes, cabbages and onions do not close it until after the plant begins to wilt. Each stomate opens into an air chamber. Carbon dioxide enters the leaf through the stomates by diffusion. It enters as a gas and goes into solution in the moisture on the cell walls and diffuses through the walls into the cytoplasm of the cell to one of the chloroplasts where, in the presence of light, it will be synthesized into carbohydrates. The apple leaf has approximately 400,000 stomates, all located on the undersurface. If each of these stomates were 1 in. in diameter, the 1 sq. ft. of leaf area assumed to be necessary for the production of a satisfactory apple would be increased to 12 acres.

The veins and midribs of the leaf are composed of conducting tissues, which supply the leaf with water and mineral elements and permit the translocation of synthesized food from the leaf blade. The leaf is attached to the shoot by the petiole.

Two important fundamental plant processes taking place in the leaf are photosynthesis and transpiration.

Photosynthesis refers to the manufacture of simple carbohydrates by the green part of the leaf and really means synthesis with the aid of light. The combination of carbon dioxide and water involves a series of chemical transformations, but the process may be illustrated by the following equation:

$$6\,CO_2 + 6\,H_2O + \text{energy} \rightarrow C_6H_{12}O_6 + 6\,O_2$$

The main product of photosynthesis is a sugar known as "glucose," and the by-product is a gas known as "oxygen."

Pruning, spraying, irrigation, addition of fertilizers and thinning of flowers, fruits or plants are a few of the principal horticultural practices associated with the maintenance of efficient photosynthetic activity. Carbohydrate synthesis in apple leaves is four times as great on bright, clear, cool days than on heavy cloudy days and 10 to 20 per cent greater in the morning than in the afternoon. Dark green leaves are three times as efficient as light green leaves. Photosynthetic activities will continue in green leaves until they are frozen. If the synthesized foods do not accumulate in the leaf blade, a mature apple tree, under favorable environmental conditions, will manufacture about 1 lb. of carbohydrates per hour utilizing the carbon dioxide from 600,000 cu. ft. of air. In general, 1 sq. ft. of leaf area, about 45 leaves, is required for the satisfactory development of one apple.

The functions of transpiration are not fully known; and although the process is often considered a necessary evil, it seems to be essential in providing a continuous supply of water from the roots to the transpiring surfaces. This water is used as the medium of transportation within the plant. The transpiration from the leaf also appears to aid in keeping the stems and leaves cool.

Flowers.—The flowers of many horticultural crops are objects of aesthetic value, but the flowers of all fruit crops and of certain vegetables are absolutely necessary for the production of the edible product.

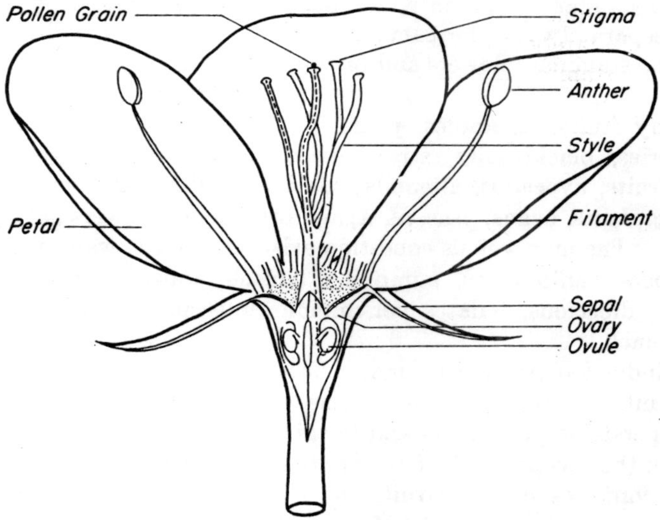

FIG. 69.—Diagram of an apple flower.

The flower is a modified stem containing the structures for sexual reproduction. The typical perfect flower of many horticultural plants has sepals, petals, stamens and pistils, which are thought to be modified leaves.

A knowledge of the parts of a flower is essential to an understanding of the processes of pollination and fertilization, both of which are necessary in the setting of fruits of the great majority of horticultural plants. Only a few abnormal varieties of fruits are seedless.

The sepals and petals are generally known as the accessory parts of the flowers. The sepals generally form the outermost cycle of the floral axis and often resemble the vegetative leaf in color and structure. Inside the sepals are the petals, which are usually colored and often possess nectary glands which attract insects.

The essential parts of the flower are the stamens, which produce the male element, and the pistil, which contains the female element. The pistil is composed of four parts. At the bottom is the ovary, in which are the ovules, and extending upward from the ovary is the style, at the top of which is the stigma. The stamen consists of two parts: the stalk and the anther. The anther, borne on the tip of the stalk, comprises one or more pollen sacs containing the male elements or pollen.

A perfect flower is one that contains both stamens and pistils. An imperfect flower is one that contains only stamens or pistils. A monoecious plant is one in which pistillate and staminate flowers are borne separately on the same plant. A dioecious condition exists when the staminate flowers and pistillate flowers are borne on separate plants.

Many fruits, as apples, pears, peaches, plums, labrusca grapes, raspberries, blackberries, strawberries (most varieties), gooseberries citrus fruits, avocados, almonds; some vegetables, as tomatoes, peppers, eggplants, beans, peas, okra; and ornamental flowers have perfect flowers. The monoecious condition exists in pecans, walnuts, filberts, cucumbers, cantaloupes, pumpkins, squash, watermelons and sweet corn; the dioecious, in dates, persimmons, muscadine grapes, asparagus and spinach.

As indicated previously, many horticultural plants are grown for their fruits or seeds. In most cases the development of fruit and seed depends on pollination and fertilization.

With the exception of a few varieties of fruits like the navel orange, certain varieties of grapefruit, the edible banana and Thompson's Seedless grape, the formation of seed is essential to the normal development of fruits. Pollination and fertilization are prerequisite to the formation of seed.

Pollination is the transfer of the pollen grain or male element from the anther to a stigma. The distance of the transfer may be very short when it occurs in the same flower or very long as when the pollen is transported from the anther of one plant to the pistil of another. The term "cross-pollination" is modified somewhat in the case of plants propagated asexually. In this case cross-pollination means that the pollen is transferred from the stamen of one horticultural variety to the pistil of a different horticultural variety. For example, if the pollen from one Jonathan apple tree is transferred to the pistil of another Jonathan apple tree, even though the trees are 10 miles apart, the process is referred to as self-pollination. On the other hand, if the pollen of one Marglobe tomato plant is transferred to the stigma of another Marglobe tomato plant 10 ft. away, it would be referred to as cross-pollination.

The transfer of pollen is generally accomplished by the agencies of wind (including gravity), insects and water. Most fruit and many vegetable plants are pollinated by insects, whereas nut fruits like pecans, filberts and walnuts are pollinated largely by wind.

The second step in the process of seed formation is known as fertilization. This process involves two steps: the growth of the pollen tube down the style of the pistil; and the uniting of the male element, or sperm, with the female element, or ovule, resulting in the first cell of a new plant.

Compatibility of the male element with the female element, resulting in the production of viable seed, varies in varieties of horticultural plants, and several terms are used to indicate the degree of fertility in horticultural plants propagated asexually. Self-fertile varieties of horticultural plants are those which set viable seed with pollen from the same variety. Self-sterile varieties are those which do not set viable seed with pollen from the same variety. Intersterile varieties are those which do not set viable seed with the pollen of one another.

Obtaining viable seed is primarily the problem of plant breeders and seed growers. Fruit producers are interested chiefly in obtaining a set of fruit. The presence of fruit is not a sure indication of the presence of viable seed. Fruitfulness is not synonomous with viable seed. As in the case of fertility, special terms are used to designate the various degrees of fruitfulness in asexually reproduced plants. Self-fruitful varieties are those which set fruit without fertilization or when fertilized by pollen of the same variety. Self-unfruitful varieties are those which do not set fruit when self-pollinated. Inter-unfruitful varieties are those which do not set fruit with their own pollen or with the pollen of certain other varieties of the same kind of plant.

A brief summary relative to the degree of fruitfulness that exists among varieties of the various kinds of asexually propagated horticultural fruits follows:

Apples.—Practically all commercial varieties of apple are benefited by cross-fertilization. They may be placed conveniently in the following groups which include but a few of the many varieties:

Self-unfruitful:

Arkansas.	Fameuse.	Stayman.
Delicious.	McIntosh.	Winesap.
Golden Delicious.	Northern Spy.	

Partly self-fruitful:

Baldwin.	Rome.	Northwestern Greening.
Ben Davis.	Wagener.	Willow Twig.
Duchess.	Wealthy.	Yellow Transparent.
Early Harvest.	Jonathan.	Yellow Newtown.
Spitzenburg.	Maiden Blush.	York.
Grimes.		

Self-unfruitful and cross-unfruitful:

Winesap.	Arkansas.	Stayman.

Pears.—The following are a few of the important varieties of pears that are apparently self-unfruitful: Anjou, Clairgeau, Clapp, Le Conte, Garber, Howell, Kieffer, Winter Nelis. A few of the many varieties that seem to set a fair crop without cross-fertilization are Bartlett, Bosc, Comice, Hardy and Seckel.

Peaches and Apricots.—Most varieties of peaches and apricots are self-fruitful. The most notable exception in the case of the peaches is the J. H. Hale variety, which produces defective pollen.

Plums and Prunes.—There are numbers of species and varieties of hybrid plums grown in the United States in which unfruitfulness exists in varying degrees. It is more simple to consider the varieties according to species type rather than individually.

Most of the varieties derived from the native American species like Wild goose, Miner, De Soto and others are self-unfruitful but will set fruit when cross-fertilized with one another. The varieties of Japanese plums (*Prunus salicina*) and their hybrids are nearly all self-unfruitful. A few of the important varieties of this group include Abundance, Burbank, Formosa, Kelsey and Wickson, and they will set fruit when planted together.

There are two groups of European plums: *P. domestica*, which is largely self-fruitful; and *P. institia*, both of which are largely self-fruitful and cross-fruitful.

A number of hardy varieties of plum have been originated by crossing American with Japanese species. These hybrid varieties are often referred to as Hansen's hybrids and Minnesota hybrids. They are both self-unfruitful and cross-unfruitful. The recommended practice is to interplant varieties of native American species, like the Miner and the Wolf, with the hybrid varieties, like Kahinta, Waneta and Winona.

Cherries.—All popular varieties of sour cherries (*Prunus cerasus*) are self-fruitful, but fruitfulness is generally improved by planting two varieties together such as Early Richmond and Montmorency.

Sweet cherries (*P. avium*) consist of many varieties that are self-unfruitful. Three of the leading varieties grown on the Pacific coast (Bing, Lambert and Napoleon) are both self-unfruitful and inter-unfruitful. The hybrids (Duke cherries) which are hybrids between sour and sweet cherries are all self-unfruitful. Varieties of sweet cherries are apparently good pollenizers for the Duke varieties.

Grapes.—Most of the commercial varieties of grapes grown in the United States are self-fruitful. A few varieties like Brighton, Barry, Lindley and Vergennes are apparently self-unfruitful and inter-unfruitful.

The muscadine grape (*Vitis rotundifolia*) grown widely in the South and the native wild grape (*Vitis vulpina*) must be provided with pollen-bearing or male vines in order to get fruit.

Raspberries, Blackberries and Dewberries.—Almost all the varieties of raspberries and blackberries are apparently self-fruitful. The Lucretia dewberry is self-fruitful, but other varieties exhibit at least partial self-unfruitfulness.

Strawberries.—Most of the important varieties of strawberries will set fruit without cross-fertilization. A few, like Gandy, produce insufficient pollen, and a few produce imperfect flowers; for a satisfactory crop these must be planted with suitable varieties.

Nut Trees.—Walnuts are generally self-fruitful, almonds mostly self-unfruitful and filberts largely self-unfruitful.

Pecans are generally divided into two groups. Varieties like Mobile and Success shed pollen when their pistils are receptive and consequently are self-fertilized. Those like Delmas, Schley and Stuart mature their pistils before the pollen is ready and must be cross-fertilized in order to set fruit.

THE ROOT

The root of the plant is chiefly below ground; it is the other principal part of the plant. Its primary functions are mechanical support for the stem, absorption and conduction of water and mineral nutrients and storage of food.

There are two well-defined types of root system in horticultural plants; the taproot and the fibrous. If the primary root tends to go straight down and develops a conspicuous structure from which laterals arise, as we have in plants like the oak and radish, it is known as a taproot system. When the primary root and the laterals from it develop more or less equally, it is known as a fibrous root system. Roots of plants like the beet and radish develop fleshy taproots, whereas any part of the sweet-potato root may become fleshy. These fleshy roots in certain vegetables constitute the part of the plant that is of economic importance to the horticulturist.

The structure of the root is similar to that of the stem. One encounters at the very tips of roots, the root cap, which is for the protec-

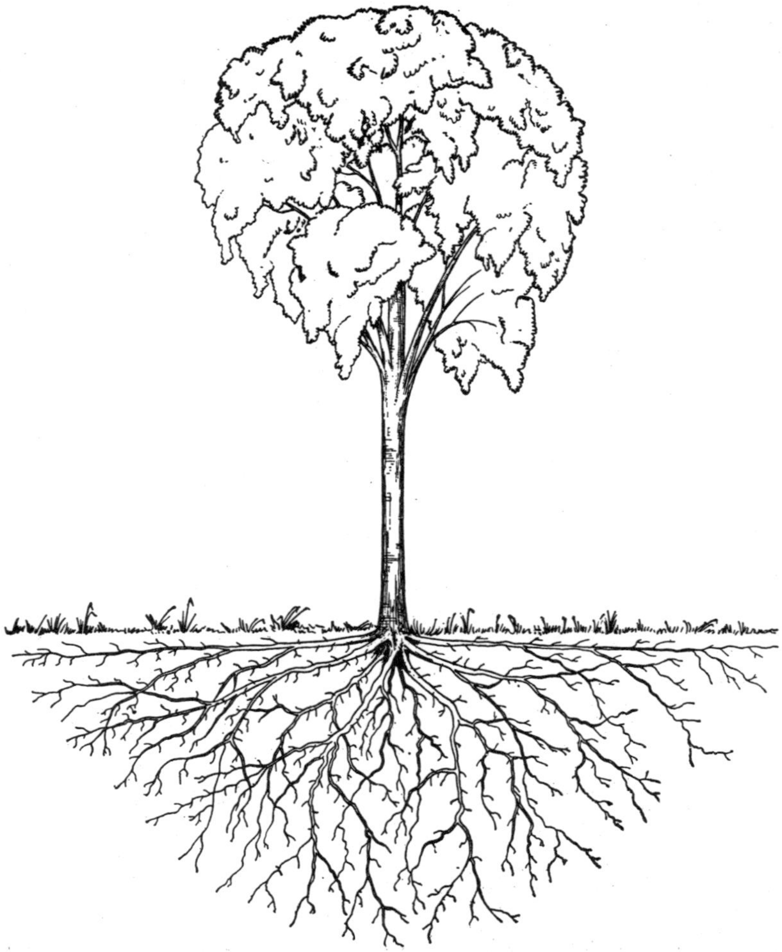

FIG. 70.—Diagrammatic sketch showing the distribution of the roots of a tree in relation to the top.

tion of the zone of actively dividing cells which is just back of it. Behind the zone of cell division one encounters the zone of cell enlargement, and back of this the zone of cell maturation. Most of the absorptive power of the root occurs a short distance back of the tip, a region in which epidermal cells push out root hairs. The absorptive power of the root is increased many fold by the presence of the root hairs. The life of a root hair is generally very short; in many cases it remains alive for only a few days. New root hairs are being formed

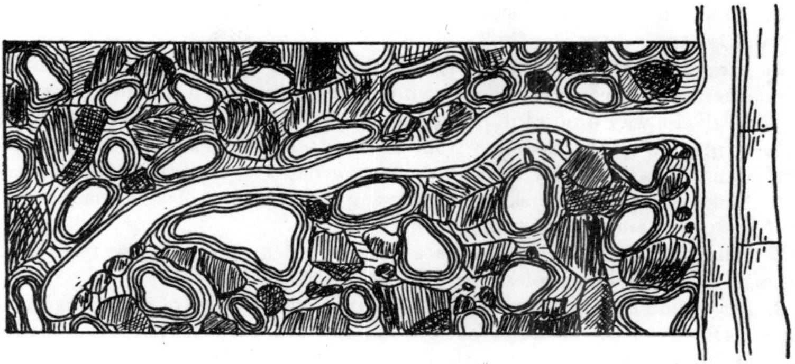

FIG. 71.—Diagram to illustrate a root hair in the soil. (*After Sachs.*)

continuously as the root elongates, however; and as the root pushes through the soil, the absorptive zone is maintained.

Review Questions

1. What is the structural unit of all horticultural plants?
2. What are the principal parts of a cell?
3. What is protoplasm?
4. Are the walls surrounding all plant cells alike?
5. What constituents of the nucleus are considered as carriers of hereditary characteristics?
6. Is there anything in the vacuoles of a cell?
7. How do cells grow?
8. What is a plant tissue?
9. What are five principal plant tissues?
10. What is the distinguishing characteristic of meristematic tissue?
11. What are the two chief types of stems of horticultural plants?
12. What are the classes of stems in an apple tree?
13. What is a bud?
14. Are there buds on stems more than one year old?
15. What are the chief functions of the leaf?
16. Where and of what materials are carbohydrates made?
17. What are the essential parts of the flower?
18. What is the essential difference between monoecious and dioecious plants?
19. What is meant by pollination?
20. What is meant by fertilization?
21. What is meant by self-fertile?
22. What is meant by self-fruitful?
23. What are the chief functions of a root?
24. Where are root hairs located?
25. What is the chief function of root hairs?

Problems

1. Make a diagrammatic sketch showing the principal constituents of a cell of an apple shoot.

2. Explain how a tomato fruit increases in size.

3. Make a diagrammatic sketch showing the principal tissues in the transverse section of an apple twig.

4. Mr. A failed to place a tree guard around the trunk of a vigorous six-year-old Jonathan apple tree. During the winter the rabbits ate a ring of "bark" about 2 in. wide all the way around it. Explain to Mr. A what may happen to the tree.

5. Explain why the distance from the surface of the soil to the center of a lateral branch on an elm tree does not become greater as the tree grows larger.

6. Make diagrammatic sketches showing the contents of an expanded flower bud of the peach and of the apple.

7. What leaf area and approximate number of leaves are necessary to produce 6 bu. of No. 1 Delicious apples with 125 apples per bushel?

8. Mr. A had two varieties of tomatoes and two varieties of onions in his home garden. He saved the seeds of both crops and much to his surprise discovered that the tomatoes came true to variety but the onions did not. Explain.

9. Mr. A is planning to plant a 40-acre apple orchard of Jonathan, Delicious and Winesap. Make a diagrammatic sketch showing a satisfactory arrangement of these varieties for best results.

10. A florist intends to grow a group of French marigold plants. He plans to grow one-half of them by broadcasting the seed in flats and, when the seedlings are showing their second leaf, to prick them out and set 1½ in. apart in flats. Later they will be transferred to 2½-in. and finally to 4-in. pots. The other half will be planted directly in 4-in. pots and grown without transplanting until ready for sale. State and explain which group of plants will be ready for sale first.

Suggested Collateral Readings

1. Coulter, M. C., "The Story of the Plant Kingdom," pp. 134–181, The University of Chicago Press, Chicago, 1935.

2. Gager, C. S., "General Botany," pp. 14–52, P. Blakiston's Son & Company, Philadelphia, 1926.

3. Holman, R. M., and W. W. Robbins, "A Textbook of General Botany," pp. 1–308, John Wiley & Sons, Inc., New York, 1939.

4. Talbert, T. J., and A. E. Murneek, "Fruit Crops," pp. 109–120, Lea & Febiger, Philadelphia, 1939.

5. Thomas, M., "Plant Physiology," pp. 1–23, 162–249, P. Blakiston's Son & Company, Philadelphia, 1935.

CHAPTER VI

GROWTH OF HORTICULTURAL PLANTS

Growth means the increase in size of the plant as a whole or one or more of its parts. As explained in Chap. V, it involves cell division, cell enlargement and cell maturation. The horticulturist is interested in the time during which the growth processes take place, because the time and manner of performing various horticultural practices are influenced by the period and stage of plant growth.

DEGREES OF PLANT ACTIVITY

Horticultural plants have a period known as the growing period or the growing season when they show visible activity of growth and a period known as the non-growing, or dormant, period when the plants show no visible activity of growth.

The Growing Period

The most obvious thing that a plant does is to grow or increase in size. The most noticeable period of growth is during the unfolding of buds into shoots, leaves and flowers. This activity generally begins, in temperate regions, in the spring when environmental conditions are favorable. Growth is slow for a short period when it is just starting, becoming rapid in the spring and early summer when conditions are most favorable and gradually decreasing in rate in the late summer and fall until finally a period arrives when apparently all growth has ceased.

The Non-growing Period

During the non-growing, or dormant, period there is no visible growth activity by the plant. There are two separate and distinct sets of conditions that operate for various periods of time and during which plants or plant parts do not grow. One period is known as the dormant period, and the other as the rest period. The two periods may—and in fact always do—overlap in time.

That period when plants show no active growth because of external or environmental conditions is referred to as the dormant period. In temperate regions this begins in the fall when the temperature becomes too low for plant growth and ends in the spring when moisture and temperature conditions are favorable for growth.

The rest period is that during which the plant or plant part shows no visible growth activity even though placed in environmental conditions favorable for active growth. For example, bulbs, tubers, buds and other growing parts of woody plants do not grow during certain

FIG. 72.—A blueberry plant; in the branch at the right the rest period was broken by exposure to cold, that at the left is still in resting period. For several weeks preceding the taking of the picture the entire plant had been exposed to growing-season temperatures. (*After Coville, from Gardner, Bradford and Hooker.*)

periods of the year. The length of the non-growing, or rest, period varies with the kind of plant, its part and its environment. If one cuts branches from apple trees in November and places them in water in a temperature of 70°F., they wili fail to grow; but if one repeats the process in February, growth will start, thereby indicating that the rest period has been completed. The plant enters the rest period gradually, and one part may be in a rest period while others of the same plant will be showing active growth. For example, axillary buds formed in May on the shoots of an apple tree usually show no activity until the following spring.

The dormant period may be compared with the rest period as follows:

Dormant period:
1. Due to external conditions.
2. Entire plant enters at the same time.
3. Begins in fall in temperate regions
4. Ends in spring in temperate regions.
5. Duration lengthened by cold temperatures.
6. All parts of a plant affected.

Rest period:
1. Due to internal conditions.
2. Plant enters gradually.
3. May begin at any time of year.
4. Ends in winter in woody plants of temperate regions.
5. Duration shortened by cold temperatures.
6. Some parts of plant not affected.

Although no external activity may be apparent when the plant is dormant, other growth processes which are internal and imperceptible are taking place. For example, the initiation of blossoms begins during the season preceding the actual appearance of the flowers, but the continued development of the floral parts may continue for the greater part of the winter. The cell walls continue to thicken after the plant has become dormant.

Apparently some plants do not have a resting period, and its length in others varies with different species and with different varieties of the same species. Apparently all parts of the plant have a resting period except the roots. The seeds of many plants have a resting period, referred to as the afterripening period, and the length of this period varies with different seeds and with the environment of the seeds. Afterripening in seeds will be discussed in more detail in Chap. XI.

The cause or causes of the resting period have never been fully determined. Apparently the resting period is associated with physical and chemical conditions within the cells of the plant, and the substances necessary for growth become unavailable.

Although the duration of the rest period varies according to the kind of plant, apparently the length of the period is fairly definite for any one variety or species of plant. Observations of woody plants indicate that if they enter the rest period late in the fall, they come out of it later than if they had entered it earlier. Conversely, if they enter the rest period early, they come out of it early.

The rest period is of considerable practical significance in certain horticultural practices. Of particular significance are some of the problems involved in breaking it.

The rest period may be shortened by natural conditions such as heat, cold or drought. It may also be broken by artificial means, as mechanical wounding, keeping at cold temperatures and the use of certain chemicals. A few examples of horticultural practices associated with breaking the rest period will help to show the significance of the condition to horticultural plants.

Southern peach growers often encourage late fall growth in an effort to postpone the beginning of the rest period, extend it longer and thus delay the trees' coming into flower. This reduces the danger of injury by spring frosts. Peach growers in sections of California have difficulty with their trees following a warm winter. Cool temperatures hasten the rate of the activities that must take place during the rest period before the plant can resume active growth. If unfavorable warm weather prevails during the rest period, the rate of the after-ripening process is slowed up considerably. This is especially true of the shoot buds. As a result of such conditions, the peach trees will be coming into bloom for a period of a month or more, and the shoot buds will still be in their rest period. The growers are attempting to solve this problem by developing a variety of peach that will pass through its rest period at a higher temperature. The Babcock variety shows some promise of being satisfactory in this respect.

The Triumph variety of Irish potato is often planted as early as September and October in southern Florida, Bermuda and the West Indies, and the seed tubers are often obtained from western Nebraska. In order to force these potatoes to grow soon after they are planted, it is necessary to break the rest period with a chemical or with heat treatment.

Often it is desirable to force certain seeds, bulbs, tubers and woody plants into growth before their normal time. Various treatments for breaking the rest period are used. One successful method consists of bringing woody and herbaceous perennials into the greenhouse after they have been kept cool (35°F.) or allowed to freeze, usually in early December; placing them in a tight box and treating them with ethylene chlorohydrin gas. Some of the woody plants, as lilacs, some types of spiraea and azalea and some of the herbaceous plants, as some types of delphinium, gaillardia and chrysanthemum, may be hastened into early flowering by this method.

FACTORS INFLUENCING PLANT GROWTH

Both favorable environment and available food are necessary for plant growth.

Environment

The principal environmental factors affecting plant growth are temperature, moisture, soil and light. The plant's growth is influenced by any one of these, and any one may become a limiting factor to growth. The growth of the plant will be influenced if the temperature is too high or too low, if the moisture is too great or too little, if the soil is too heavy or too light or too fertile or too infertile and if the light is too intense or too diffused for the optimum requirements of the particular plant or part of plant. If any one of these factors becomes too pronounced, as the absence of moisture, it becomes a limiting factor. The retarding effect of the environment on the growth of a plant may be compared with the retarding effect of certain conditions on the speed of an automobile. The speed of the car is influenced by the construction and mechanical efficiency of the car, by the skill of the driver, by the type of the highway, by the amount and kind of traffic and by the regulatory laws. Any one of these, for example, the absence of gasoline or an impassable road, may become the limiting rather than an influencing factor, but growth of the plant and movement of the car will again proceed when the limiting factor is remedied. Optimum growth of the plant and speed of the car will occur only when all influencing factors are mutually favorable.

Food Supply

Many scientists compare a living plant to a man-made machine. A steam engine converts the energy that was in coal into the energy of work. Similarly, a living plant converts the food that it makes from raw materials into energy which it utilizes in performing its functions. In order for plants to obtain the energy stored in the synthesized food, a process known as "respiration" takes place which liberates the energy stored during the photosynthetic process. This process occurs in all living cells at all times oxygen is used, food is broken down, carbon dioxide is given off and energy is released. The simple chemical reaction may be represented by the following equation:

$$C_6H_{12}O_6 + 6\ O_2 \rightarrow \text{energy} + 6\ CO_2 + 6\ H_2O$$

The student will notice that this is the reverse of the equation for photosynthesis.

Foods in general may be placed in one of three categories: carbohydrates, fats and proteins. Although fats play some part in the growth of horticultural plants, the horticulturist is particularly concerned in the relationship between the carbohydrates and proteins.

Carbohydrates.—It was learned earlier that one of the fundamental plant processes was making carbohydrates in those cells containing chlorophyll. The leaves of plants are favorably designed to perform this function. Carbohydrates are exceedingly variable, ranging from simple glucose to complex cellulose and starches. Glucose is the first simple sugar made in the photosynthetic process. It may be used or stored in place, transported to other parts of the plant where it is used or stored but most likely transformed into more complex carbohydrates, such as cellulose, which is an important constituent of cell walls, and starch, the most compact form in which carbohydrates are stored.

Proteins.—Proteins are exceedingly variable and complex products synthesized by living plants. A protein is formed when certain carbohydrates, which contain carbon, hydrogen and oxygen, are combined in the living plant with nitrogen, usually sulfur, and often phosphorus. The mineral elements necessary for protein formation are obtained from the soil. Although the making of carbohydrates occurs only in cells having chlorophyll in the presence of light, it appears that proteins can be synthesized by any living cell at any time that the raw products are available and the conditions favorable.

PHASES OF PLANT GROWTH

In general, there are two phases of plant growth: the vegetative phase, which is associated with the development of stems and leaves, and the reproductive phase, which is associated with the development of flowers, fruits, seeds and fleshy roots and stems. When the vegetative phase is dominant, carbohydrates are utilized; conversely, when the reproductive phase is dominant, carbohydrates accumulate.

The two phases of growth exhibited by a plant occur in cycles. The cycles of carbohydrate utilization and carbohydrate accumulation are not the same for all plants or for the same plants under different conditions. Since these phases are influenced by environmental and nutritional conditions, the horticulturist has an opportunity to modify, to a limited extent, the type of growth made by plants by manipulating the environment. A plant may be in the vegetative phase for all of the season or for many seasons or in the vegetative phase for one part of a season and the reproductive phase during another part of the same season. The raspberry cane will grow an entire season and not flower until the next season; the young apple tree may grow for six or seven years without flowering; and the snap bean will make rapid vegetative growth for the first few weeks and then produce flowers and fruit. A single organ of a plant, as a flowering bud, first goes through a stage when the vegetative phase is dominant and then through a stage when the reproductive phase is dominant.

The carbohydrate-nitrogen hypothesis advances the idea that the amount and kind of growth made by a plant are influenced largely by the carbohydrates and proteins available for the use of the plant during various stages of its life cycle. On this hypothesis, which presupposes the availability of adequate water and mineral elements other than nitrogen, plants can be separated into four distinct groups or, rather, into four phases of growth. There are no fixed lines of demarcation separating these phases of growth, and one phase will blend imperceptibly into the adjacent phase or phases. Future investigations may prove, disprove or modify this hypothesis; but at present it serves, in its broad sense, as a suitable means of discussing certain phases of plant growth.

Group 1 is known as the highly vegetative, non-flowering group and may be represented by C/N+. Plants in this group have an adequate supply of carbohydrates and an abundant supply of nitrogen. The abundant nitrogen supply makes it possible for the plant to utilize the carbohydrates as they are synthesized and thus prevents their accumulation. (A college freshman with good health, a surplus supply of money and a good automobile will not store up much reserve.)

Group 2 is known as the moderately vegetative, flowering and fruiting group and may be represented by C/N. Plants in this group have an adequate supply of carbohydrates combined with an adequate amount of nitrogen; consequently, only moderate vegetative growth occurs, and carbohydrates accumulate in the vicinity of the developing buds and flower buds are initiated. (A college freshman with good health and sufficient funds for all necessary activities does good scholastic work.)

Group 3 is known as the flowering but non-fruiting group and may be represented by C/N−. Plants in this group have an adequate supply of carbohydrates but are deficient in nitrogen; consequently, only fair vegetative growth occurs, but flower buds are formed. Flowers and fruits either drop while small or do not develop fully because of the lack of sufficient nitrogen. (A college freshman with fair health, a small amount of money in reserve and a part-time job that was not continued after Christmas has scholastic difficulties.)

Group 4 is known as the weakly vegetative, non-flowering and non-fruiting group and may be represented by C−/N−. Plants in this group are deficient in both carbohydrates and nitrogen; consequently, the vegetative growth is weak; the leaves are few in number, small in size and yellowish green in color; and no flower buds are formed. (A college student with poor health, no money and unable to get temporary employment fails in scholastic achievement.)

During its complete life cycle every plant passes through these four stages. The young plant has a relatively large, well-exposed leaf area and an adequate, uncrowded root system in a fertile soil. Under such conditions there is an adequate supply of carbohydrates and an abundant supply of nitrogen which are expended in vegetative growth —the C/N+ phase. As the plant becomes older and larger, the number of growing points increases rapidly; the leaves begin to shade one another somewhat; and the root system fills its allotted space more fully and begins to meet competition from the roots of other plants. The leaf area supplies adequate carbohydrates, but the greatly increased number of growing points (buds) has resulted in a decrease in the relative amount of nitrogen available for each bud. The lessening in the amount of nitrogen checks the utilization of carbohydrates in vegetative growth, and the carbohydrates accumulate. This adequate supply of nitrogen combined with more carbohydrates than are needed for the activities of the plant will result in the initiation of flowers or the formation of reproductive organs—the C/N phase. As the perennial plant continues to grow, the number of growing points will continue to increase, competition among the leaves for water and light will become more pronounced, a decrease in the photosynthetic efficiency of the individual leaves will occur, the amount of carbohydrates for both the top and roots will be decreased and the nitrogen supply to each growing point will continue to decrease. For a time there will be sufficient carbohydrates to cause the formation of flowers, but the early growth of the plant the following spring, which is made chiefly from the stored food, will make such a heavy demand on the already low nitrogen reserve that the supply will soon be exhausted, the plant will be weakened and the flowers or young fruits will fall off—the C/N− phase. With increased age and size this condition becomes more pronounced. The decreased carbohydrate supply to the roots further retards the growth of a root system already unable to supply adequate nitrogen to the top. Thus the relative nitrogen supply is decreased further, and this results in a decrease in the size and number of leaves with an accompanying decrease in the quantity of carbohydrates synthesized. The amount of growth is very little; the foliage is scant in amount, small in size and yellowish green in color; and flower buds are not formed—the C−/N− phase. Unless this condition is remedied, the plant will die.

The object of the horticulturist is to get plants into the productive (C/N) stage as soon as desirable and to keep them there as long as it is financially profitable. He does this by various practices which influence the type of growth made by the plant. The type of growth is a response to the relationship existing between the carbohydrates and nitrogen within the plant. It is a response to the utilization or accumulation of reserves—chiefly carbohydrates.

When nitrogen is dominant, the plant is vegetative, and the carbohydrates are utilized in making more stems, leaves and roots. When the carbohydrates are dominant, the carbohydrates accumulate, with the resultant lessening in vegetative growth, maturation of cells, formation of flower buds and formation of storage organs usually associated with reproduction.

Cell Maturation

When woody plants begin to slow down in vegetative activity, carbohydrates begin to accumulate. One says that the plant is preparing for the low temperatures of winter. The accumulation of carbohydrates brings about the maturation of the live cells. The cell walls are thickened; and largely by the deposition of various carbohydrates, such as cellulose, the moisture in the cell protoplasm decreases, and various food products accumulate in the cells. If environmental conditions, as a warm moist fall, or cultural practices, as a late cultivation, stimulate late vegetative growth and delay the time at which this maturation of the tissues takes place, the plant is likely to be injured by cold during the dormant season.

Flower-bud Formation

Apparently in all plants that have mixed flower buds, each growing shoot point or bud is potentially a flower bud. During the early development of these buds they appear to be identical in structure, but gradually certain microscopically morphological changes take place in some of them. These changes are the formation of flowers. Chemical differences within the different buds must initiate and bring about these morphological changes which result in the formation of flower parts rather than the continuation of vegetative growth. The initiation of flowers in growing buds appears to be associated with the presence of an adequate supply of nitrogen and the presence of more carbohydrates than are necessary to carry on the vegetative activities of that particular bud or stem point.

Differentiation of flower parts apparently occurs in the life of an actively growing bud during certain stages in the growth of the bud, provided that the adequate supply of nitrogen and the reserve supply of carbohydrates are present during the time the bud is passing through this stage. Flower buds are formed at different times in different kinds of plants and at different times in various buds of the same plant.

At Ames, Iowa, the great majority of terminal buds on spurs of mature Jonathan apple trees that formed flowers did so during the period between June 15 and July 1, but flowers were formed on Dunlap strawberry plants between September 1 and 20. On the same Jonathan apple tree, however, there were some flowers that formed terminally on shoots 6 in. or more long as well as on the short spurs. Since the shoot continued to elongate until late July, the terminal bud was not formed, and flower formation could not take place until after appreciable terminal elongation of the shoot had ceased.

It is important to remember that flower formation is associated with an internal chemical composition and that it does not necessarily occur at any specific date. Carbohydrates will accumulate when the vegetative activity of a plant or any part of a plant is retarded, without a corresponding check in the photosynthetic activity. The retardation in vegetative growth may be due to inherent characteristics of the plant; to natural environmental conditions, as unfavorable temperature or moisture; or to some practice performed by the grower. The mechanical operation of ringing (cutting through the "bark" or removing a ring of bark from a stem) is sometimes resorted to by horticulturists to induce flowering or fruiting by particular plants or kinds of plants. Filler apple trees and excessively vigorous nonflowering individual trees are often ringed to promote bearing. Removing a narrow ring of bark from the trunk of the Corinth grape—the dried currant of commerce—induces the setting of the seedless

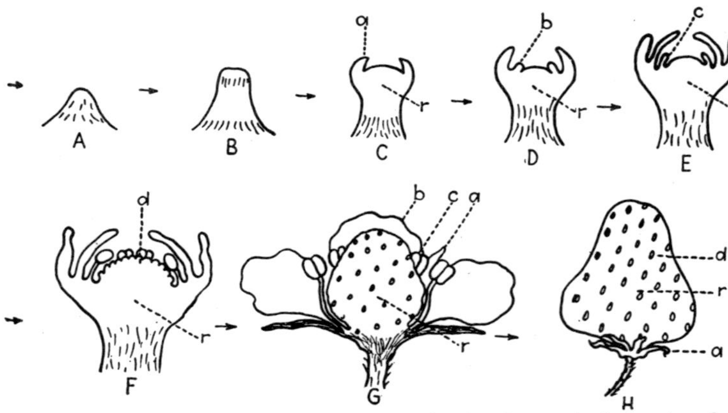

Fig. 73.—Diagrammatic sketches showing the development of the strawberry: *A*, rounded growing point; *B*, flattened growing point, first indication of a fruit bud formation; *C-a*, first indication of sepals; *D-b*, first indication of petals; *E-c*, first indication of stamens; *F-d*, first indication of pistils; *G*, fruit (*a*) sepals, (*b*) petals, (*c*) stamen, (*d*) pistil, (*r*) receptacle; *H*, (*a*) sepal (*d*) seed, (*r*) receptacle.

fruits which fail to set without such treatment. Ringing severs the tissues of the phloem and prevents the flow of the carbohydrates from the top to the roots until the wound has healed and new passages are formed. During this period carbohydrates accumulated in the portion of the plant above the ring and were instrumental in bringing about the changes in growth.

The carbohydrate-nitrogen relationship in plants may be represented by four sliding weights on the level beam of a balance. The highly vegetative, non-flowering phase (C/N+) is on the left; the moderately vegetative, flowering and fruiting or productive phase (C/N) near the center; the weakly vegetative, flowering but non-fruiting phase (C/N−) next and the weakly vegetative, non-flowering phase (C−/N−) on the right.

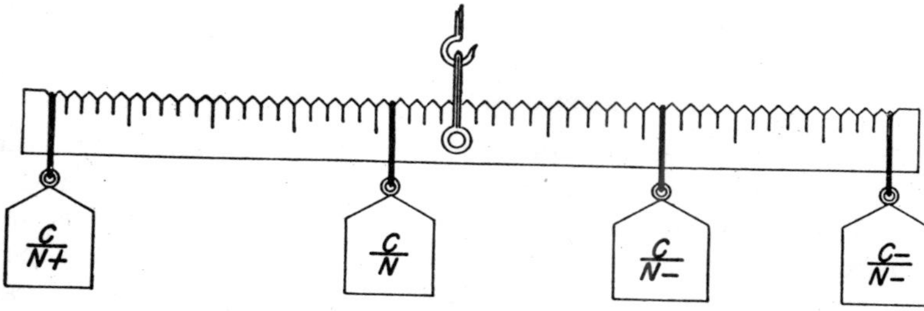

The horticulturist endeavors to keep the plant or plants on the rather wide C/N part of the beam after they have passed through their earlier vigorously vegetative C/N+ stage and before they pass into the C/N− and C−/N− stages in their normal progress towards final death and destruction.

Let us assume that a plant is in the C/N+ condition and the grower applies a quickly available nitrogenous fertilizer. Since he has added N where there is already an oversupply of nitrogen, the C/N+ weight is pushed farther to the left, and fruiting is delayed. If the fertilizer had been applied to the plants in the C/N− condition, the plus nitrogen would have canceled the minus nitrogen, as the C/N− weight would have been moved to the left, and the C/N, or moderately vegetative and productive stage, would have been attained. If the highly vegetative, non-flowering C/N+ plant had been root pruned, the nitrogen-absorbing area would have been decreased, the nitrogen absorbed by the plant would have been less or nitrogen would be subtracted from the C/N+ condition. The C/N+ weight would move to the right, and the plant would enter the C/N stage of growth.

A mature asparagus planting passes through the following cycle annually. In the spring there is a period of carbohydrate utilization when the new spears that are growing from the carbohydrates, stored the preceding year, are being removed. After the grower stops harvesting the spears and allows the stems and leaves to grow, there is a period of carbohydrate utilization as the new tops are forming, followed by a period of carbohydrate accumulation and storage in the roots when the new tops have developed to such a size that they manufacture more carbohydrates than are utilized by the activities of the plants. The successful asparagus grower will cease cutting spears sufficiently early to allow the plants ample time to develop an adequate leaf area to produce an abundant supply of reserve carbohydrates for next season's crop of spears.

The grower should know the growth of his plants sufficiently well to determine their position on the balance beam and adjust his operations to maintain the plants in their optimum state of growth and fruitfulness.

Development of Storage Organs

A plant utilizes carbohydrates in carrying on its various activities, and the carbohydrates that are not required for these activities accumulate in various parts of the plants or in specialized storage organs, as fruits, tubers and bulbs, or in fleshy leaves, stems and roots. Environmental factors often influence the development of these storage organs through their effects on growth responses. Maximum tuber formation of Irish potatoes takes place when they are growing at a temperature of 63°F. As temperature increases, tuberization decreases until a temperature of 84°F. is reached, when no tuberization occurs. At the higher temperature the carbohydrates are apparently used as they are synthesized for top growth; but at the lower temperatures the growth of the top is retarded, and the carbohydrates accumulate and are stored in the tubers. Asparagus is not grown successfully in Florida because the tops of the plants continue to grow and thus utilize the carbohydrates as they are synthesized. In more favorable locations, the top growth is retarded after a time by cooler temperatures, and the carbohydrates synthesized thereafter are largely available for storage in the roots.

Growing, Flowering and Fruiting Habits

The growing, flowering and fruiting habits of plants vary. In many kinds of plants the shoots grow only at the end, and although lateral buds are formed in the axes of the leaves, lateral shoots are not produced. In many other kinds of plants the shoots give rise to lateral shoots, and in some plants the lateral shoots may continue to produce side branches themselves. A one-year-old plant of one kind may be as elaborately branched as a four-year-old plant of another habit of growth. Not only do plants vary in their habits of growth, but they vary even more widely in their habits of flowering and fruiting, and even the same kind of plant will vary but to a lesser degree. Intelligent training and pruning of plants is based upon a knowledge of the growing, flowering and fruiting habits of the plants. An intelligent grower can read the history of the past performance of his plants,

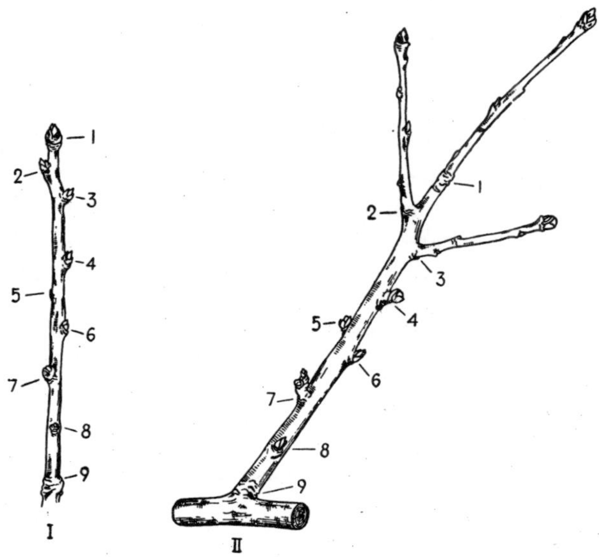

FIG. 74.—Diagrammatic sketches illustrating growth habit of apples. I, twig showing nine buds; II, two-year branch showing possible developments from the buds in twig I. Buds 1 and 2 developed into vigorous twigs terminated by shoot buds, and bud 3 developed into a less vigorous twig terminated by a fruit bud. Bud 4 developed into a vigorous spur terminated by a fruit bud. Bud 5 developed into a less vigorous spur terminated by a fruit bud. Bud 6 developed into a weakly spur terminated by a shoot bud. Bud 7, being a flower bud, produced fruit and a terminal shoot bud. Bud 8 remained latent. Bud 9 died.

predict the future and govern his pruning and other practices accordingly.

Flowering and fruiting habits of horticultural plants are very diverse and quite complex. It will be advisable, however, to discuss a few of the simpler and more common fruiting habits of woody plants. Fruits are borne terminally or laterally on wood that was produced during the previous season or the current season and singly or in clusters of two or more. The peach bears fruit laterally on one-year-old twigs from simple buds that contain but one flower each. Raspberries are produced terminally in clusters on current season's wood. The grape produces fruit in clusters near the base of current season's shoots which generally arise from the one-year-old canes. Cherries produce laterally, singly or in small clusters on one-year-old twigs and on the one-year-old growth of spurs. In horticultural usage a spur is considered as a slow-growing twig or branch that, under normal conditions, did not make over 2 in. of growth the previous season. The apple, with its mixed fruit bud, exhibits several of these types of fruiting habits. It produces terminally on one-year-old twigs, laterally on twigs but mostly terminally on spurs that are one year old or older. The spur-bearing habit is so pronounced in some varieties of apples, as Duchess, Jonathan, Wealthy, Rhode Island Greening and York Imperial, that they are known as "spur bearers." This habit of bearing may lead to pronounced biennial bearing. Some apple varieties, as Rome Beauty, Ben Davis, Gano, Golden Delicious and Yellow Transparent, especially while young, produce fruit on the ends of relatively long one-year-old twigs. More rarely, but under very favorable conditions, varieties such as the Wagener will produce fruit laterally on the one-year-old twigs.

The woody ornamentals possess similar flowering habits. The lilac flowers terminally and near the ends of the one-year-old twigs; forsythia laterally on twigs; the spiraeas and viburnums terminally on current season's shoots.

Review Questions

1. What is growth?
2. What are the two periods of plant activity?
3. What is the duration of the growing period?
4. What is meant by the dormant period?
5. What is meant by the rest period?
6. Do the dormant periods and rest periods coincide in length of time?
7. How can the duration of the dormant period be altered?
8. How can the duration of the rest period be altered?
9. What two principal factors influence plant growth?
10. What are the four principal environmental factors that influence plant growth?
11. Under what conditions might an influencing factor of plant growth become a limiting factor?
12. How is a carbohydrate within the plant made available for use by the plant?
13. What are the two principal groups of food synthesized and used by plants?
14. What are the two principal phases of plant growth?
15. What is the carbohydrate-nitrogen hypothesis?
16. When carbohydrates accumulate, what three activities occur in the plant?
17. What is flower-bud formation?
18. What causes flower-bud formation?

19. When does flower-bud formation occur?

20. Where are flower buds formed on an apple tree?

Problems

1. There is a peach tree growing just outside the south wall of the greenhouse. On Oct. 1 an opening was made in the greenhouse, and branch *A*, while still attached to the tree, was brought into the house and kept under conditions favorable for growth. On the first of the following February a similar branch, *B*, was brought into the greenhouse in a corresponding fashion. Branch *C*, similar to branches *A* and *B*, was left outside the greenhouse. Give the order in which growth started in these three similar branches. Explain.

2. You have an old apple tree that for the past two years has made very little growth, has blossomed profusely each spring but has not produced any fruit although several years ago it produced satisfactory crops. Explain the cause of blossoming, and state what might be done to produce fruit.

3. A month before Mother's Day a carnation grower noticed that the plants in one house were growing vigorously but would not be in bloom until ten days or two weeks after Mother's Day, at which time he planned for a full crop. State a possible cause, and suggest and explain a remedy.

4. A 40-acre, nine-year-old, vigorously growing apple orchard was given a heavy pruning during March. State and explain the results on the following crop.

5. Make a diagrammatic sketch of an apple branch showing the locations of fruit buds.

Suggested Collateral Readings

1. Gourley, J. H., "Textbook of Pomology," pp. 19–73, The Macmillan Company, New York, 1922.

2. Holman, R. M., and W. W. Robbins, "A Textbook of General Botany," pp. 309–334, John Wiley & Sons, Inc., New York, 1939.

3. Kraus, E. J., Sources and Cycles of the Nutritive Elements, "U.S. Department of Agriculture Yearbook," pp. 405–417, 1939.

4. Maximov, N. A., "Plant Physiology," McGraw-Hill Book Company, Inc., New York, 1938.

CHAPTER VII

THE HORTICULTURAL PLANT IN RELATION TO TEMPERATURE

The behavior of a complicated organism such as a plant is governed indirectly by such fundamental factors as temperature, moisture, light and soil. These function together in the activities of the plant, and it can hardly be said that one is more important than another. At various times during the plant's life, however, one of the factors may be classed as a limiting factor. The individual processes of a plant are so interdependent that difficulty is often experienced in determining whether a particular process is influenced more by one factor or another or by a combination of two or more factors.

FAVORABLE TEMPERATURES

There is a minimum and maximum temperature below or above which a plant does not grow, and for each kind of plant there is an optimum temperature at which it grows or functions best. The minimum temperature suitable for growth for most horticultural plants is around 40 to 43°F., and the maximum from 85 to 114°F., whereas the optimum ranges around 75 to 85°F. At the optimum temperature the plant makes satisfactory growth of vegetative, reproductive and storage parts; at temperatures slightly above, the optimum, but not injurious, carbohydrates are utilized, and vegetative growth is rapid; and at temperatures slightly below the optimum, vegetative growth is slow, and carbohydrates accumulate.

Most people have noticed that all the different kinds of plants in the same locality do not start to grow at the same time in the spring, continue to grow at the same rates during the warmer temperatures of summer or stop growing at the same time as the temperature becomes cooler in the fall. As one travels from north to south in the United States, he notes that the types of plants change. As one passes into glasshouses in which florists and vegetable growers are producing different types of plants, one cannot fail to note that the temperatures of the various houses are different. Seeds of corn, tomatoes, squash, peas, radishes and lettuce are planted at different seasons because some germinate satisfactorily only at low temperatures, and others germinate satisfactorily only at high temperatures. Certain kinds of plants grow only when the temperature is cool and others grow better; and still others grow when the temperature is warm. According to their temperature requirements horticultural plants may be grouped as cool-season or warm-season crops.

Cool-season Crops

Cool-season crops are those which make their most satisfactory growth at relatively cool temperatures. They may grow during the cool temperatures of spring, the cool temperatures of fall or during both the periods. This makes it possible in some sections to produce both a spring and a fall crop of certain kinds of plants. Some varieties of the same kind of plant are better adapted to cool temperatures than others. The McIntosh apple grows well and attains a high quality under the cool growing conditions of New England but is unsuited to the warmer sections farther south.

Fruits.—Of those fruits which grow successfully in the temperate zone, cranberries, raspberries, dewberries, blueberries, currants, gooseberries, strawberries, plums, labrusca grapes, apples and pears do best in localities with relatively cool growing seasons. Some of these fruits require cooler temperature than others, and their successful production is restricted to certain localities, but a cool though short-season crop, as the strawberry, produces its fruit in the winter in Florida and in June in Iowa. Even within the same kind of cool-season fruit there are varieties better adapted to one temperature than they are to another during the growing season. The Fameuse and McIntosh varieties of apples grow satisfactorily and attain their high quality only when grown under the cool temperatures of Michigan, New York and New England, becoming coarse textured and of poor flavor when grown in the warmer areas farther south. The Wealthy apple grown in Michigan, Minnesota and northern Iowa is better flavored than when grown in southern Iowa and Missouri. The Delicious and Winesap, however, do better when the temperatures are relatively high. The Duchess apple develops the best quality when the mean temperature from Mar. 1 to Sept. 1 is about 52°F., whereas the Grimes Golden develops the best quality when the temperature for the same period averages about 62°F. The Bartlett pear grown in the Santa Clara Valley in California is a noticeably different pear from the Bartlett produced in the much warmer Sacramento Valley about 100 miles distant.

Vegetables.—Cool-season vegetable crops may be divided into three general groups.

The first consists of short-season crops that are unable to withstand the heat of summer but that can be planted in the open early enough to attain full development before the temperature becomes too high. Crops in this group as a whole are planted early in the spring, and a second crop of many of them can be grown again in the fall. This is especially true in the South where two crops of the same kind of vegetable are produced annually. Representative crops of this group include kohlrabi, leaf lettuce, mustard, pea, radish, spinach and turnip.

A second group of cool-season vegetable crops is composed of those requiring not only a cool season but a longer period for growth than required by the kinds of vegetables in the first group. It includes early cabbage, early cauliflower and head lettuce. In some sections there is insufficient time to mature these crops from seed planted in the open before high temperatures occur; consequently, the seeds are started under glass, and the seedlings transplanted to the open field as early as temperature conditions permit. In other cases the crops are planted after the heat of the summer and grown during the cool fall weather. The long cool growing seasons during the fall in the northern part of the United States are especially favorable for the production of such cool-season crops as late cauliflower, late cabbage, Brussels sprouts and celery.

A third group of cool-season vegetable crops includes those relatively long-season crops which require cool weather during the earlier stages of development but will withstand considerable heat after becoming established. The most common representatives of this group are beets, carrots, kale, onions, parsley, parsnips and early potatoes. In general, seeds of these crops are planted in the open early in the spring. The two perennial plants asparagus and rhubarb belong in this group.

Ornamentals.—Flowers and other ornamental horticultural plants also have certain temperatures at which they make their optimum growth. The maintenance of proper temperature for specific crops grown under glass is of great importance to the florist. The temperatures are controlled, within certain limits, by manipulating the heating and ventilating systems and by shading. Although it is possible to grow in the same greenhouse some plants that do best at cool temperatures and some that do best at warm temperatures, neither type of plant will do its best, and under such conditions the grower usually compromises and maintains a temperature somewhere between the two optima. He may, however, favor one group of plants if that returns the greater profit. The leading cool-season crops of the florist are carnation, chrysanthemum, snapdragon, sweet pea, pansy and cyclamen. The bulbous plants, as tulips and narcissi, are forced at low temperatures. The deciduous shrubs, as forsythia, redbud, *Spiraea prunifolia*, flowering almond and lilac, that start growth and bloom early in the spring may be considered as cool-season plants, although they do survive the heat of the following summer.

Warm-season Crops

Warm-season crops are those which make their most satisfactory growth at relatively warm temperatures. In some localities where such temperatures are not normal, special practices are performed in growing certain crops. The most common practice is that of growing flowers and vegetables out of season in greenhouses. In certain cool sections of Europe the peach is grown successfully by being trained against south walls where the temperature is higher because of the reflected heat. Grapes grown on the terraced southern slopes in Germany, France, Spain and Italy make high-quality wines. In England the temperature never becomes high enough for the production of cucumbers, and consequently this vegetable is grown in heated greenhouses.

Fruits.—Fruits of the temperate zone, as peaches, nectarines, apricots and vinifera grapes, require more heat for their best development than does the apple. Apricots growing in certain sections of the Sacramento River Valley grow more slowly and ripen as much as three weeks later than apricots growing in other sections of the valley not subject to cool nights. Citrus fruits, like the orange, grapefruit and lemon, require more heat than the peach. Tropical fruits like the banana and pineapple require higher temperatures than do citrus fruits.

Vegetables.—There are two groups of warm-season vegetable crops. The first consists of those which will complete their growing season and perfect their products in temperate climates during the normal period of weather favorable for their development. These crops can be planted after the soil is sufficiently warm in the spring and will mature before the first killing frost in the fall. Among them are string beans, lima beans, sweet corn, cucumbers, muskmelons, watermelons, okra, squashes and pumpkins. A second group of warm-season vegetable crops comprises those which require a longer period for growth than the normal period favorable for their development. In temperate climates this longer growing period is obtained by starting the plants under glass and transplanting to the field when temperature conditions are suitable. Eggplants, peppers, sweet potatoes and tomatoes are the most common vegetables that belong to this group.

Ornamentals.—The florist groups the various plants that he is growing in different houses and regulates the temperatures according to the different kinds of plants. Some of the leading floral plants that do best when grown in warm houses at temperatures of 55°F. and higher are roses, gardenias, poinsettias, orchids, callas and begonias. Many of the ornamental shrubs, as the oleander and hibiscus, that are not hardy in cool climates grow satisfactorily at high temperatures.

UNFAVORABLE TEMPERATURES

The horticulturist is particularly concerned with the occurrence of unfavorable temperatures and the resultant influence on the behavior of plants. Either injurious high or low temperatures may occur at any time of the year and cause damage, ranging from a slight checking of growth through the destruction of the crop to the killing of the entire plant.

Effects of High Temperatures

High temperatures during either the growing or the dormant season result in various difficulties.

During the Growing Season.—Excessive heat during the growing season often results in the burning of leaves and fruit. Growers often observe that leaf injury on apple trees is particularly severe if abnormally high temperatures occur shortly after the plants have been sprayed with lime-sulfur. High temperatures cause peas and corn to pass their best stages of maturity very quickly, which results in starchiness and toughness. Irish potatoes are inclined to produce vegetative growth at the expense of tuber formation when the temperature goes above 64°F. Associated with the effects of unfavorable high temperatures during the growing season are branching and unpleasant flavor in asparagus, flower dropping in snap beans and flower stalks developing in head lettuce. Temperatures above 65°F. in glasshouses result in lack of differentiation of flower buds in stocks and many other floral plants.

During the Dormant Season.—High temperatures during the dormant season are indirectly responsible for certain kinds of damage. Abnormally high temperatures may induce cambial activity and the opening of buds early in the season. When this activity is followed shortly by freezing temperatures, damage to wood and flowers is often quite destructive. The temperature during the last three weeks of February and the first three weeks of March is very important in determining the amount of frost damage to apples in the eastern part of the United States. If the temperature during this period is above the average, the trees will blossom early, with a great likelihood of being injured by frosts.

Effects of Low Temperatures

Low temperatures occurring either during the rest period or afterward cause damage to roots, stems and buds of horticultural plants. Low temperatures occurring after the rest period is completed and growth has started in the spring often destroy the flowers and new shoot growth of woody plants and kill herbaceous plants entirely.

The importance of the great economic losses has stimulated scientists to investigate the cause of death of plant tissues by low temperature. In general, the capacity of a plant to retain moisture against the

Fig. 75.—Havoc caused by a severe winter. Old Baldwin apple trees are pulled out in Wayne County, New York.

extracting forces of freezing is associated with its ability to survive cold. Death from freezing is believed to be due to desiccation. Ice forms in the intercellular spaces and extracts water from the protoplasm. If the ice formation proceeds beyond critical limits, the protoplasm of the cells dies from loss of water. Thus, plant tissues holding greater quantities of water against freezing should be more hardy to cold than those with less capacity for water retention. The water held without freezing is spoken of as "bound water"; the water frozen, as "free water." Those plants which have a higher percentage of free water in proportion to the amount of bound water are usually less hardy than plants with a smaller ratio between free water and bound water. The amount of bound water will vary in the same tissues of the same plant under different conditions

One plausible explanation of the injury caused by low temperatures suggests that winter hardiness is dependent upon high carbohydrate and other food reserve supply in the tissues and that hardiness against low temperature is the result of structural differentiation of the protoplasm which makes it more resistant to loss of moisture. Such differentiation may be dependent upon and in part initiated by high sugar concentration in the tissues. Plants may differ in their genetic ability to use the food reserves in building a protoplasm that is resistant to low temperature.

Woody plants, which may be hardy when subjected to low temperatures if the wood is mature, may be severely injured when exposed to the same conditions if the wood is immature. This is explained on the basis of the chemical composition and physical structure of the protoplasm at different stages of maturity in the cells. In general, those horticultural practices, such as late cultivation, irrigation and the application of nitrogenous fertilizers, which tend to stimulate late fall growth result in the depletion of reserve carbohydrates, the development of soft immature shoots that are low in their water-retaining capacity and resistance to low temperatures. Such tissues may be severely injured or killed even in mild winters when properly matured tissues would be entirely unharmed.

During the Rest Period.—In a previous discussion it was learned that most deciduous trees and shrubs have a rest period during which no appreciable growth takes place even though the environmental conditions are favorable for growth. This internal condition of the plant apparently has some bearing on its ability to withstand low temperatures during the winter. The more deeply a plant or plant part is in the rest period the more resistant it is to cold. A very low temperature in the early part of the rest period or in the latter part of it will cause more injury than if such temperature came during the time when the plant was in the middle or deepest part of the period.

The root of the plant evidently does not have a rest period; consequently, one would expect more damage to roots from low temperatures than to the tops. The soil temperature does not become so low or fluctuate so rapidly and widely as the air temperature surrounding the top. Soil temperatures remain much higher than atmospheric temperatures; and in addition mulches of snow or vegetation, as that of a cover crop, give added protection to the roots by maintaining higher soil temperatures.

After the Rest Period.—The effect of low temperatures after the rest period is completed varies in relation to a number of different factors, among the most important being (1) relative hardiness of different kinds of plants and of varieties of the same kind of plant, (2) relative hardiness of the different parts of the same plant, (3) the time of occurrence of the low temperature, (4) the degree and duration of low temperature and (5) the rapidity of temperature changes.

The discussion of horticultural practices that are used to protect plants from damage caused by low temperatures will indicate the relative hardiness of the more important horticultural plants. Suffice it to say that not only do various species of plants vary in their resistance to low temperatures but also some varieties of the same species show evidence of being more resistant than other varieties. The various parts of the same plant also differ in relative hardiness to cold. The fact that trees often fail to blossom following severe winters indicates that shoot buds are more resistant to low temperatures than the flower buds. The pistil of the flower is more tender than are other parts of the same flower. The stem when considered in cross section shows that the cambium is the most resistant to low temperature, followed in turn by the cortex, the sapwood and the pith.

The time at which low temperatures occur results in different effects on the plant. The advent of early winter cold will result in the killing of the terminals of the immature shoots. In some cases the sapwood of the plant is injured and turns brown, resulting in a condition known as "black heart." Since the cambium is often uninjured, a new layer of sapwood may be formed covering the injured wood, and little damage results. Badly black hearted young nursery trees seldom survive transplanting in the spring following the damage. At Ames, Iowa, on Oct. 30, 1925, while the trees were in full leaf and active growth, the temperature dropped to −7°F. The result was the killing of the sapwood in many of the branches of three-year-old apple trees. The next spring these trees were allowed to retain all the leaf area possible by resorting to practically no pruning, and in time they outgrew the effect of the injury. Low temperatures during the winter season also cause such injuries known as "bark splitting," "trunk splitting," "branch splitting," "sunscald," "crotch injury" and "collar rot." The explanation offered for the crown and crotch injury is that these are the last parts of the plant to cease cambial activity and to mature their wood and consequently are more tender. Sunscald usually occurs on the side of the tree that is exposed to the direct and reflected rays of the sun. Most often this injury occurs on the south side of the tree which the plant is subjected to alternating changes of temperature. During the day the tissues on the branches exposed to the sun become warm and thaw out only to freeze again at night. This alternate freezing and thawing with its resultant expansion and contraction results in the final death of the tissues. Owing to the destruction of these tissues, the bark peels off.

Frost or feathery crystals of ice form on the ground and on other exposed surfaces when there is sufficient moisture in the air and the temperature of the exposed surfaces falls to 32°F. This temperature may cause considerable injury to horticultural plants. Frost does most damage to the newly opened blossoms of fruit trees and flowering plants and to young growths and newly set plants. Untimely frosts often cause immense losses to vegetables in Florida, Texas and other southern sections that grow early vegetables for the northern markets. The Irish potato crop is often decreased in the northern producing areas because early fall frosts destroy the tops before they have had sufficient time to synthesize enough reserves to cause the development of large tubers.

Subjecting seedlings of biennial plants, as celery, beets, cabbages and carrots, to a temperature of 50°F. or lower for a period of five or six weeks will cause them to produce seed stalks the first season instead of producing the characteristic edible product.

The amount of damage done to horticultural plants by unfavorable low temperatures varies with the kind of plant and part of plant, with season of the year, with the state of growth or activity of the plant, with the lowness of temperature attained, with the rate and extent of the drop in temperature and with the duration of time for which the plant is subjected to the unfavorable temperature.

PROTECTION FROM TEMPERATURES

The horticulturist resorts to various practices to protect plants from unfavorable temperatures.

Selection of Location

Location refers to the general area in which one might establish a horticultural enterprise. It is obvious that if one desired to establish an orange orchard, he would confine the location to subtropical regions. Unfortunately, the proper locations within this general region are often determined only after severe losses have demonstrated the unsuitability of other locations. With such plants as the perennial long-lived fruit trees, errors in selecting favorable locations are especially costly. After the severe freeze of 1893–1894, the orange orchards in Florida were pushed southward. The fruit belts in Michigan and New York are rather sharply fixed by the extent of influence of the adjacent bodies of water on the temperature.

Selection of Site

After a suitable location has been determined upon, the next consideration is the selection of a proper site. This is important in all horticultural plantings but probably of greater significance in fruit growing than in the production of vegetables or flowers because of the greater permanency of the fruit crops. In all localities the site should provide for adequate air and water drainage. Cold air drains away from the higher elevations and collects in the valleys. This condition leads to serious losses from late spring frosts. Excess water should also be able to drain away, for plants will not grow satisfactorily if the soil is waterlogged. For these reasons orchards are often planted on gently rolling to hilly land, but level land is just as satisfactory, or even more so, provided there is lower land adjacent which permits suitable air and water drainage. In some localities the particular slope or exposure is of importance. The south and west slopes are warmer and drier; and the north and east, cooler and more moist. In the Middle West there is more winter injury on the trunks of trees on a southwest exposure. In planting certain flowers or other plants on the home grounds it is often advisable to select the exposures best adapted to the particular plants.

Selection of Plants

The selection of plants and of varieties of those plants which are inherently resistant to unfavorable temperatures is one of the first considerations in overcoming the effects of unfavorable temperatures. Most of the state agricultural experimental stations publish lists of varieties that are adapted to the various sections of their respective states. The prospective planter should become well acquainted with the varieties that are growing satisfactorily in his community and also consult these lists for other kinds that he may wish to plant.

Horticultural plants are often classified on the basis of hardiness which means chiefly the ability of the plants to resist low temperatures. There is often a wide range between the least hardy and most hardy horticultural variety of the same kind of plant. The hardy types of fruits grown in temperate regions include the cranberry, blueberry, gooseberry, currant, red raspberry, American plum, strawberry, sour cherry and apple. The intermediate types include the dewberry, black raspberry, blackberry, damson plum, pear and sweet cherry. Fruits grown in temperate regions and generally considered as tender, or lacking resistance to low temperatures, are the European plum, Japanese plum, apricot, peach and European grape.

Fruits of the subtropical regions are relatively tender when compared to those of temperate regions. Among subtropical fruits the orange is the hardiest, followed in order by grapefruit, lemon and citron.

Vegetables also differ in their ability to withstand frost and low temperatures. Some hardy vegetables may be planted two to four weeks before the last killing frosts in the spring; half-hardy vegetables, about the time of the last killing frost; tender and very tender vegetables which are injured by light frosts and whose seed germinate poorly at low temperature must be planted or transplanted to the field after all danger of late frosts is over. The principal vegetable crops may be grouped roughly as follows in order of relative hardiness:

Hardy	Half hardy	Tender	Very tender
Asparagus	Artichoke	New Zealand spinach	Cucumber
Broccoli	Beet	Snap bean	Eggplant
Brussels sprouts	Carrot	Sweet corn	Lima bean
Cabbage	Cauliflower	Tomato	Muskmelon
Chinese cabbage	Celery		Okra

The ornamental trees, shrubs, vines and herbaceous perennials can be grouped roughly into those which are hardy and those which are

HARDY

Trees	Shrubs	Vines	Herbaceous perennials
Oaks	Viburnums	Englemann creeper	Peony
Maple	Spiraeas	Clematis	Iris
Elm	Lilacs	Boston ivy	Tulip
Norway spruce	Honeysuckle		Phlox
Junipers	Philadelphus		Daffodil
Firs	Snowberry		

TENDER

Trees	Shrubs	Vines	Herbaceous perennials
Magnolia	Elaeagnus	English ivy	Dahlia
Holly	Cape jasmine	Vinca	Gladiolus
Mimosa	Abelia	Bougainvillaea	Canna
Willow oak	Azalea		
Water oak	Crepe myrtle		
	Oleander		
	Hibiscus		

tender. In grouping them, the word "hardy" is used to signify those which are resistant to low temperatures in central and northern latitudes. The tender group is grown without protection in southerly latitudes of the United States but can be grown in more northerly latitudes only under special conditions of protection.

Special Practices

The horticulturist resorts to a number of special practices in overcoming the effects of unfavorable temperatures. Among the most important are (1) use of screens, (2) mulches, (3) hardening, (4) pruning, (5) irrigation, (6) top-working and (7) artificial heating.

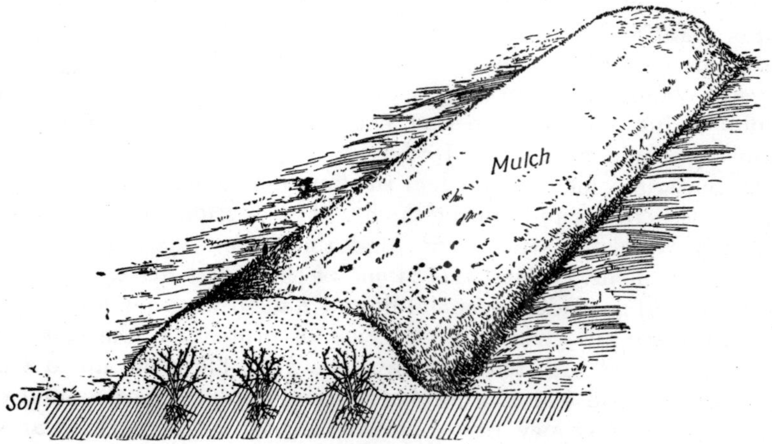

Fig. 76.—Diagrammatic sketch showing a method of protecting tender plants in cold climates.

Screening.—Plants may be protected from both high and low temperatures by screens made of laths, cloth, cut branches and growing plants. Seedlings and small transplants in the nursery are often protected from the sun by use of slat or brush covers which are raised to various heights above the seed or transplant bed. Frames constructed of lath are often built over broadleaf evergreens, particularly types like boxwood, to protect them from injury by the winter sun.

Commercial florists grow certain plants during the summer in "cloth houses." A lightweight wooden frame is covered with light muslin or "tobacco cloth." The cloth permits the entrance of sufficient light to grow such plants as roses, chrysanthemums, dahlias and asters. Such enclosures offer protection from insects as well as from the hot rays of the sun during midsummer.

Mulching.—Mulches of all types afford some protection from damage caused by low atmospheric humidity and low winter temperatures. They decrease the loss of water, prevent the soil from cooling to as low a degree and to as great a depth and prevent rapid changes in the temperature of the surface of the soil. Thus the "heaving" of plants which is caused by the alternate freezing and thawing of the surface of the soil is avoided. The early spring growth of the plants or plant parts covered by the mulch is retarded by keeping them cooler than the air above the mulch. Artificial mulches are used extensively on seedling and propagation beds; tender plants, as hybrid tea roses; many herbaceous perennials; and universally on strawberry plantings.

The time of applying the protective mulch is influenced by the kind of plant and the locality in which it is growing. Various materials are used, of which leaves, peat moss, straw, soil and glass wool are the most common. Snow is an excellent mulch if it remains on the ground during the entire period necessary. Soft-crown plants, as foxgloves, should first be protected by a light covering of twigs or similar material which will prevent the finer mulch material from matting over the crowns and causing them to die. Various straws are used frequently to mulch strawberries and the more widely grown herbaceous perennials. Glass wool is one of the newer materials, being used for covering tender perennial flowers and to wrap tender shrubs. Although roses are given various types of protection, the common method employed with tender hybrid teas is to hill them up to 8 to 9 in. with soil and cut off the plant extending above the mound of soil. Manure or strawy refuse may be placed between the plants. Canes of climbing roses, raspberries and even grapes are often removed from the trellis or support and covered with several inches of soil in those sections where damage from low winter temperatures is likely to occur.

Hardening.—Hardening of plants like cabbage, kale and celery is a practice often followed by vegetable growers to make them better able to withstand adverse conditions such as frost, drying winds, shortage of water and high temperatures which they may encounter in the field after they have been taken from the cold frame. Withholding water or lowering the temperature while the plants are still in the cold frame are the means generally used to harden a vegetable plant. Briefly, when the rate of growth is reduced, carbohydrates accumulate, because they are not used in making growth. This, in turn, is associated with the internal conditions of the plant and increases the water-retaining power of the protoplasm. Tender plants like the tomato apparently cannot be made frost-resistant.

Pruning.—Judicious pruning will lessen damage to horticultural plants by low winter temperatures. In regions of severe winter temperatures, considerable killing of tissues is likely to occur about pruning wounds made before the severe cold weather is over. This type of injury is especially noticeable on young trees during the period of training. Twigs are often headed back to particular buds which are killed by low temperatures. In regions where the trunks of trees are subject to sunscald, it is well to keep the trees low-headed, as this shades the trunk in winter. The plants should be pruned only enough to maintain the optimum amount of vegetative growth and fruitfulness. Any practice that stimulates succulent vegetative growth or prevents the plants from maturing satisfactorily will increase the probability of winter injury.

Irrigation.—Low-growing crops, as the strawberry, when grown under irrigation, have been saved from frost injury by the application of irrigation water. Water, in comparison with air, contains a large amount of latent heat. The irrigation water, on cooling, may take up so much of the cold from the air or release so much heat that the temperature of the air is kept above the frost point. Frost injury to vegetables growing on muck soil is sometimes avoided in a somewhat similar fashion. It is well known that crops on muck or peat soils are much more subject to frost than are similar crops grown on mineral soils, because mineral soils are warmer and liberate heat more rapidly. Raising the water table of muck or peat land may help to protect a crop on such soils from a light frost. The water is raised in the drainage ditch by means of dams and forced back into the drain tiles in the peat beds, thus raising the water table of the soil. Since the soil releases its latent heat to the water more readily than it does to air, the movement of the latent heat from the soil to the air is thus speeded up by passing through the water. This increased rate of liberation of latent heat of the soil combined with the liberation of the latent heat of the water itself is often sufficiently rapid to avoid injury by frost.

Top-working.—A practice that is becoming of greater importance in the protection of fruit trees from low winter temperatures is that of double-working, or top-working. This is being used particularly with apple trees. Apple trees are especially subject to the effect of low temperature at the collar and the main crotches. Apple varieties such as the Virginia Crab and Hibernal are more resistant to damage from low temperature than are such varieties as Jonathan and Grimes Golden. Top-working is discussed in Chap. XI.

Artificial Heating.—Heat as a means of protecting horticultural plants from injury by low temperatures includes the use of cold frames, hotbeds, glasshouses and heaters.

A cold frame is a structure that utilizes only the natural heat from the sun to furnish plants with a warmer environment. A simple type has the back or north side 6 in. higher than the south side. Since the

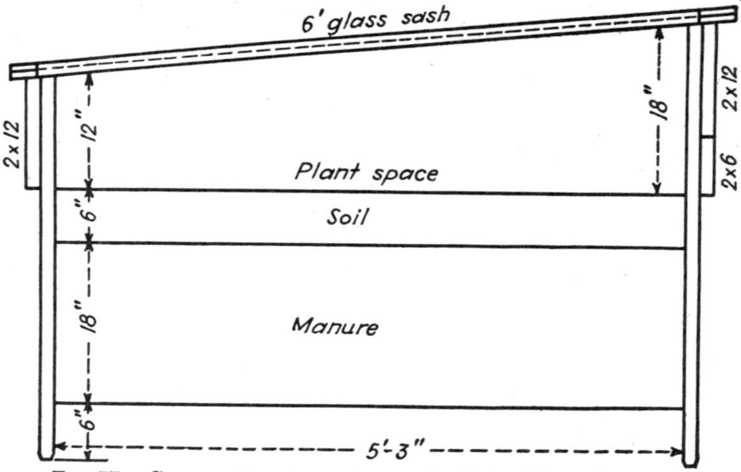

FIG. 77.—Cross section of a permanent hotbed. (*Iowa Exp. Sta.*)

FIG. 78.—Orchard heating. (*Gulf Fertilizer Co.*)

standard sash is 3 ft. wide and 6 ft. long, the cold frame is generally 6 ft. wide, with the length determined by the number of sash to be used and the economy of handling. Although cold frames are generally covered with glass sash, cloth covers are widely used in the South. Cold frames are used for growing certain early crops like lettuce and radishes to maturity, for starting early plants that are to be transplanted to the field, for hardening of plants before transplanting and for overwintering certain plants.

A hotbed differs from a cold frame in that heat, in addition to that from the sun, is supplied by the fermentation of horse manure or other organic matter, by hot-water or steam pipes or by hot-air flues or by electric cables. Hotbeds are generally used for growing early crops of quick growing vegetables, to start plants that are to be transplanted and for the propagation of many kinds of plants.

Glasshouses vary greatly in size and structure. They are heated artificially and are used principally for the production of cut flowers and pot plants out of season. Both vegetables and fruits are produced under glass, but the area devoted to these crops under glass in the United States is relatively very small when compared to the areas in the open.

Heating of orchards to prevent damage from frost is an established practice in some regions. It is extensive in the citrus orchards of California. The temperature is raised artificially by the use of great numbers of oil-burning heaters distributed throughout the orchard. These produce a warm blanket of air which covers the orchard.

Review Questions

1. What are the meanings of minimum, maximum and optimum temperatures for plant growth?

2. What is meant by a cool-season crop?

3. Give two examples of fruits, vegetables and ornamental plants that are considered as cool-season crops.

4. What is meant by a warm-season crop?

5. Give two examples of fruits, vegetables and ornamental plants that are considered as warm-season crops.

6. What kinds of temperatures are injurious to plant growth?

7. At what seasons of the year may high temperatures be injurious?

8. What is the effect of freezing temperatures on the moisture of protoplasm?

9. Why are some plants less injured than others by low temperatures?

10. At what period in its annual life cycle is a plant most resistant to low temperatures?

11. What factors influence the degree of injury of a plant from low temperatures after the rest period?

12. How may plants be protected from possible injury by low temperatures?

13. Are all kinds of plants equally resistant to unfavorable low temperatures?

14. Are all horticultural varieties of the same kind of plant equally resistant to low temperatures?

15. How do mulches provide protection from injury by low temperatures?

16. Name the methods used by the horticulturist in supplying heat to plants to avoid injury by low temperatures.

17. How does the application of irrigation water to a strawberry planting lessen the chance of injury by a mild late frost?

Problems

1. A florist has a 50- by 120-ft. glasshouse, half of which is devoted to Briarcliff hybrid tea roses and half to Red Spectrum carnations. The roses are growing satisfactorily, but the carnations have weak spindly stems with many of the flowers splitting. What is the cause of this condition, and what factors will govern the remedial measures adopted?

2. You are a county agent in a truck-crop community and have been asked to advise the planting date for transplanting cabbage and tomatoes to the field.

3. Mrs. A is growing peonies, iris, foxglove, tulips and hybrid tea roses. How should she provide suitable winter protection for them?

4. In examining a mature apple tree in April you noted that the bark only on the south side of the trunk was split in strips and appeared to be loosely attached. Diagnose the probable cause, and recommend a practical treatment.

Suggested Collateral Readings

1. Auchter, E. C., and H. B. Knapp, "Orchard and Small Fruit Culture," pp. 467–484, John Wiley & Sons, Inc., New York, 1937.

2. Gardner, V. R., *et al.*, "Fundamentals of Fruit Production," pp. 265–431, McGraw-Hill Book Company, Inc., New York, 1939.

3. Knott, J. E., "Vegetable Growing," pp. 37–50, Lea & Febiger, Philadelphia, 1935.

4. Lloyd, J. W., "Productive Vegetable Gardening," pp. 43–47, J. B. Lippincott Company, Philadelphia, 1930.

5. Weaver, J. E., and F. E. Clements, "Plant Ecology," pp. 356–379, McGraw-Hill Book Company, Inc., New York, 1938.

6. White, E. A., "The Florist Business," pp. 63–76, The Macmillan Company, New York, 1933.

7. Young, F. D., Frost and the Prevention of Frost Damage, *U.S. Dept. Agr. Farmers' Bul.* 1588: 1–62, 1935.

CHAPTER VIII

THE HORTICULTURAL PLANT IN RELATION TO MOISTURE

If any one factor might be said to be the most essential for plant growth, it is moisture. The importance of moisture is indicated when one notes that it constitutes the greater part of living plants; that it is an essential ingredient of carbohydrates and other chemical products of the plant; that it is required to maintain the turgidity of living cells, a condition necessary for fundamental physiological activities of the plant; that it is a solvent for mineral elements that enter the plant through the roots and for the gases that enter through the leaves; and that it serves as a means by which substances are transported within the plant.

WATER IN THE PLANT

A plant has been defined as a "supported column of water." This definition would indicate that all its physiological processes occur in solutions and take place when large quantities of water are present. The active plant contains water in two different states, which may be referred to as bound water and free water. The free water in the plant absorbs and gives off heat, evaporates at 212°F., freezes at 32°F. and moves freely from one part of the plant to another. The bound water is practically a constituent of chemical substances in the plant; it does not give off or regulate heat at 212°F.; it does not freeze easily, if it freezes at all; and it acts more like a solid than like water. Free water may become bound water, and bound water may become free water.

Obtaining Water

Most horticultural plants obtain water from the soil through minute openings in the walls of the root hairs by a process known as "diffusion." The molecules of water are in constant rapid motion. If one of them strikes an opening in the wall of the root hair, it will pass through the opening if the hole is large enough to permit its passage. When there are more molecules of water on the outside of the root hair than on the inside of the same root hair, there will be a greater number of chances for molecules on the outside to hit openings than there are chances for those on the inside to hit openings. Consequently, as long as this water condition is maintained on the two sides of the root-hair membrane, more water will pass from the soil into the plant than passes from the plant into the soil. In a soil containing adequate water for satisfactory plant growth the amount of water inside is always less than the amount outside the root hair. Under such conditions, water continues to pass into the plant from the soil. The rate of the movement of water into the plant will vary according to the difference between the amount of water on the inside of the root hair and the amount on the outside, as the two amounts will attempt to reach a state of equilibrium. The rate, therefore, will be influenced by the condi-

FIG. 79.—Irrigation in ancient Assyria as depicted in the palace at Koyunjik. An aqueduct is shown at the right and laterals in foreground. (*Gardner, Bradford and Hooker.*)

tions of the plant and its environment, both of which may be regulated, to some extent, by the grower.

USING WATER

Different kinds of plants and the same kind in different environments vary in the amount of water that they take in and in the amount that they give off. The total amount used by a plant and the amount used for each of its many activities is never constant. Optimum nutritive conditions provide for the most economical use of water.

FUNCTIONS OF WATER

Water has certain functions that it performs or for which it is used by the plant. The relative amounts used for the various functions will vary according to the activities of the plant and the amount of water available. When the water supply is limited, the vegetative parts of the plant seem to be supplied with water before the fruits that may be growing on that plant. In cases of extreme drought, water passes from the developing fruits to the leaves, and the fruits may wither and drop. An abundant water supply is necessary for satisfactory growth of leafy plants like lettuce, spinach and lawn grasses.

Free water normally is a constituent of all plant tissues. It comprises from 50 to 75 per cent of the green weight of all leaves and twigs, from 60 to 85 per cent of the weight of roots, from 70 to 90 per cent of the weight of herbaceous plants and 85 per cent or more of the weight of fleshy fruits and vegetables. It comprises 85 per cent of the fruit of the apple, 90 per cent of the fruit of the strawberry, 90 per cent of the head of cabbage and 90 per cent of spinach.

Water is indispensable as a plant nutrient. It is a component of protoplasm and, with carbon dioxide, is essential in synthesizing various plant foods that are utilized or stored by the plant. Water keeps the living cells turgid, and only turgid cells carry on their physiological activities satisfactorily. It is the solvent in which all the nutrients essential to green plants are dissolved. The elements must be in solution before they can enter the plant. Carbon dioxide is dissolved in the moisture film on the cells of the stomatal cavity. Water serves as a means by which materials are transported within the plant. The elements that enter the plant and many of the various products manufactured by it must be moved from one part of the plant to another. The movement of these materials is independent of the movement of the water, but the materials must have water as a medium in which to move. Water as a carrier is indispensable to the plant. Its loss by transpiration may be of some importance in raising water in the plant. Transpiration may assist to keep the plant cool, as the water within cells must absorb some of the heat produced by the various activities within the plant.

The Plant's Water Requirements

The term "water requirement" is used to designate the units of water required for the production of one unit of dry matter. The water requirement of different kinds of plants varies widely, ranging from 50 lb. for each pound of dry matter produced by a pine tree to a 1,000 lb. for each pound of dry matter produced by such succulent plants as the alfalfa. The water requirements of the same kinds of plants vary with the conditions under which the plants are growing.

An apple tree uses about 500 lb. of water for each pound of dry matter produced. A mature bearing apple tree will produce about 35 lb. of dry matter per year in stems, roots and foliage at an expenditure of 17,500 lb., or about 2,000 gal., of water. An added 7.2 lb. of dry matter will be produced in each bushel of apples which will require an additional 3,600 lb. of water for each bushel. Such an apple tree will transpire water at the rate of about 25 gal. per day. Under certain conditions, as high as 50 gal., or 1 bbl., per day will be transpired by the mature bearing apple tree. The water requirement of an acre of mature apple trees in bearing will be about 10 or 11 acre-in. of water. Because of surface runoff, percolation, evaporation and other losses, only about one-third of the rainfall is available for the use of the apple trees so that an annual precipitation of 30 to 33 in. would be required for the orchard.

Optimum nutritive conditions provide for the most economical use of water. The amount used by a vigorously growing plant is greater than that used by a weakly growing plant of the same variety. The water requirement, or the amount of water required, for the production of 1 lb. of dry matter is less in the vigorously growing plant than in the less vigorously growing plant. All factors tending to increase the nutrition and growth of plants will tend to decrease their water requirements, and all tending to decrease the nutrition and growth will increase their water requirements.

FACTORS INFLUENCING THE USE OF WATER

The amount of water that must be supplied from the soil to maintain the plant in an optimum functioning condition varies throughout the season, from day to day and even during different periods of the same day. The major factors that influence the amount of water used by a plant are the rate at which water is supplied by the soil and the rate at which it is lost by the plant.

Rate at Which Water Is Supplied by a Soil

The rate at which water is supplied to a plant by the soil is influenced (1) by the amount of water in the soil, (2) by the availability of the water as influenced by the texture and structure of the soil and (3) by the water-absorbing area of the root hairs. These factors are interdependent and are influenced by soil and atmospheric temperature. When there is ample water in the soil to supply the needs of the plant the addition of more water may only prove injurious. The injurious effects may be due to excessive leaching and to an actual decrease in the amounts of available plant food elements. The absence of an adequate supply of oxygen decreases oxidation and retards the activities of the beneficial soil organisms which need oxygen to carry on their processes of liberating plant food elements.

A relatively low water content, provided there is enough to insure adequate growth, results in a greater root development and consequently an increased absorbing area, whereas, too much water results in the development of a smaller and more shallow root system. If a high concentration of mineral salts is in solution, the amount of water available to the plant is decreased, and some of the dissolved salts themselves may be absorbed by the plants and have toxic effects on the tissues. This injuriously high concentration of dissolved salts may occur in cases of too heavy applications of fertilizers and in irrigated soils. Some soils of this nature are spoken of as "alkali soils."

The water in the soil surrounds the soil particles and occupies more or less of the spaces between the soil particles. Of two volumes of soil, the one composed of the finer particles will have the greater surface area and can consequently hold more water. If the amount of water in a given soil volume is limited, the film surrounding the soil particle may be so thin that the suction power of the root hair is less than the force of adhesion existing between the soil particle and the film of water, and the plant cannot obtain sufficient water. Soil particles of very small size or of varying sizes may be packed so closely together that aeration, as well as water movement, is greatly restricted.

Water enters or is absorbed by the plant through openings in the root hairs. Those conditions of nutrition, moisture, temperature and aeration within the plant and soil that favor the growth of root hairs will therefore increase the area by which the plant obtains water from the films surrounding the soil particles.

Rate at Which Water Is Lost by a Plant

The rate at which water is lost by a plant to the atmosphere is influenced (1) by the kind of plant and the amount of transpiring area, (2) by the relationship that exists between the free water in the plant and the humidity of the atmosphere surrounding the plant, (3) by the temperature of the atmosphere, (4) by the rate of movement of the air and (5) by the exposure to light.

The amount of water lost from plants by transpiration is influenced greatly by the presence, amount and kind of hairs and cuticle on the foliage and stems; by the depression of stomates; by the presence of stomates on but one surface of the leaf and the amount of stomatal opening for a given leaf area and by the position, size, number or even absence of leaves.

A plant that contains an abundant supply of water will lose more by transpiration to a dry atmosphere than to a moist atmosphere. The moisture in the humid stomatal cavity diffuses through the stomatal opening into the less humid atmosphere. The higher the atmospheric humidity the nearer the state of equilibrium between it and the humidity of the stomatal cavity and the less the amount of water lost by the plant to the surrounding atmosphere. In nurseries and greenhouses it is a common practice to maintain a high atmospheric humidity and thus keep plants from wilting by syringing them and the structures in which they are growing.

The amount of water lost by transpiration increases with a rise in temperature and an increase in the velocity of the wind. The higher temperature increases the activities of the plants, and the water content of the air increases as the temperature rises. When a quiet layer of air surrounds the leaf of a plant, the layer of air becomes almost saturated with water vapor because of the slowness with which the moisture diffuses away from the leaf. This moist blanket of air surrounding the leaf checks transpiration. Movement of air sweeps away this blanket of moist air and brings other and drier air into contact with the surface of the leaf. Plants lose the greatest amount of water on dry, hot, windy days.

The opening and closing of stomates seems to be influenced somewhat by light, as the stomates are wider open in light than in darkness. This effect, however, is often more than counterbalanced by the effects of temperature and humidity during the daytime. The temperature is higher in sunlight, and the humidity is less, and both conditions tend to increase the loss of moisture by transpiration and cause a closing of the stomates. Shading plants consequently decreases transpiration.

During days of even moderately high transpiration, water is not supplied to fruit trees sufficiently rapidly by the roots to maintain maximum efficiency. Under such conditions stomates usually do not remain open later than noon on bright clear days, but on cloudy or rainy days most of them may remain open throughout the entire day. A reduction of as much as 10 per cent may occur in the moisture content of the leaves between sunrise and sunset. Fruit enlargement slows down and may cease entirely during a period of the afternoon. This is especially noticeable with citrus fruits which may actually shrink in the afternoons, and consequently many lemon growers limit fruit picking to the morning hours.

In general, with other factors equal, the larger the area of the leaf surface and the greater the number of stomates the greater would be the outgo of water. Although the outgo of water would be greater, the amount used per pound of dry matter produced would not necessarily be greater. The horticulturist generally cuts off a portion of the top of newly set nursery plants in recognition of the fact that the leaf area should be reduced in order to decrease transpiration. The reason for doing this is based upon the fact that a part of the root system was destroyed when the plant was dug, and consequently the reduced root system may be insufficient to supply the top with the required amount of water. If the plant has a sufficient root system, as is the case with evergreens that are balled and burlapped, there is no need to cut out any of the branches when moisture is the only factor to consider.

WATER IN THE SOIL

The soil may be considered as a reservoir in which water is stored. The water supply of the soil is maintained by natural precipitation and by irrigation. Water is lost from the soil to the use of the plant by surface runoff, by percolation through the soil and by evaporation from the soil. It is lost from the plant by transpiration. Soils that are fully occupied by plants lose approximately ten times as much moisture by transpiration through the leaves of the plants as by direct evaporation from the ground. Obviously, deciduous plants need the greatest amount of water during the active growing season.

Conditions of Water in Soil

Water is held in the soil with varying degrees of force depending on its condition there. The usual classification of soil water is based upon the effects produced by gravitational and capillary forces. When gravitational forces predominate, which is the case in soils nearly filled with water, water percolates downward first from the larger pore spaces where capillary forces are weakest. This gravitational water is over and above that required to satisfy the forces of capillarity. The downward movement will continue until the capillary forces equal or exceed the downward gravitational pull. The water remaining in the soil particles held by capillary attraction is known as "capillary water" and is the water available to the plant. It is free to move but always moves in the direction in which capillary tension is greatest. As the water content of the soil is decreased further, the water will no longer move under capillary forces. This water, which is held so tenaciously as very thin films about the soil particles and as small wedges at points of contact of soil particles, is known as "hygroscopic water" and is unavailable to plants. The greater part of it is associated with the colloidal part of the soil. If the hygroscopic water evaporates, a point is reached at which no more will be lost at ordinary temperatures. If the soil is heated red-hot, the last water, known as "combined water," held by chemical rather than physical forces, will be driven off.

Amount of Water in Soil

Correct interpretation of irrigation practices and correct selection of soils for horticultural crops is based fundamentally upon the relationship between the plant and its environment and the soil and its water content.

With good drainage conditions any particular soil will take up and hold against the forces of gravity only a certain rather definite amount of water. This amount is termed "field capacity" and varies widely with soils differing in texture and structure. In general, the finer textured soils have the higher total water-holding, or field capacity, but much of the water may be unavailable to the plant.

A certain part of the water held in any soil remains unavailable for plant usage. The moisture content of a soil at which a plant wilts when growing under conditions of limited transpiration, and at which it does not recover turgidity without the addition of water, is termed the "wilting percentage" of that soil. This is the state of equilibrium that exists between the suction forces of the plant and the adhesive forces of the soil. This wilting percentage also varies with soil type but apparently does not vary appreciably with the kind of plant growing in the soil. It is apparently determined largely by forces within the soil itself rather than by forces within the plant.

The availability of the soil moisture to the plant is, obviously, more important than the total amount of water in the soil. The volume of soil taken up by the soil particles themselves varies with the texture and structure of the particular soil. Sandy and sandy loam soils, when saturated, hold from 2.5 to 3 in. of water per foot of depth, of which 1.25 to 1.75 in. is available to the plant after the removal of gravitational water. Clay loams and clays hold from 3.5 to 4.5 in. of water to the foot when completely saturated, of which 2 to 2.5 in. is available to the plant. In general, the amount of moisture at the wilting percentage is higher in the fine-textured and lower in the coarser soils. Sticky clays may show the wilting percentage as half of the field capacity, whereas with the coarser soils the wilting percentage is, in general, less than half the field capacity. Soils may become waterlogged at any season. This condition will prevent growth and may cause the death of the roots due to lack of aeration. It often causes the failure of orchards that require deep, well-drained soils.

Movement of Water in Soil

All the water that falls on the soil does not enter it, as some is lost to the use of plants by surface runoff. It is this water that is largely instrumental in soil erosion. The amount of runoff is influenced by the topographical features of the area, the amount and duration of the rainfall, the characteristics of the soil, the conditions of the soil and the soil cover. Under similar conditions the runoff from sod and cover-crop areas is decidedly less than that from cultivated areas.

Gravitational water moves downward until it reaches an impervious layer in the soil. This causes it to collect and form a saturated area, the upper layer of which is known as the "water table." The water in this saturated area moves slowly by the force of gravity into drainage channels, springs, streams or lakes. The rate of this movement is influenced by the steepness of the slope of the impervious layer that stopped its downward movement and the character of the soil or material through which it must pass. A rain will moisten the top layer of soil, and the films of water surrounding the soil particles will become thicker. Gravitational forces will exceed capillary forces, and the water will move downward until these two forces are equalized.

The capillary movement of water is always in the direction of the greatest capillary tension, or toward the thinner film. It is over but very short distances, and the lateral movement practically negligible. Capillary water may move for a distance of 1 or 2 ft. at a moderately rapid rate in moist soils. This provides for the movement of water to the roots in the immediate vicinity. Roots grow only in moist soils. With an available water supply, the amount of water absorbed from a given area is about directly proportional to the absorbing roots in that area. The distribution and ramification of the root system, although modified by soil fertility, is influenced largely by the availability of a water supply.

EFFECTS OF ABNORMAL AMOUNTS OF WATER

Horticultural plants should have an adequate and steady supply of water. The presence of too much or the absence of sufficient water will result in injury to the plants.

Excess of Water

When the income of water is greater than the outgo, the plant may show certain unfavorable effects. It may be supplied with abundant moisture, and the evaporation be fairly low, resulting in an increase of water pressure within the plant, increasing cell size and often the bursting of cells. An excess of water suddenly following a prolonged dry period often results in the bursting of cabbage heads and the cracking of apples, plums, cherries, carrots, beets and tomatoes. The development of these products was retarded during a dry period, and the epidermis and outer layers of tissue lost the ability to expand rapidly when they again received sufficient moisture. When onions are grown under irrigation, the soil must be kept moist in order that the plants may continue to grow. A temporary check in growth because of water shortage results in the maturing of the outer scales; and when water is added again, the inner scales resume growth, splitting the outer ones. Secondary growth of shoots of many trees and shrubs occurs when heavy late summer rains follow a long dry period.

Shortage of Water

When the outgo of water is greater than the income, the plant shows unfavorable effects. Under conditions of acute shortage of moisture, the stomates of the leaves close, and the leaves wilt and curl. The length of time that the stomates remain closed and the leaves wilted and curled depends upon the acuteness of the shortage and the length of the period of drought. Under conditions of prolonged water shortage there is a marked decrease in total carbohydrate synthesis which is correlated with the period of stomatal opening and shortage of water.

Early defoliation of plants is often associated with a shortage of water. Early defoliation of woody plants like fruit trees reduces the manufacture of carbohydrates late in the season, and this may result in the tree's being more susceptible to winter injury and in less growth the following spring. As mentioned earlier, water shortage results in a reduction of the growth rate of fruit. The size of fruit will be reduced in approximate proportion to the length of time during which growth is restricted. Drought spot of apples is attributed to a shortage of water.

Review Questions

1. What are the important functions of moisture to the plant?
2. In what states does water exist within the plant?
3. How does a plant obtain water from the soil?
4. Can water that has entered the root of the plant return to the soil through the root?
5. Do the foliage or the developing fruits have first demand on a limited water supply?
6. About what percentage of the green weight of a plant is composed of water?
7. Do the mineral elements within the plant move with or independently of the movement of the water within the plant?
8. What is meant by the plant's water requirement?
9. What are the major factors that influence the amount of water used by a plant?
10. What factors influence the rate at which water is supplied to the plant by the soil?

11. What factors influence the rate at which water is lost by a plant?

12. How is water lost from the soil?

13. What is gravitational water?

14. What is capillary water?

15. What is hygroscopic water?

16. What is combined water?

17. How is the water that is available to a plant held in the soil?

18. What are some of the effects on the plant of an excess supply of water in the soil?

19. What are some of the effects on the plant of a shortage of water in the soil?

Problems

1. Mr. A has an apple orchard heavily loaded with fruit. The season has become unusually dry in July, and he cannot irrigate but is considering removing some of the small fruit with the idea of conserving water. State and justify your advice to Mr. A.

2. Of two adjacent tomato plantings *A* and *B*, the plants in plot *A* are making 25 per cent more growth and producing 20 per cent more crop than those in plot *B*. State and explain which group of plants is using the most water and which group has the higher water requirement.

3. What would be the water requirement in acre-inches under the following conditions? Mature apple trees are planted 36 ft. apart by the square system and are producing 35 lb. of dry matter in stems, leaves and roots per year in addition to 10 bu. of apples. A gallon of water occupies 231 cu. in. and weighs 8.33 lb. Assuming that only 33⅓ per cent of the precipitation was available to the plant, what should be the annual precipitation to provide adequate moisture for this orchard?

4. Mr. A has decided to save some money on labor in irrigation and used two irrigation furrows between his rows of fruit trees 30 ft. apart instead of six furrows which he has used formerly. State and explain the probable results.

Suggested Collateral Readings

1. Gager, C. S., "General Botany," pp. 29–130, P. Blakiston's Son & Company, Philadelphia, 1926.

2. Gardner, V. R., *et al.*, "Fundamentals of Fruit Production," pp. 1–35, McGraw-Hill Book Company, Inc., New York, 1939.

3. Lyon, T. L., and H. O. Buckman, "The Nature and Property of Soil," pp. 137–165, The Macmillan Company, New York, 1939.

4. Thomas, M., "Plant Physiology," pp. 82–141, P. Blakiston's Son & Company, Philadelphia, 1935.

5. Weaver, J. E., and F. E. Clements, "Plant Ecology," pp. 424–453, McGraw-Hill Book Company, Inc., New York, 1938.

6. White, E. A., "The Florist Business," pp. 101–104, The Macmillan Company, New York, 1923.

CHAPTER IX

THE HORTICULTURAL PLANT IN RELATION TO LIGHT

Everyone has noticed that most plants will bend toward light and grow better in light than in shade; that when grown in darkness plants either lose or never develop a green color and have elongated, slender and weak stems and leaves which remain small and rudimentary; and that some plants, as the tulip and certain water lilies, open their flowers in the light and close them in darkness. It is not so commonly known that the blooming times of plants are influenced largely by the total number of hours of light rather than its intensity. For example, plants that bloom in the spring and fall do so because in their natural environment they receive about 8 hr. of sunlight. Those which flower in the middle of the summer do so because they have a total of 16 hr. or more of light. To illustrate: During the winter of 1936 it was noticed that the poinsettias (*Euphorbia pulcherrima*) in the horticultural greenhouses at Iowa State College were not coming along satisfactorily and would not bloom by Christmas. Since these plants are used exclusively for Christmas trade, it is important that they bloom at that time. That year they had been treated the same as in previous years, and they had always blossomed satisfactorily. Investigation showed that the reason that they would not bloom by Christmas was because the janitor had inadvertently exposed them to too much light. The poinsettias happened to be on a bench directly under some strong electric lights which were lighted from 4 to 6 hr. each evening. They would not bloom with this long daily light exposure but flowered normally when the practice of turning on the electric lights was discontinued. Obviously the effect of the daily light period on the ability of plants to flower and reproduce themselves by seeds exerts a tremendous influence on the distribution of different kinds of plants on the surface of the earth. All the energy on the earth comes directly or indirectly from the sun. It is used by the green parts of the plants in manufacturing carbohydrates by a process known as "photosynthesis," discussed in Chap. VII. The photosynthetic process is carried on only in the presence of light; and furthermore, in the majority of cases, light is necessary for the formation of chlorophyll. All green plants require it for normal growth.

Light is that part of radiant energy which is visible to the human eye. It is, however, but a small part of the great electromagnetic spectrum. The wave lengths in the visible spectrum, together with a small portion in the shorter ultraviolet and in the longer infrared sections, comprise the wave lengths that are significant in the life of plants. Light has an effect on the natural distribution of plants and upon the amount and kind of growth made by them. It must be remembered, however, that light is but one of the factors of the external environment influencing the activities of plants and that these various factors are interdependent. Similar responses can be obtained by altering either light or temperature. For the optimum growth of any plant with a given temperature, mineral nutrients, water supply and other environmental factors there are certain optimum quality, intensity and duration of light. Light acts indirectly on carbon assimilation by raising the temperature of the leaf and by stimulating the guard cells of the stomates to open. The growth of the plant is influenced by the kind of light, its intensity and the duration of daily exposure to it.

INFLUENCE OF THE KIND OR QUALITY OF LIGHT

The visible spectrum is made up of lights of different kinds or different qualities. The shortest waves produce the violet, and the longest the red color. Different kinds or qualities of light, occupying different areas in the visible spectrum, have various effects on green plants. The short ultraviolet rays are distinctly harmful to plants, but under natural conditions many of these rays are filtered out by the ozone in the outer layers of the atmosphere and consequently never reach the surface of the earth. Experiments have shown that all kinds of the light of the visible spectrum are necessary for normal, vigorous plant growth but that different regions of the spectrum influence plant activities in different ways. Photosynthesis proceeds more rapidly in the red than in the blue-violet end of the spectrum, but the blue-violet and ultraviolet region have the most influence on the development of red pigment in apples. If apples while still green are covered with red cellophane bags which exclude the rays from the blue-violet end of the spectrum, they will not take on their characteristic red color as they mature on the tree. This is the reason that apples develop a redder color in regions with a clear dry atmosphere, because more blue-violet rays reach the fruit than in humid regions where some of these rays are absorbed by moisture in the atmosphere. Concord grapes, however, can be bagged while green but will develop their characteristic purple color.

The distribution of green, blue-green, brown and red algae at different depths in the oceans is apparently in partial response, at least, to the ability of the various algae to utilize the particular light rays that are able to penetrate to the different depths in the water.

INFLUENCE OF LIGHT INTENSITY

It is readily noted that some kinds of plants grow better in less light than others and that some apparently are injured by the full noonday sunlight. Less than 2 per cent of full sunlight is used for photosynthesis. Supplementing the daily light period of sunlight with electric lights of approximately one-thousandth of the intensity of the noonday sunlight is sufficient to bring about the flowering response of long-day plants and cause short-day plants to continue vigorous vegetative growth and fail to bloom. Some plants will grow over a wide range of light intensities; others will bloom over a very limited range.

Certain kinds of plants are able to follow one another in succession under natural conditions, because some are able to stand the partial shade of other plants. In general, those species whose light requirement for survival is comparatively little are able to exist for a long time in the shade; and when more favorable light conditions are presented, these shade-tolerant seedlings are in a position to replace their non-tolerant competitors.

The amount of light reaching the interior of dense-foliaged trees may be reduced to 10 per cent or even as low as 1 per cent of that reaching the exposed area. Under such conditions the leaves become small and yellow, and photosynthesis is greatly reduced. With this reduced photosynthesis, the growth becomes slender; and in the case of apple trees the fruit fails to develop the best flavor, remaining green in color and small in size.

Partial shade is often beneficial in growing plants. Many seedlings, cuttings, grafted plants and new transplants in the nursery or field are protected from too much heat and too rapid transpiration by various shading devices. The quality of some horticultural plants is improved when they are grown in partial shade. The reduction in light is also accompanied by a lowering of the temperature and a lessening in the air movement, thus decreasing transpiration.

Reduction in light intensity tends to increase the length of stem and to promote the growth of broad, thin leaves with loose, open structure. The total absence of light results in the production of plants with weak stems with little mechanical tissues. Partial shade induces succulence and tender structure in asparagus, cauliflower, celery and lettuce. Many vegetable crops grown for the vegetative parts, as potato, carrot, turnip, beet and cabbage, yield best where there is a high percentage of cloudy days. Half shade is employed in forcing rhubarb, and under such conditions it does not develop toughness. In Formosa and to some extent in the Philippines sunlight is excessive for the satisfactory production of the Smooth Cayenne pineapple, so that this variety is grown in the partial shade of trees. The best tea leaves are produced from plants growing in partial shade under trees, and coffee likewise is produced most satisfactorily in subtropical regions when the plants are partially shaded by forest trees. The leaves of tobacco plants grown under shade are large, broad and thin, with poorly developed veins and abundant spongy parenchyma. Such leaves are used for cigar wrappers.

For a given leaf area with a certain chlorophyll content, there appears to be a maximum rate for photosynthesis for any particular light intensity; and for each intensity there is only one temperature at which the photosynthetic process of the plant works most efficiently.

INFLUENCE OF THE DURATION OF LIGHT

In Chap. VI it was noted that the type and amount of plant growth were closely associated with carbohydrate utilization and carbohydrate accumulation. Since the manufacture of carbohydrates by the green plant is influenced by light, it is known that, other conditions being suitable and similar, the green plant that is exposed to favorable light for the longest period of time will synthesize the most carbohydrates. Through the action of light in the formation of various pigments, carbohydrates, etc., profound influences are brought to bear not only upon the processes of growth but also upon the processes of differentiation of cells and organs of the plant. Thus the daily duration of the exposure of green plants to adequate light apparently has an influence in determining whether the plant will develop only vegetatively or will produce flowers. This is probably the morphological or physical response to an internal chemical condition of the plant which was influenced, at least in part, by the duration of the daily exposure to light and the influence of the light on the utilization and accumulation of carbohydrates. The daily period of illumination seems not only to influence the quantity of photosynthetic material that is formed but also may determine the use to which the carbohydrates are put.

Light exerts a marked influence upon the different organs of a plant. Many plants, as the Irish potato, produce tubers only in comparatively short days. Typical tuber formation takes place when the daily exposure to light is decreased below the optimum for stem elongation. Bulb formation in the onion will take place when the daily light period is increased above the optimum for stem growth. Evidence indicates that the yearly variations in the chemical composition of grapes from

FIG. 80.—Method of shading with black cloth. (*Courtesy E. C. Volz.*)

FIG. 81.—Effect of 4 hr. additional electric light (background) for 40 days on Feverfew. Check is in foreground. (*Courtesy F. A. Volz.*)

the same vineyard or from vineyards of the same variety located at different areas are due in great measure to differences in the amount of sunlight received during the growing season. Many plants will flower only when they are exposed to rather definite daily periods of light, whereas the flowering of others does not appear to be influenced by exposure. Those plants which blossom only when exposed to a daily light exposure of 12 hr. or less are known as "short-day plants"; those which require 16 hr. or more are known as "long-day plants." Those which flower irrespective of the number of hours of light are known as "neutral plants." The term "photoperiodism" has been coined to

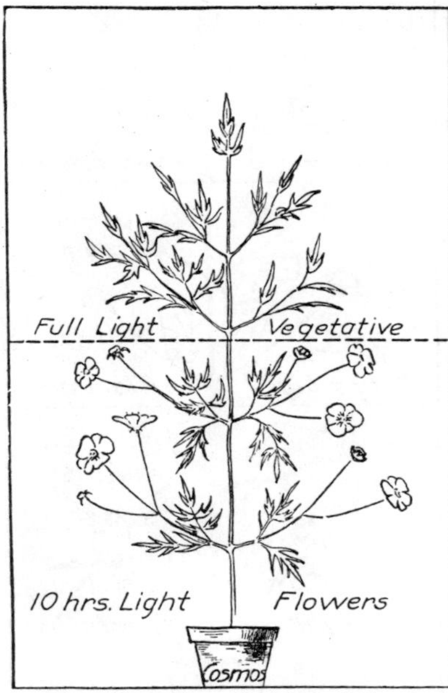

FIG. 82.—The upper part of the plant was exposed to the full length of daylight (about 16 hr.) and the lower part of the same plant received but 10 hr. of light daily. The upper part of the plant remained vegetative and the lower part flowered.

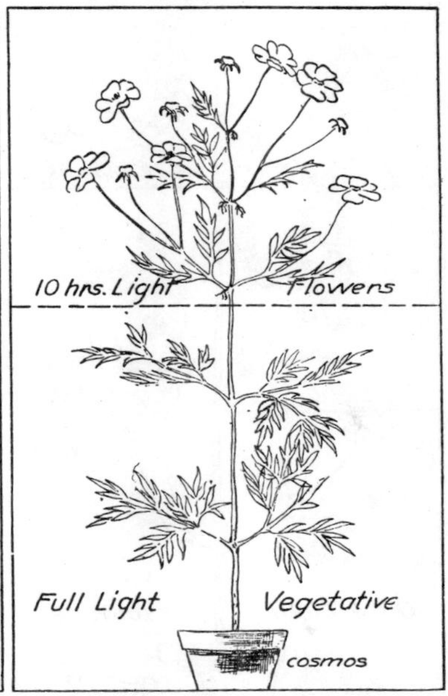

FIG. 83.—When the treatment was reversed and the upper part of the plant received but 10 hr. of light daily, the lower part of the plant remained vegetative and the upper part flowered.

designate the responses that plants make to the length of daily exposure to light.

An experiment was conducted to determine the length of time that a herbaceous plant might be maintained in a vegetative condition without flowering by controlling the daily exposure of light. Two species of Sedum were used in the experiment. These are all long-day plants, but some were exposed to daily light periods of 8, 10 and 12 hr., and the check plants received the full daily period of light. Throughout a period of eight years for *S. woodwardi* and nine for *S. spectabile*, the plants remained healthy and vigorous. All plants that received 12 hr. or less of light each day from March to October did not flower but made only vegetative growth. The check plants which received the full daily period of light blossomed normally each year. Beginning in June of the tenth year, the *S. spectabile* plants that had been growing

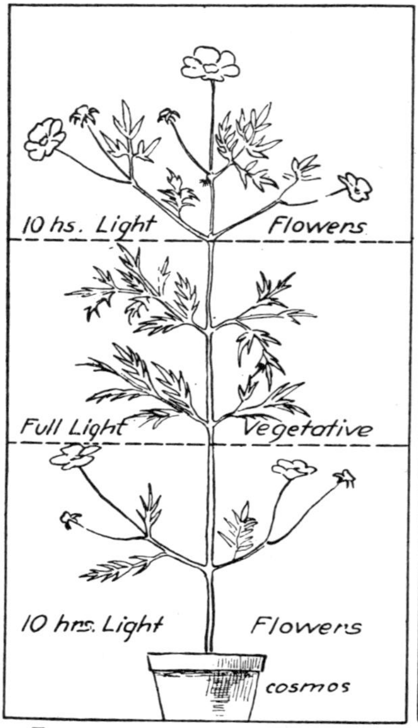

FIG. 84.—After the middle of July, the central portion of the same cosmos plant received the full length of daylight, but the upper and lower portions of the plant received but 10 hr. of light daily. The central portion remained vegetative but the upper and lower portions responded in the characteristic manner of a short-day plant and produced flowers.

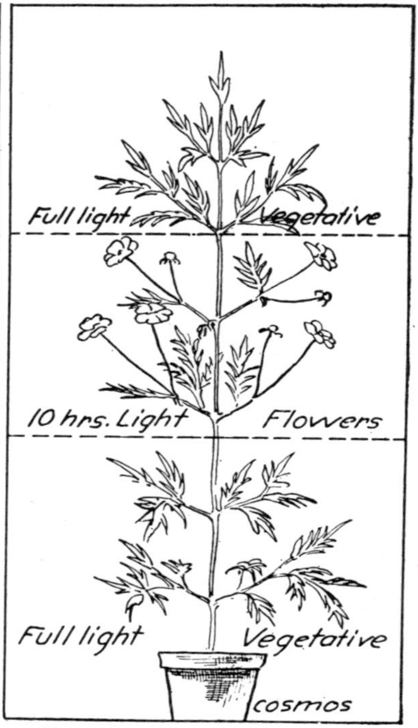

FIG. 85.—When the treatment was reversed and the upper and lower portions of the plant received the full length of daylight and the central portions of the plant received but 10 hr. of light daily, the upper and lower portions remained vegetative and the central portion produced flowers.

for nine years in shortened light periods were exposed to the light during the full summer day. Although they had not flowered for the nine previous years, they flowered in August at the same time as the check plants that had flowered each of the previous nine years. Short days apparently prevented flowering in these plants.

Cosmos sulphureus, a short-day plant which flowers when exposed to a daily light period of 10 to 12 hr., shows not only the response of a plant to different periods of daily exposure to light but also rather strikingly the localization of the responses to restricted parts of the same plant.

Florists are growing flowers out of season by altering the daily exposure to light. They are increasing the daily light period in the winter season on long-day plants, as feverfew (*Chrysanthemum parthenium*), with satisfactory results. Adequate lighting is obtained with 100-watt reflector lights placed about 18 in. above the plants. One light is sufficient for an area 4 ft. square. Some short-day plants, as chrysanthemum (*indicum* x *morifolium*) and stevia (*Piqueria trinervia*), are being produced commercially out of season by shading with black cloth and thus reducing the daily period of exposure to light to 8 or 10 hr., depending upon the requirements of the particular plant being grown. Standard varieties of chrysanthemums are brought into flowering earlier by shortening the daily exposure to light. The treatment is started about six weeks after planting or when the plants are 18 to 24 in. high. Experiments have shown that an 11-hr. day is the most satisfactory for blossom formation and that the most satisfactory method of controlling the light period is to place a heavy black cloth about the plants at five o'clock in the afternoon and remove it at seven the following morning. Treatment is continued for 30 to 35 days or until the flower bud is from ¾ to 1 in. in diameter.

Review Questions

1. What is light?

2. Is light required for the photosynthetic process?

3. Is any kind of light injurious to the plant?

4. Are there different kinds or qualities of light?

5. Does the green plant require all the kinds of light of the visible spectrum for normal development?

6. Do the different kinds of light have the same effect on the green plant?

7. Does the plant require all the light of full sunlight for photosynthetic activity?

8. Are plants or leaves of plants grown in shade different from the same kinds of plants or leaves grown in the sunlight?

9. Do all kinds of plants require the same intensity of light?

10. Are any plants intentionally shaded during growth?

11. Does the duration of the daily exposure to light appear to have any influence on the kind of growth of horticultural plants?

12. Will different parts of the same plant respond characteristically to different daily exposures to light?

13. What is meant by a short-day plant?

14. What is meant by a long-day plant?

15. What is meant by a neutral plant?

16. What might be a plausible explanation for the two types of growth—vegetative and flowering—made by plants?

Problems

1. An apple grower has noticed that the Jonathan apples in the tops of the trees are always more highly colored than those on the lower branches of the same trees. He would like to know the cause of this difference in color and if he could do anything to have all the apples the same high color.

2. A nurseryman has always shaded his green-wood cuttings but has been told that sunlight is necessary for photosynthetic activity and that the cuttings would do better if they were not shaded. What is your advice?

3. A commercial florist has a planting of chrysanthemums (*C. indicum* x *morifolium*) that he wishes to bring into bloom a couple of weeks sooner than similar plants in another house. He wants to know if this can be done and, if so, how he should proceed.

4. A vegetable grower wishes to know why the young tomato plants growing closely together in flats are taller and more slender than similar plants that had been transplanted from flats to 4-in. pots.

Suggested Collateral Readings

1. ARTHUR, J. M., *et al.*, Artificial Climate, pp. 416–482, *Amer. Jour. Bot.*, Vol. 17, 1930.

2. LAURIE, A., and G. H. POESCH, Photoperiodism. The Value of Supplementary Illumination and Reduction of Light on Flowering Plants in the Greenhouse, *Ohio Agr. Expt. Sta. Bul.* 512: 1–42, 1932.

3. LAURIE, A., and G. H. POESCH, "Commercial Flower Forcing," pp. 45–79, P. Blakiston's Son & Company, Inc., Philadelphia, 1939.

4. WEAVER, J. E., and F. E. CLEMENTS, "Plant Ecology," pp. 380–417, McGraw-Hill Book Company, Inc., New York, 1938.

CHAPTER X

THE HORTICULTURAL PLANT IN RELATION TO SOIL

Soil is the product resulting from the various stages of disintegration and decay of rocks and plant and animal materials in which plants can grow. The resulting product varies widely in physical, chemical and biological characteristics and consequently in its ability to support

Fig. 86.—Fruit orchards in the world famous Hood River Valley of Oregon. Compare with Fig. 17. (*Union Pacific Railroad.*)

plant growth. In general, soils may be classified as mineral or organic. The geologic, or parent, materials from which mineral soils developed originated chiefly through the disintegration and decay of consolidated rock, whereas the parent materials from which organic soils developed originated chiefly through the decomposition of plant and animal materials. The average content of organic matter in the topsoils of the mineral soils of the United States is about 2 per cent, and that of their subsoils is about 0.8 per cent. Peat and muck soils, which often contain more than 70 per cent of organic matter, are important for growing certain horticultural crops. Soils must have a suitable topography, a satisfactory texture and structure and adequate fertility for the successful and profitable production of horticultural crops.

SOIL TOPOGRAPHY

The importance of soil topography in relation to air and water drainage was discussed in Chap. VII. Certain soil topographies are more suitable than others for the satisfactory and economical production of horticultural crops because of certain cultural operations and the influence of necessary tillage practices on soil erosion. Flowers and vegetables and a few fruits, as strawberries and raspberries, are generally confined to level or slightly rolling land, as the necessity for tillage in such crops would lead to injury by erosion on steeply sloping land. Tree fruits and grapes may be planted on steeper slopes, and the system of soil management adjusted accordingly. Many of the orchards in the Appalachian Mountains are planted on steep hilly land, whereas in the irrigated sections of the West the orchards occupy the level land in the valleys. The degree of slope that is suitable from the standpoint of cultural practices, soil erosion and operating costs varies widely with different localities.

SOIL FERTILITY

The fertility of a soil is a measure of its ability to support plant life when provided with favorable temperature, moisture and light. The growth of the plant is an index of the relative amounts of essential elements in the soil that are available for plant use. Soils exhibit all stages of fertility from sterile sand to excessively fertile loams. A fertile soil in one locality might be considered only a moderately fertile to infertile soil when compared to soil in another locality. The degree of fertility will vary also with the kind of plant being produced. A very acid soil would be unproductive and considered infertile if used for growing asparagus, but it would be considered quite fertile if devoted to the production of blueberries.

The fertility of the soil is determined by the interaction of physical, chemical and biological forces. No one of these forces operates independently. A change in the physical structure of the soil alters the temperature, moisture and air conditions in the soil, and these influence the kind and amount of oxidation and biological activities, which in turn have their influence on the chemical condition of the soil. For example, organic matter may be added to improve the physical texture of a soil, but with this improved condition the organisms responsible for decay of organic matter soon destroy the organic matter that was added and liberate for use of the growing plants the chemicals bound up in the organic matter that was added to the soil. Organic matter is also an important source of much of the food supply of the micro-organisms in the soil and consequently may cause their rapid increase.

Many factors, as soil texture, soil aeration, soil temperature, moisture supply, organic supply, the presence and availability of plant nutrients, the kind and amount of soil organisms and their activity and the soil reaction, are associated with the problem of soil fertility.

Nutrient Supply

The plant nutrient supply in the soil exists in both inorganic and organic forms. Physical, chemical and biological processes are continually at work breaking down the complex chemical substances into their elements and making the nutrients available to the plant. All the processes involved in making organic and inorganic nutrients available for plant use are not fully understood. It is known, however, that there are several reactions occurring in the soil that produce weak acids and that these acids liberate unavailable nutrients. Certain bacteria and other microorganisms as well as larger animals aid in the decomposition of the mineral and organic matter of the soil and reduce the nutrients to forms in which they are soluble and available to the plant.

Soil Organisms

The soil is teeming with both plant and animal organisms, most of which are only microscopic in size. These organisms are responsible for the biochemical processes that reduce plant and animal remains to the carbon dioxide, ammonia and minerals from which they were made. There are hundreds of kinds of microorganisms in the soil carrying on the processes of decay and decomposition.

In the breakdown of organic matter most of the carbon dioxide escapes from the soil into the air, but some is combined with soil water. The ammonia, however, is absorbed by the soil, and very little is lost. That absorbed is rapidly changed to nitrite, and this to nitrate, by nitrifying bacteria. Azotobacter are a type of soil-living bacteria that can use the free nitrogen of the air in building up the proteins in their own bodies and, by continued growth, reproduction and death, increase the nitrogen supply of the soil. Several different groups of bacteria are capable of producing nodules on the roots of certain leguminous plants. These nodules are caused by bacteria that penetrate the roots and stimulate the plant to produce an enlarged growth, or nodule, at that point. The bacteria grow and reproduce inside the nodules, getting their carbohydrate and mineral food from the plant and their nitrogen from the air to form proteins which are released to the plant. Legumes, therefore, are able to grow normally in soil poor in nitrogen, provided the soil contains suitable nodule-forming bacteria and other conditions are favorable. Such legumes if left in the soil increase its nitrogen content.

Soil Reaction

The growers of vegetables, small fruits and ornamentals in areas along the Atlantic Coastal Plain and Piedmont soil provinces have become concerned comparatively recently with the acuteness of the problem of soil acidity. The use of large quantities of distinctly acid-forming fertilizers since about 1925 has increased the degree of acidity in many sections so that some crops are apparently severely injured by the soil conditions. In the important commercial vegetable area on the eastern shore of Virginia and Maryland, a test of several thousand soil samples showed an acidity much too high for the successful and economical production of many vegetable crops. Because of the economic importance of the vegetable crops, the problem of soil acidity is particularly acute in areas along the Atlantic Coastal Plain but is also of importance in other areas and under other situations. Soils are acid, neutral or alkaline in their reaction.

Plants vary greatly in their requirement with reference to the degree of acidity or alkalinity of the soil. Asparagus, spinach, lettuce, currants and gooseberries grow best when the soil is only slightly acid, neutral or slightly alkaline; potatoes, watermelons and strawberries prefer slightly acid soils; blueberries, cranberries and azaleas grow best when the soil is highly acid. The degree of acidity or alkalinity is expressed by the hydrogen-ion, or pH, scale. A pH value of 7.0 represents neutrality. Values higher than 7.0, such as 7.2, 8.0, 9.0, 9.5 and 10.0, denote alkalinity, the degree of alkalinity increasing with the pH. Values lower than 7.0, as 6.8, 4.0, 2.5, denote acidity, the degree of acidity increasing as the pH decreases. The scale of pH values is logarithmic, which means that a soil with a reaction of pH 4.5 is ten times as acid as one of pH 5.5.

It is easier to lower the acidity or increase the alkalinity of a soil than it is to increase its acidity or lower its alkalinity. It is also easier to change the reaction in a light soil than in a heavy soil. A sandy soil with a pH of 5.0 can be changed to a pH of 6.0 by the addition of about 1,000 lb. of hydrated lime per acre. For heavier soils, larger quantities of hydrated lime are necessary to obtain a corresponding reduction in acidity. Plowing under animal manures, green manures or large amounts of organic matter is the cheapest way of increasing the acidity of a soil, but the change takes place very slowly. The use of acid to increase the pH range is too expensive for com-

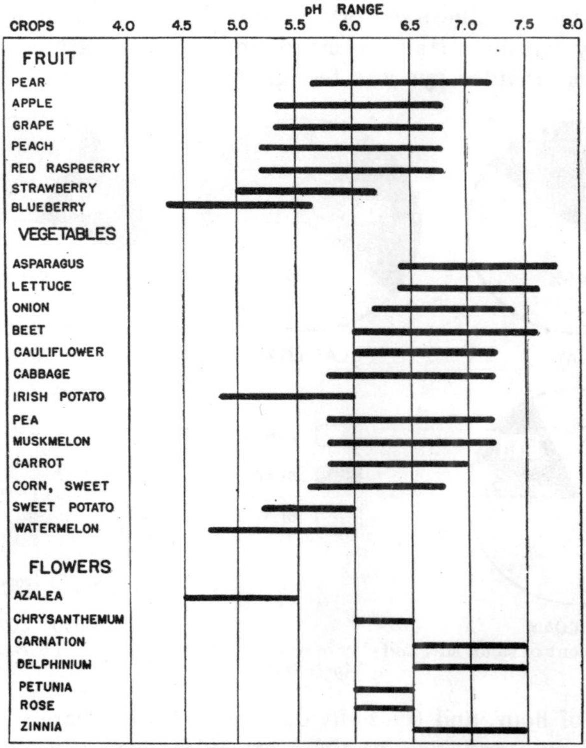

FIG. 87.—Desirable pH range for selected horticultural crops.

mercial practice. Aluminum sulfate is often used to increase the acidity of small quantities of soil.

The pH of soils is measured in various ways, ranging from the rather crude litmus test through the fairly accurate colorometric methods to the highly accurate electrometric determinations.

PHYSICAL CONDITION OF SOIL

The maintenance of a suitable physical condition of the soil is fully as important in the production of satisfactory crops as is the maintenance of suitable fertility. The physical condition of the soii is determined by the texture, which refers to the size of the individual soil particles, and the structure, which refers to the arrangement of these soil particles within the soil mass. The principal soil groups based on the size of soil particles are sand, silt and clay. Varying proportions of these soil particles of different sizes will arrange themselves differently and produce radically different physical conditions in the soil mass. The sizes of the particles in each of these three groups vary within certain specified limits. Sand grains are easily visible to the unaided eye and feel gritty to the fingers; silt has the

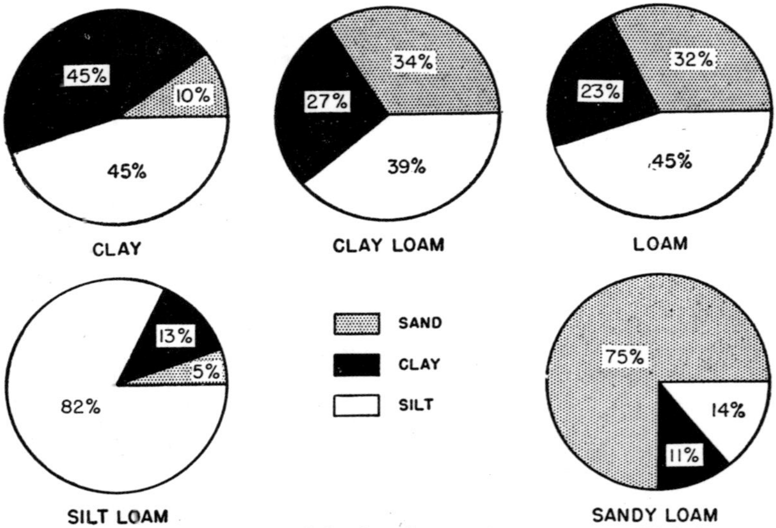

FIG. 88.—Per cent of sand, silt, and clay in soils of various textures. (*U.S. Department of Agriculture.*)

appearance of flour, and the individual particles are barely visible to the unaided eye; and clay particles are microscopic or less in size, some of the particles actually becoming colloidal. Soils containing various proportions of different-sized particles have been classified according to texture and given specific names which designate the relative amounts of the different-sized particles present in the soil mass. On the basis of their porosity, soils are considered as heavy, medium and light.

HEAVY-TEXTURED SOILS

Clay and clay loam soils are termed "heavy soils." Heavy soils are very fine textured. The compactness of such soils inhibits the ready absorption of moisture, prevents adequate percolation and aeration, causes excessive surface runoff and holds large quantities of water so tenaciously that it is often too slowly available for the plant. These soils are likely to be fertile but are cold and relatively unyielding to the growth of roots. Heavy soils are difficult to handle and must be worked at the proper time to prevent baking and the formation of hard clods. Heavy soils puddle after rains and form a crust which often makes it difficult or impossible for young seedlings to break through.

Light-textured Soils

Sandy soils and sandy loams are light soils. Light-textured soils are too loose and absorb water readily, allowing it to percolate away rapidly carrying with it the soluble elements. Light soils, consequently, are likely to be relatively infertile. They are warm, well aerated, well drained and offer little resistance to the penetration of roots. Sandy loam soils are perhaps more widely used for vegetable growing than any other type of soil. Sandy loams are easily worked into fine seedbeds and warm up early in the spring. Sandy loams underlaid with a fairly heavy subsoil that drains well but is also retentive of moisture are the best type for a large number of vegetable crops.

Medium-textured Soils

Loam and silt loam soils are generally considered medium textured. An attempt is made to combine the good features of the heavy soil with the good features of the light soil and to eliminate, in so far as possible, the undesirable features of both types. The texture of the soil will be best when it is composed of suitable quantities of various sized particles of sand, silt and clay intermixed with decaying organic matter. Such soils furnish the most desirable medium for the growth of roots of most horticultural plants. The medium-textured soils often appear cloddy when dry, but the lumps are broken easily, and the soil becomes soft and floury. Silt soils warm slowly and are generally fertile.

IDEAL SOILS FOR SELECTED HORTICULTURAL CROPS

Because of the fact that horticultural plants have different soil requirements and that the soil requirements for the same kind of plant vary with climatic conditions, definite specifications for the ideal soil for all horticultural crops for the United States as a whole would be difficult to formulate. There are, however, qualities that characterize a soil adapted for fruits, for vegetables or for ornamentals.

Fruits

The correct selection of the proper soil for tree fruits is perhaps of more importance than that entailed in the selection of soil for any other group of crops. This is true because an orchard represents a long-time and costly investment, and the plants occupy the same area for many years. The discovery that the soil is unsuited for tree fruits often occurs only after the orchard has been growing for a number of years. This results in great economic loss, because to improve the soil is impossible or highly expensive and to retain the orchard means only poor to moderate yields. The idea that land too infertile for

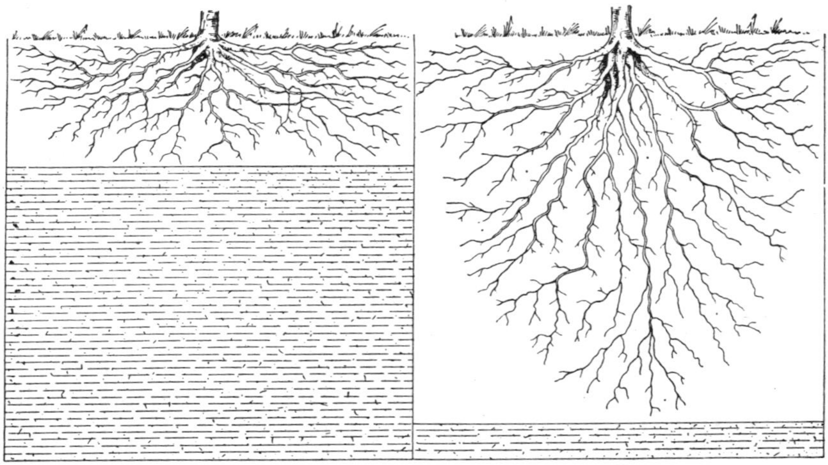

A B

FIG. 89.—Diagrammatic sketches showing the distribution of the root system of a tree in: *A*, poorly drained, and *B*, well-drained soil.

farm crops might well be used for an orchard is a result of the observation that trees often lived and bore fruit on soil that did not produce a profitable grain crop and that trees planted on more fertile soils were more subject to winter injury and blight and frequently were unreliable in bearing. For profitable production, fruit trees must be planted on moderately fertile soil of suitable texture.

The texture of both surface and subsoil is important in the success of an orchard. Fruit trees have relatively wide and deep root systems which occupy practically the same area for many years. The texture of the subsoil is very important in the development and growth of the root system. The texture and fertility of the topsoil can be changed fairly readily, but very little change can be made in the fertility and practically none can be made in the texture of the subsoil. Consequently, in selecting a soil suitable for tree fruits the subsoil should receive more consideration than the topsoil. The ideal texture of the soil for orchards is one that never becomes waterlogged but stores a large amount of available water. In general, the silt loams combine excellent available moisture storage with good aeration and drainage. The moderate clays and clay loams are excellent when they are underlaid at a depth of several feet with gravel or sand which permits suitable drainage.

Orchard soils should be sufficiently deep to supply the needs of the trees. This depth will vary in different sections according to the soil type and climatic conditions. In general, a deep soil means a topsoil of 12 to 18 in. which is underlaid with a fairly porous subsoil which gives proper drainage and aeration and thus will permit the penetration of roots to a depth of from 4 to 5 ft. to as much as 8 to 10 ft. A soil 5 ft. deep is generally recommended as the minimum in irrigated sections. In West Virginia, where the precipitation is fairly high and evenly distributed over the year, a soil 3 ft. deep is satisfactory for apples. In Nebraska, where apple orchards are planted on loess soils and where protracted summer droughts occur, a depth of 25 ft. is being recommended. In New York, apple trees live longer and produce more profitably on well-drained, even-textured, sandy loam soils which permit root penetration to a depth of 8 to 9 ft. than on more heavy-textured soils in which the tree roots penetrate only about 3 ft.

Regardless of the depth recommended, it is important to emphasize that in no case should one plant an orchard where a bedrock, hardpan or impervious layer in the subsoil will cause waterlogging and poor root growth. Often a fertile topsoil may cover a hardpan of clay and be quite deceiving to the prospective orchardists. The accumulation of water in spots below the surface often goes unnoticed until the trees begin to show signs of weakened growth. A high water table at any time of the year may result in the killing of roots during the wet weather and consequently be as fatal to the trees as a permanent high water table.

Soil fertility that produces satisfactory yields of the general farm crops is considered sufficiently fertile for orchards. The fertility can be increased by adding fertilizers and by turning under leguminous crops, but it is better to start with a suitably fertile soil rather than plan on making a fertile out of an infertile soil.

The deciduous tree fruits have a wide tolerance to soil reaction, and no optimum has been determined for any of them. The soil reaction should be considered, however, in reference to the planting of certain cover crops. Chlorosis of the leaves often occurs on fruit trees growing in very alkaline soils. This condition results in the weakening and often death of the trees.

Small fruits are grown on a great variety of soils and require soils that are moderately fertile, well drained, easily cultivated and of good moisture-holding capacity.

Apple.—The characteristics mentioned previously concerning the requisites for tree fruits in general apply to the type of soil favorable for apple trees. The loams are more suitable than any other of the many different types of soils that are producing apples successfully throughout the United States. There are some special varietal considerations. For example, Rhode Island Greening does better on rich heavy soil than the Baldwin does.

Peach.—Peaches are grown on soils of somewhat lighter texture than that considered most suitable for apples. Types of soil existing in some of the principal peach sections of the United States include gravelly and sandy loams in New York, sandy soils in North Carolina, sandy loam in Georgia and fine sandy loam in California.

Strawberry.—Sandy loam soils are the most desirable type for growing strawberries, but other types are suitable provided they are well drained but retentive of moisture and fairly fertile. The reaction of the soil is of minor importance, provided it does not reach extremes. Experiments indicate that growth is obtained on soils with a range of pH from 5.0 to 7.0 when the organic content is relatively high. With lower organic content the range for best growth is less, being from 5.7 to 6.5 or less. Strawberries are producing satisfactory crops in Delaware and Virginia on sandy loam soils, in Louisiana on silt loam and in Arkansas and Missouri on shale loam.

Vegetables

Vegetable crops are grown most extensively on deep, well-drained, friable soils ranging in texture from sands to clay loams. A high level of fertility and a good supply of decomposing organic matter are essential. Soil reaction varies considerably, but the majority of vegetables are intolerant of strong acidity or alkalinity. Liming is a common practice in the vegetable areas of the East and South.

The quickly warming sandy soils are well suited to the early short-season crops and for those vegetables like sweet potatoes, cucumbers and tomatoes which require high temperatures for their best development. Vegetables grown in clay loams are generally slower in developing than when grown in lighter soils, and the root-crops are poorly formed and have numerous coarse side roots.

Muck and peat soils are of importance in some areas for growing such cool-season vegetables as potatoes, celery, onions and cabbage. Peat and muck have developed from the decomposition of vegetation in marshes, bogs and swamp forests. Muck is more fully decomposed than peat, but the type of peat or muck depends upon the plant material from which it was derived. Peat soils that resulted from the decomposition of evergreens are usually unsatisfactory owing to resin deposits. Peat and muck soils may be high or low in lime, low in phosphorous and potash and high in nitrogen.

Irish Potatoes.—Sandy loams and peat soil are generally used for commercial plantings of Irish potatoes. These soils are loose, deep, well drained, friable and fertile. In the light, well-aerated soils the tubers are smooth and of typical varietal shape. Potatoes grow best on acid soils, and potato scab is less prevalent than in neutral or alkaline soils.

Tomato.—Good crops of tomatoes are produced on a wide variety of soil types. In Florida they are grown on well-drained sandy land, on marl and on muck; in Texas in the lower Rio Grande Valley they are grown on sandy loams and clay loams. The "quick" sandy soils are selected for the early crop, but heavier, more fertile and more highly productive soils are selected for the late market and canning crops. Muck and peat are seldom used for tomatoes because they are high in nitrogen and moisture, and this combination is conducive to excessive growth of vines and less fruit.

Onion.—Sandy loams, silt loams and muck soils are the preferred types for onions. Much of the onion crop in the North and in California is grown in muck soils. These soils generally have a high water-holding capacity and, as a rule, are loose and fertile, a condition that favors the development of a smooth bulb. The best reaction of the soil for onions is one that is neutral or slightly alkaline.

Ornamentals

The wide variety of ornamental trees, shrubs and herbaceous annuals and perennials grow on many diverse types of soil. There are ornamentals adapted to soils that vary from sterile, infertile sands of the dunes to the deep and fertile soils of the prairie. Some ornamentals prefer cool, moist soils; others require warm, well-drained soils. With all this variation, however, there are a few rather definite soil specifications common to a large number of ornamental plants.

The texture of the subsoil for trees and shrubs whose roots penetrate several feet should be similar to that suggested for tree fruits. The texture of the best all-round garden soil for herbaceous flowering plants is a sandy loam in the topsoil underlaid with a subsoil that is retentive of moisture but drains well.

The soil reaction is important for herbaceous flower plants. The majority of cultivated shrubs and herbaceous flowers thrive best when growing in soil that is neutral to slightly alkaline. The broadleaf evergreens—azaleas, laurels, rhododendrons—the heathers and a large number of native woodland flowers, including arbutus and lady's-slippers, require acid soils.

Rose.—Well-drained, heavy clay loam soils that are retentive of moisture are well suited to all kinds of roses. Many excellent roses are grown in a slightly acid or neutral, fertile topsoil of only 10 in., provided it is over a well-drained subsoil; but better results will be obtained if the topsoil is 18 in. deep. Although the soil characteristics suggested are considered the best, roses will thrive in fertile, well-drained garden soils of textures that vary from heavy clay loams to light sandy loams.

Rhododendron.—Soil acidity is the watchword among those who grow rhododendrons and azaleas. The amount of acidity needed for best results ranges between pH 4.5 and 5.0. Where plenty of organic matter, such as peat, is present, the pH can go up to 5.5 without injury.

A proper garden soil for rhododendrons and azaleas contains a high percentage of peat, half-rotted oak leaves or other acid-forming fibrous organic materials. This type of soil possesses both the physical and the chemical requirements of a soil suited for rhododendrons. The acidity and organic matter should be maintained by the annual application of an acid-forming mulch such as oak leaves or peat. Some idea of the percentage of organic matter needed may be noted from the fact that where sharp, quartz, neutral or acid sand is used it should be mixed with 50 per cent or more of peat by volume.

Tulip.—Tulips grow well and produce satisfactory bulbs on any fertile, friable, well-drained soil.

ARTIFICIAL ROOTING MEDIUMS

Recently there has been considerable interest in growing plants in nutrient solutions. The plants are placed in sand, gravel, crushed rock or cinders in wide, shallow, watertight benches. The required nutrient solution is pumped into the bottom of the bench until the gravel or other rooting medium is just flooded. The solution is then allowed to drain away from the bench. This operation is repeated three or four times each 24 hr. The system requires knowledge in selecting a suitable rooting medium, careful attention in preparing and maintaining a suitable nutrient solution, skill in maintaining a suitable pH condition in the nutrient solution and judgment in providing suitable aeration.

Review Questions

1. What is soil?

2. What is mineral soil?

3. What is an organic soil?

4. Why are orchards generally planted on rolling land in the East and level land in the West?

5. What is meant by a fertile soil?

6. What three interacting forces determine the fertility of a soil?

7. In what forms do the plant food elements exist in the soil?

8. How are these elements made available to the plant?

9. Is the soil only an inert substance?

10. What is meant by pH 7?

11. What will a soil solution testing pH 5 do to blue litmus paper?

12. What will a soil solution testing pH 8 do to blue litmus paper?

13. What determines the physical condition of a soil?

14. Is the type of subsoil of more significance in growing fruit trees or vegetables?

15. Characterize the ideal soil for an apple orchard.

16. Characterize the ideal soil for growing Irish potatoes.

17. Characterize the ideal soil for growing rhododendron.

Problems

1. Using different-sized circles to represent sand, silt and clay, make a diagrammatic sketch showing the appearance of a section of soil containing only sand particles of one size. Make a second diagram showing the same section of sand to which has been added 14 per cent silt. Make a third diagrammatic sketch showing the same section of sand and silt but to which 11 per cent clay has been added.

2. State and explain (1) the effects of the foregoing on surface area and (2) its relation to water-holding capacity and feeding area of roots.

3. A florist is using the local city water for a large house of azaleas. The leaves are turning yellow and dropping. What is your procedure in diagnosing the trouble, and what are your suggestions for correcting the difficulty?

4. State and justify your opinion on the possibility of the production of vegetables and flowers in artificial rooting mediums replacing their culture in soil.

Suggested Collateral Readings

1. Emerson, Paul, "Principles of Soil Technology," The Macmillan Company, New York, 1930.

2. Morgan, M. F., *et al.*, The Soil Requirements of Economic Plants, "U.S. Department of Agriculture Yearbook," pp. 753–776, 1938.

3. Thompson, H. C., "Vegetable Crops," pp. 24–35, McGraw-Hill Book Company, Inc., New York, 1939.

4. Weaver, J. E., and F. E. Clements, "Plant Ecology," pp. 173–232, McGraw-Hill Book Company, Inc., New York, 1938.

5. Weir, W. W., "Soil Science," J. B. Lippincott Company, Philadelphia, 1936.

CHAPTER XI

PROPAGATION OF HORTICULTURAL PLANTS

The earth is covered, the waters are filled, the atmosphere is permeated and the soil is teeming with plants. These plants vary from the single-celled, ephemeral microscopic bacterium that is liberating nitrogen in the soil or from the fungus spore of apple scab or brown rot that is floating in the air to the more diversified, complex mosses and liverworts through increasingly complex ferns, until the variations in size and complexity of structure have been climaxed in such plants as the giant redwoods, asters and orchids. Even under adverse circumstances most of the different kinds of plants are able to perpetuate themselves, and under more favorable conditions they increase in numbers. Such perpetuation or increase in numbers is known as "plant propagation."

Even the casual observer will note the many different varieties of the same kind of horticultural plant. For example, the products of apple trees vary in color from red to yellow and in taste from sweet to sour. Similarly, there are many different varieties of vegetables and of ornamental plants. Many of these varieties have been known for two or three centuries, whereas others have been known for only a few years. One of the problems of the horticulturist which is associated with the propagation or perpetuation of a particular variety is that of obtaining better varieties. The attempt to improve plants by obtaining new characteristics or combining in one individual characteristics that are now present in two or more individuals has resulted in hundreds of new varieties of horticultural plants. Probably 25 per cent of the varieties of vegetables and flowers that are grown today were unknown twenty-five years ago. Hybridizers have created a startling array of many new kinds of plants. The garlion is a cross between the garlic and onion, the topeppo combines the tomato and the pepper, and the tangelo is a cross between the grapefruit and the tangerine. New kinds and varieties of flowers in whole rainbows of colors are being produced. The plantsman attempts to improve the present plants by crossing or hybridizing plants with different characteristics which he hopes to combine in one individual and by discovering and perpetuating fortuitous changes or mutations that he observes. Both these methods are proving very satisfactory in improving horticultural plants. The present horticultural plants have arisen largely without the intentional assistance of man.

The Delicious apple is an unknown seedling which originated in Iowa, but the Starking arose as a mutation on a Delicious tree growing in New Jersey. The McIntosh apple is a chance seedling which arose in Ontario, Can.; the Ben Davis arose as an unknown seedling, probably somewhere in the Shenandoah Valley, but the Cortland, credited with being the best apple introduced in the twentieth century, is a

FIG. 90.—Improving horticultural plants, each screen contains a single plant selected because of its desirable characteristics. Screened plants are protected from undesirable pollen, and this controlled pollination makes it possible to combine desirable characteristics of known parents in the new individual. (*Courtesy of Ferry-Morse.*)

seedling of a handmade cross of the Ben Davis by McIntosh, which was made at the New York State Agricultural Experiment Station at Geneva, N. Y. Innumerable similar examples could be given of other named varieties of horticultural plants that have arisen in all these ways.

More horticultural plants in the future will be the results of well-planned hybridizing and the selection and propagation of desirable fortuitous or caused mutations. In the future, plant breeders think that every climatic zone will have a much wider variety of plants due to the creation of new varieties which will flourish away from their original environment.

With such great differences in plants as have been noted, one would expect corresponding variations in the methods by which these diverse plants are propagated. Great variations in the manner of propagation do exist, but a detailed study reveals a surprising similarity in the fundamental principles underlying the propagation of all plants.

METHODS OF PROPAGATING PLANTS

With all this great diversity in plants there are but two methods by which they can be propagated: by sexual and by asexual means. Some kinds of plants can be propagated only by one method, whereas others can be propagated by both.

SEXUAL PROPAGATION OF PLANTS

Some of the lower plants are propagated by sexual spores, and some of the higher plants by seeds. The formation of such spores and seeds is brought about by the fusing of two cells commonly known as "male" and "female gametes," or "sex cells," or as "pollen" and "ovule." The fusing of the male and female cells results in the formation of a single new cell which is the first cell of a new plant. If this new individual is a member of one of the lower divisions of the plant kingdom, it is known as a "spore"; but if it belongs to the highest division of the plant kingdom, that which produces flowers, it is known as a "seed." The spore remains in the single-cell stage through a resting period after which it starts to grow. The first cell of the seed multiplies and forms a small plant. A seed may be defined therefore as an embryonic plant with or without a reserve food supply all of which may be enclosed in a seed coat.

Seeds are used extensively by the horticulturists to propagate plants. Since there are certain limitations to their use and requirements for their successful germination, it will be advisable to consider them further.

Seed Formation.—Seeds are formed, in flowering plants, as the result of the fusing of two sex cells, the male and female gametes, or the sperm and egg cells, which are produced in different parts of the same plant or different parts of different plants. The formation of these sex cells; their transfer from the anther to the pistil, known as "pollination"; and the fusing of the sperm and egg, known as "fertilization," were explained in the discussion of the flower in Chap. V.

Seed Production.—The production of seeds to produce horticultural plants varies in method and extent from those of the home gardener who saves a few bean, corn, tomato, pea, zinnia, nasturtium and other vegetable and flower seeds, to the amateur or commercial collector who gathers large quantities of seeds from the native habitats of the plants, to the commercial grower of seeds who devotes large acreages to the production of a single kind or variety of vegetable or flower seed. These seed-producing areas are located in regions especially suited to the production of seed of that particular plant.

Cleaning Seed.—The method followed in cleaning seeds will vary according to whether the seed are in fleshy or dry fruits.

The manner in which fleshy fruits are handled is influenced by the amount and kind of pulp, the time and condition of the fruit when harvested and the time and manner of storing the seed. Some fleshy fruits that contain but a relatively small amount of pulp or but a single seed to the fruit are spread in thin layers and dried. The dried pulp may be removed or planted with the seed. Fleshy fruits with a small amount of pulp and several or many seed are ground and allowed to ferment in a small amount of water for a few days. The pulp is then

FIG. 91.—Drying onion seed, seed are spread on huge canvas sheets and are raked every day until thoroughly dry. (*Courtesy of Ferry-Morse.*)

raked or decanted off and the seed washed from the remaining pulp over screens. The seed are then dried and stored or mixed with moist sand or peat and stored. In some cases the ground pulp is not fermented but is dried with the seed, and the seed and pulp are stored dry or added to moist sand and peat and stored.

The pulp is usually removed from very fleshy fruits, as the peach and plum. The fruit is spread in a layer or placed in small piles and left until the flesh begins to decay. The seed or pits are then separated from the pulp by masceration and washing. Seeds of the apple and pear are obtained from the mascerated pulp at cider mills.

Seeds are cleaned of foreign matter and graded to size by means of screens of various sizes and by being subjected to a current of air.

Storage of Seed.—Various seeds are stored in different ways. Some require special storage conditions to insure satisfactory germination. Most seeds are stored dry at about room temperature. Since the germination of some kinds of seeds is retarded or prevented by dry storage and room temperatures, such seeds are mixed with a moist medium, as sand or peat, and stored moist at a temperature slightly above freezing. This is known as "stratification." In the temperate regions these storage conditions are simulated for many of this kind of seeds by planting in the fall. Usually such seeds will germinate the following spring.

The longevity, or ability of a seed to live, is influenced by the kind of seed, the size, the composition and structure and the conditions under which the seed is stored. Many seeds over one year old are not suitable for planting. There are records of seeds two hundred years old that germinated satisfactorily, but most seeds have lost much of their viability after three or four years.

Seed Germination.—For germination and growth of seedlings, viable seeds, which are embryonic plants, require suitable conditions of moisture, temperature and oxygen. A very small group of seeds seems to require an exposure to light before they will germinate. In Chap. VI it was noted that many plants pass through a period of dormancy, or afterripening. Many seeds also have a dormancy period during which they will not germinate even though placed in the optimum conditions for germination. The duration of this period varies with the kind of seed. Dormancy in seeds may be due to the seed covering or to dormancy in the embryo itself. The embryo may be capable of growing but may be unable to grow because of the encasing coverings surrounding the seed. This covering may prevent growth of the embryo by the complete inhibition of water absorption, by interfering with or preventing the absorption of oxygen or the elimination of carbon dioxide and by a mechanical resistance to the expansion of the embryo. The seed may fail to grow because of the fact that the embryo is not yet fully developed or, if fully developed, the embryo or some part of it is in a state of dormancy itself through which it must pass, thus bringing about certain chemical and physical changes before germination can take place.

Various means have been found to hasten the germination of seeds. If the delayed germination is due to the encasing structures, or seed coat, it can be hastened by injuring or breaking the coverings. This is often known as "scarifying" the seeds. The period of dormancy in the embryo itself can often be shortened by storing the seeds at a temperature slightly above freezing. Apparently the necessary changes that take place during the afterripening process proceed more rapidly at this than at higher or lower temperatures.

Asexual Propagation of Plants

Asexual, vegetative or clonal propagation of plants is that type in which a vegetative part of the plant, as a leaf, stem or root, is placed in such an environment that it develops into a new plant. The new plant is merely a portion of a single parent plant and therefore is similar to that parent plant. Even without the intervention of man, many plants under natural conditions reproduce by some vegetative means. Many of the plants that produce sexual spores also produce asexual spores during the rapid-growing phase of their life cycle. These asexual spores are single cells which are part of but a single parent plant. This will be discussed more fully in Chap. XV in the section dealing with fungous diseases.

Requirements for Asexual Propagation of Plants.—It is quite generally believed that all horticultural plants could be propagated asexually if one knew exactly what should be done. One basic requirement must be possessed by any vegetative portion of the plant before it will produce a new plant. That is that the vegetative portion of the plant used for propagation must either possess or have the ability to produce meristematic tissues that are able to develop growth centers that will produce shoots and roots. The leaf of the common geranium will develop roots but will not produce a stem, but a leaf from the peperomia or gloxinia will produce not only roots but also a growing point which develops into a stem.

In addition to possessing this ability to produce roots and stems, the vegetative portion must be placed in a suitable environment if it is to produce a new plant. The chief factors to be considered in the environment are temperature, moisture, light and air, or oxygen supply. These are important in the asexual propagation of all plants but will differ in degree according to the particular plant being propagated.

Reasons for Propagating Plants Asexually.—There are several reasons why the horticulturist finds it necessary or advisable to propagate plants asexually.

One is to perpetuate plants that either never possessed or have lost their ability to produce seeds or that develop seeds only under special conditions. In this group might be listed the Smyrna fig, the edible banana, the Washington navel orange, the Eureka lemon, Marsh's Seedless grapefruit, dried currant, Thompson's Seedless grape and many other seedless fruits as well as many of the ornamental plants.

Another important reason is to perpetuate a plant that will not come true from seed. Earlier in this chapter it was noted that the embryo, or young plant in the seed, was the progeny of two parents and possessed characteristics of each. If the parents were unlike in some characteristics, the seedling plant would exhibit some characteristics of one parent and some characteristics of the other. The new individual would not be the same as either parent. Since in asexual propagation the new plant is merely a part of one plant, it will be as near like that one plant as it is possible for two plants to be alike. None of the common tree fruits, as the apple, pear, peach, plum, cherry, orange, fig, pecan, etc., will produce true from seed. The same is true for many ornamental trees, shrubs and flowers. A group of American elm trees propagated from seed will show wide variation. Because of these variations, nurserymen are now propagating the American elm, as well as many other ornamental trees, by asexual means. The same is true for many ornamental shrubs and flowers.

Another reason for propagating plants asexually is to perpetuate their particular form. Some plants appear different when young or in their juvenile form, and this young, or juvenile, form can be perpetuated by propagating new plants from vegetative parts of the young plant. The new plant, even when fully grown, will have the appearance of the original young plant from which it was produced. A condition somewhat similar to this is obtained when prostrate or creeping forms, as *Catalpa bungei*, Camperdown elm and weeping mulberry are placed on upright stems.

In some cases plants are propagated with greater ease and more speedily by vegetative means than by seeds. With the date, for example, the offset, or cutting, is already a large plant when taken.

In many cases plants are propagated asexually to increase their resistance or to develop their immunity to a particular pest. A good example is the grafting of the European grape, *Vitis vinifera*, on rootstocks of one of the American species of grape. The fleshy root of the European grape is attacked by an insect known as "phylloxera," although the insect does very little or no damage to the more fibrous roots of the various American species of grapes. A similar case exists with the apple. In Australia the woolly aphis does considerable damage to the roots of the apple; but as the roots of the Northern Spy are resistant to the attacks of this insect, many of the apples are grown on this resistant rootstock.

The adaptability of a plant to a particular location or habitat can often be increased by asexual means of propagation. In growing citrus fruits, grapes, bench roses and many other kinds of plants, rather extensive use is made of certain rootstocks that, because of particular habits of growth, are especially adapted to certain locations or soils. In certain sections of the Mississippi Valley certain varieties of apples are better able to withstand the adverse climate when grown on intermediate stocks, as Hibernal and Virginia Crab, or some other varieties.

Another reason for propagating plants asexually is to change the nature or amount of growth made by the plant. This can be done in some cases by using certain stocks with certain scions. The rate and amount of growth made by a plant is determined by both the root and the top. A small or slow-growing root system will delay and decrease the growth of the top, and a small slow-growing top will react in a similar fashion on the root system. The commercial pear tree is dwarfed in size by being grown on the root of the quince which has a much smaller root system than that of the pear. A corresponding result takes place when the standard apple is placed on the inherently small root system of the Doucin and Paradise apple stocks. The only influence that the stock has on the top or the top has on the stock seems to be of a nutritional nature. The one part of the plant can grow only as fast and as much as the size of the other part of the plant will allow.

Methods of Propagating Plants Asexually.—Asexually propagated plants are propagated on their own roots or on the roots of other plants. The portion of the plant used for propagation containing the growing point (bud) or the tissues capable of producing such a point may develop its own roots, or it may be placed on the root of another plant. The method used varies with the kind of plant and the results desired.

Plants Propagated on Their Own Roots.—Vegetative parts of plants that are to be used for propagation must possess the bud or the stem-growing point or the ability to produce such a growing point and also must be able to produce roots. During the development of the stems of many plants "root initials" are formed in the stems. These are generally most abundant at or near the nodes. Under favorable conditions they will start to grow and form roots. Many plants that do not normally possess such root initials appear to be capable of producing them under favorable conditions. As was noted before, the development of roots appears to be possible even though the development of a bud or growing stem point is impossible. When plants are reproduced on their own roots, the vegetative portion of the plant used for propagation purposes is either attached to the mother plant or detached from the mother plant. With some kinds of plants it appears to make very little difference, as both methods are used successfully.

PARTS NOT DETACHED BEFORE ROOTING.—The most common ways in which the new plant is not detached from the mother plant

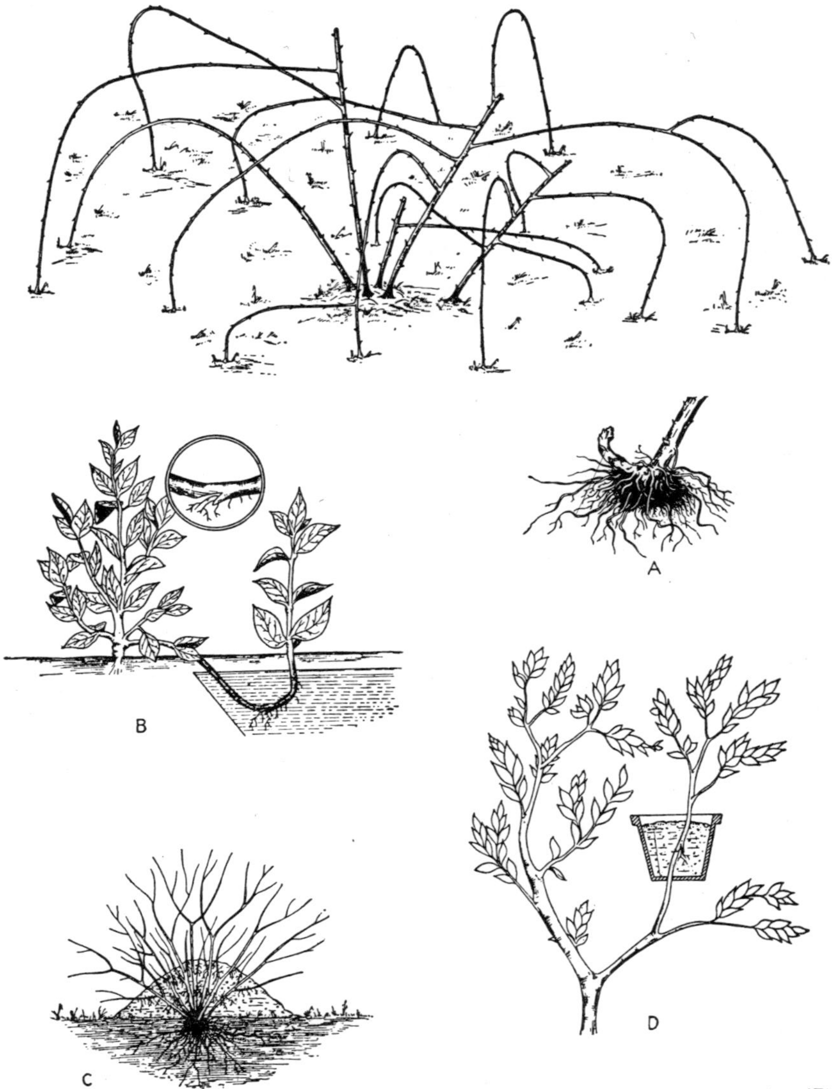

FIG. 92.—Diagrammatic representation of several forms of layerage, (*A*) tip, (*B*) simple, (*C*) mound, (*D*) air.

until it is established on its own root system are by runners, or stolons, and by layerage.

Runners, or stolons, and rhizomes: A runner, or stolon, is a stem that runs horizontally along the surface of the ground. It forms roots and new shoots at the various nodes. The strawberry is a common example of a plant that propagates naturally in this fashion. Creeping bent grass reproduces and spreads by stolons, and the procumbent stems of the crab grass root readily at the nodes. Rhizomes are similar to stolons except that they are stems that run horizontally beneath the surface of the soil. Lily of the valley propagates naturally by rhizomes. This habit of propagation makes quack grass a very difficult weed to control after it once gains entrance in a lawn.

Layerage: Layerage is the method of propagating plants while they are still attached to the mother plant. It is accomplished by naturally or artificially bringing a portion of the stem into contact with or beneath the soil, thus placing it in a position to favor the formation of roots. The formation of roots can be hastened in many cases by ringing, notching, twisting, tonguing or in any other way injuring the portion of the stem in contact with the soil or other rooting medium. There are several different types of layerage differing chiefly in the method used to produce suitable environment about the stem for the production of roots. The type used is adapted to the kind of plant to be propagated.

Tip layerage occurs naturally with the black raspberry. Toward the close of the growing season the ends or tips of the shoots that come and remain in contact with the soil or are buried 1 or 2 in. below its surface become enlarged, send out a mass of fibrous roots and form an upright growing point or bud.

In simple layerage a bent and usually wounded portion of the stem of the plant is buried 1 or 2 in. below the surface of the soil. Roots will develop from the buried portion, and the rooted portion can then be cut from the mother plant. Many kinds of plants, as cornus, hydrangea, rose, grape, carnation and spiraea are propagated by simple layerage. The time at which the layering is done will be influenced by the kind of plant and the locality. Shoots of deciduous shrubs are usually layered during the latter part of the growing season, or the twigs are layered before growth starts the following spring. The length of time required for rooting will be influenced by the kind of plant, the time at which the layering is done, the condition of the twig used, the environmental conditions, etc.

Mound layerage is practiced on some plants, the stems of which cannot readily be bent to the ground. Preparatory to mounding, the plant is usually pruned severely to cause the development of vigorous lateral shoots near the ground. In the fall or the following spring, soil is mounded and packed about the plant so that it covers the bases of the vigorous twigs. Usually satisfactory rooting will take place on the twigs within one year so that the mound of soil can be removed and the rooted twigs cut from the parent plant. The gooseberry and the Paradise and other apples are propagated in this fashion.

Air layerage, pot layerage or Chinese layerage is that type of layerage in which the soil or rooting medium is placed about the stems of the plants that are high in the air. It is practiced on those plants

FIG. 93.—Root cutting of sweet potato.

which cannot be bent to the ground for simple layerage and on those with which mounding would be impracticable. Since a desirable portion of the plant cannot be brought to the soil or rooting medium, the latter is taken to a suitable part of the plant. The twig to be layered is usually ringed or notched. A flowerpot is cut in two longitudinally and wired together about the stem, covering the wounded area. The pot is then filled with a suitable rooting medium which is kept moist. After a time roots form from the wounded stem in the pot. The stem is then cut off just below the pot containing the newly rooted plant. In some cases balls of sphagnum moss or other suitable rooting mediums are tied about the wounded stem, and the pot is not used. This form of layerage is used to propagate crotons and rubber plants. It is used quite extensively in China for many plants, one of which is the litchi. Rather large plants can be obtained fairly quickly in this fashion.

PARTS ALWAYS DETACHED BEFORE ROOTING.—In many cases it is necessary or advisable to separate the portion of the plant to be used for propagation from the parent plant. This type of asexual propagation goes under the general classification of cuttage. Roots, stems and leaves are used of the various plants. With certain plants only one part, as the stem, is used; but with other plants, stems, roots and leaves may all be used satisfactorily.

FIG. 94.—Rooted herbaceous cutting of geranium.

Root cuttings: A root cutting is a portion of the root used to propagate the plant. Since roots do not possess buds, it is necessary that the roots of plants that can be used for propagation possess the ability to produce a bud or a growing stem point. Only the roots of certain plants possess this ability. Both plants with fleshy roots, as the sweet potato, and those with fibrous roots, as blackberry, red raspberry, horseradish and phlox, can be propagated by root cuttings. The method of handling varies with the plant being propagated, but the general practices are similar. In order to prevent decay the fleshy roots are not cut into smaller pieces, but the fibrous roots are cut into small pieces and planted much as one would plant seeds. After a time shoots will develop from the pieces of roots, and a new plant will be obtained.

Stem cuttings: A stem cutting is a portion of the stem used to propagate the plant. The stem portion must contain one or more nodes, or buds. The bud will expand into a stem. Roots may develop on the old stem, on the new stem or on both the old and the new stems. Stem cuttings can be made from herbaceous stems and from green wood and hard or mature stems. The line of demarcation between the types of stems is not always clear.

Herbaceous stem cuttings are made of the soft succulent stems with part or all of the leaves attached, as geranium, coleus, phlox, chrysanthemum and carnation. A portion of the stem, usually the tip, about 3 in. long is cut; the basal leaves are removed and possibly portions of the remaining leaves. The basal end of the cutting is then placed about 1 in. deep in sand or some other suitable rooting medium, which is packed about the stem; and the temperature, moisture and light of the cutting bed are regulated according to the requirements of the particular plant.

Fig. 95.—Rooted hardwood cutting of grape.

Green-wood or softwood cuttings are made of the growing shoots of many woody plants, as the rose, privet and hibiscus. They are made at some time during the growing season, and part of the leaves are left attached. They are handled in much the same fashion as are the herbaceous stem cuttings.

Hardwood cuttings are made of the matured one-year-old twigs of many woody plants. Instead of taking the growing shoot, the latter is allowed to mature and drop its foliage. The cuttings are taken any time after the foliage has dropped. If taken before spring they are cut about 6 in. long or so that they contain at least two nodes, tied in bundles with the bases of the cuttings in the same plane, and stored over winter in sand, sphagnum moss or some other moist medium at a temperature of about 40°F. Under such conditions the cuttings will not be injured by adverse winter conditions, the rest period will be completed, callus will form, the primordia of roots will form and the roots may grow some before conditions are suitable for planting out-of-doors in the spring. Stems of various evergreens are used as cuttings. Some will root readily from cuttings, whereas others root with difficulty or not at all. Most root more slowly than hardwood cuttings and are more particular relative to the season of the year when the cuttings are made.

Leaf cuttings: Many herbaceous-like plants with thick leaves, as bryophyllum, sansevieria and sedum, or leaves with heavy veins, as rex begonia, if placed under favorable conditions will form stem-growing points and roots and thus produce new plants. The method employed

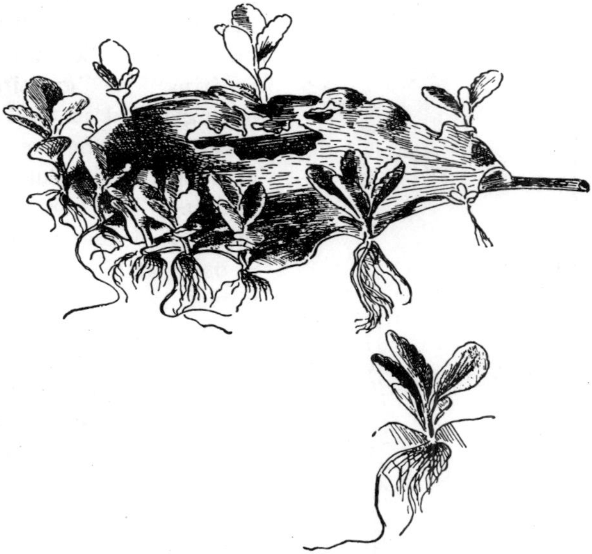

Fig. 96.—Leaf cutting of *Bryophyllum crenatum.*

varies with the kind of plant, but in all cases the leaf or portion of leaf is placed in a cutting bed or suitable medium, and the temperature, moisture and light are regulated to meet the requirements of the plant being propagated.

Parts Generally Detached from the Parent Plant before Rooting.—As would be expected, there is a group of plants lying between those which are propagated while still attached to the parent plant—layerage—and those always detached from the parent plant—cuttage. Members of this group may or may not be detached from the parent plant before rooting. There are two main divisions in this group, although the two divisions grade imperceptibly into one another. Those which are propagated by separation are those which fall apart naturally, as plants propagated by bulbs and corms. Those which are propagated by division are those in which it is generally necessary to cut or break the parent plant into pieces. Obviously in certain plants as they become old, as the iris, the parent plant might fall apart naturally, whereas the same plant when young would have to be cut into parts. Other examples are peony, rhubarb, asparagus, lily of the valley, canna, echeveria, pineapple and date.

Buds, or Growing Points, on the Roots of Other Plants.—In an operation known as "graftage" one or more buds of a plant are attached to the root system of some other plant. If a short twig containing one or more buds is used, the operation is known as "scion grafting," or "grafting"; but if a single bud is used with very little or none of the twig attached, it is known as "bud grafting," or "budding." This method of asexual propagation is used with such plants as the apple, pear, peach and many others that cannot be propagated readily on their own roots but seem to require a nurse root to become established. It is also used better to adapt plants to particular soils and conditions, as hybrid tea roses on Manetti rose stock in greenhouse benches; to adapt the plant better to climatic conditions, as double-working Grimes Golden apple on trunks more resistant to winter injury; to enable the plant better to withstand the attacks of a pest, as grafting the European grape on American types of roots; and to modify the growth of the plant, as dwarfing the pear tree by growing it on a quince root.

FIG. 97.—Root graft of apple, (*a*) cion, (*b*) root, (*c*) completed graft.

For successful union of the stock and scion it is essential that the cambium or growing tissues of the stock and scion be in contact or very close to one another. Not all kinds of plants can be grafted on to all stocks or roots. There must be certain inherent characteristics common to the two plants before a satisfactory union will take place. These characteristics are generally spoken of as "botanical relationships." Within the limits of possible union there are some plants that form a more congenial or satisfactory union than others. Certain apple or cherry stocks will be found to make more satisfactory unions with particular varieties of apple and cherry than with other varieties of the same fruits. There are many different forms of grafting adapted to particular plants or specific purposes. In scion grafting the scion may be placed directly on the root, in which case it is known as "root grafting"; or the scion may be placed on the trunk or branches of the tree when it is given a specific name generally descriptive of the graft. "Double-working" is merely a special form in which the scion is first grafted on to the root and later, as the plant becomes larger, scions or buds are placed in the trunk or main branches of the plant; so the plant is grafted at two different times, or double-worked. This is used to grow tender varieties on more hardy trunks and to grow more than one variety on a single trunk. In bud grafting or budding a single bud with a small amount of "bark" is removed from the growing twig and inserted into a suitable incision made on the stock so that the cambial regions of the stock and scion are in contact. There are many forms of budding adapted to particular purposes or kinds of plants and given descriptive names, but the fundamental essentials are the same in all forms.

VARIABLE FACTORS INFLUENCING THE SUCCESSFUL PROPAGATION OF PLANTS

Besides the inherent characteristics of the different kinds of plants, a number of variable factors contribute to success or failure in propagating plants. These influencing factors are of varying degrees of importance with different plants and even with the same plant at different seasons of the year.

Temperature

The rate of growth of all plants, and even of different parts of the same plant, is influenced by temperature. The germination of seeds and the development of roots will be influenced by the temperature of the soil or rooting mediums in which the seeds or other plant parts are placed, and the rate of top growth will be influenced by the temperature of the air in which the tops of the plants are growing. Just as certain seeds will germinate at different temperatures, so will certain plants form roots at various temperatures. A soil temperature suitable for the germination of pea seed will result in the decay of corn seed. Roots will develop and grow at temperatures lower than will the tops. In propagating plants by cuttings an attempt is made to maintain an optimum temperature of the rooting mediums to encourage rapid and successful root formation and growth while the temperature of the atmosphere is kept lower than that required for the optimum growth of the top of the plant. These conditions result in a more rapid formation and growth of the roots and a lessening or retarding of top growth. The energies of the plant are directed toward the production of roots on the cutting so that the roots can supply the needs of the top for water and mineral elements.

Moisture

Moisture is required for plant growth. The newly planted seed must absorb it in order to germinate; and the vegetative portion of the plant, used for propagation, must be furnished a suitable water

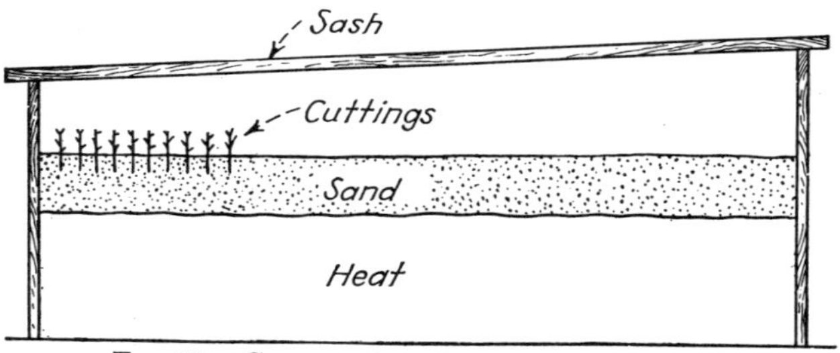

Fig. 98.—Cross section of a propagation frame.

supply. Too much water in the rooting medium will prevent satisfactory aeration and result in the destruction of the seed or plant part; too little will result in desiccation and death. Different kinds of plants require different amounts of water in the rooting medium for best results. The atmospheric humidity is important, especially with plant parts possessing leaves. Water is being lost by the leaves to the atmosphere, and the rate of this loss will be influenced largely by the humidity. If the atmosphere is dry and the cutting has few or no roots, the leaves may lose water more rapidly than it can be replaced from the rooting medium, with the result that the cutting dries out and dies. The moisture in the rooting mediums is regulated by using various mediums with different moisture-retaining capacities, and the atmospheric moisture is regulated by enclosing the propagation bench with glass, muslin or other suitable material and by spraying moisture into the enclosed area. The high atmospheric humidity checks transpiration from the plant. Such conditions, however, are very favorable for certain destructive diseases, so constant attention is necessary to note the first appearance of such diseases and check them by lowering the humidity.

Light

Light is essential for germination of the seeds of a few plants, as *Veronica longifolia;* it is detrimental to the germination of the seeds of a small number of plants, as *Nigella arvensis;* but it is of no significance to the germination of the seeds of most plants. In propagating horticultural plants one is interested in reducing the amount of light on the propagating bed because of the influence of shade in decreasing the loss of moisture from the cutting and from the soil. Too much light will cause newly set cuttings to lose water more rapidly than it can be replaced; and too much light, combined with air movement, will cause the surface soil to dry and form a hard crust which will interfere with the germination of small seeds and the emergence of delicate seedlings. At the same time, sufficient light must be provided to allow for adequate photosynthetic activity and to control certain fungous diseases. Growing shoots or portions of shoots in darkness so that they are etiolated has improved the rooting properties of some stems, when the etiolated portion comprised the basal part of the stem cutting.

Oxygen Supply

Oxygen is necessary for plant growth, and aeration is essential for the formation of satisfactory roots on cuttings. Since about 20 per cent of the atmosphere is oxygen, which is three to four times as much as the plant can use, all the rooting mediums that are used in seed beds and cutting beds are satisfactory from the standpoint of aeration.

Time of Performing the Work

The time at which the work is done has a great influence on the probable success in propagating many plants. As has been noted previously, most seeds and plants have a period of rest, or afterripening, through which they must pass before they will resume active growth. Many seeds that are placed in the soil in the fall will not germinate until the following spring. The tops of the Irish potato plant will wither and die, but the tuber remaining in the ground will not sprout for several weeks; however, the same tuber sprouts quickly under the same conditions the following spring. The period of rest influences the time at which the sexually and asexually propagated plant starts to grow. If propagated at an unfavorable time, the seed or other plant part might be destroyed before or shortly after it started to grow.

Plants and plant parts that exhibit dormancy enter that state slowly and emerge from it slowly. Plant growth appears to take place in cycles or waves, not being at the same rate at all times. Other conditions being favorable, one would start cuttings, make grafts, etc., when this cyclic growth is in the stage to produce the most satisfactory results. Scion grafting should be done, and green and herbaceous cuttings should be made just as the plants are entering an active growing phase; budding, in which it might be desired that only union should take place, to be followed by growth the following spring, should be performed just as the plant is entering one of its low periods of growth. If it is desired that the inserted bud should develop into a shoot the same season, then the budding should be done much earlier in the growing season.

Seeds that have a long afterripening period should be collected and stored under conditions that will be favorable for afterripening and the retarding of germination until conditions are suitable for planting. The manner of storage will vary with the kind of seed.

In regions where there is danger of winter injury to the twigs, scions for grafting and twigs for hardwood cuttings should be collected and stored before the advent of unfavorable weather. Many of the greenwood shoots root better when the shoots are in their earlier stages of growth. For this reason such green-wood cuttings are taken at rather definite periods. For example, better rooting is obtained with greenwood cuttings of the lilac when they are taken close to the time when the terminal bud is forming. In the United States the best success is obtained with some of the narrow-leaved evergreens when cuttings are taken during November through March.

Size of Seed or Vegetative Part of Plant Used

The first growth of the new plant, whether it be from a seed or from a vegetative part of the plant, as a stem or leaf, must be made from the food supply stored within the portion of the plant used for propagation purposes. Consequently, with other factors being equal, the most satisfactory early growth is made from the seeds and vegetative parts of the plant that possess the largest supply of available food that can be used in plant growth. Generally the size of the seed and size of the vegetative part of the plant used for asexual propagation is associated with the amount of food reserves available for the growth of the new plant. Consequently the size of the seed and the size of the vegetative portion used will have an influence on the speed and success of propagation. With starchy seeds, the medium-sized to large seeds, which are also those of the highest specific gravity, are generally the most satisfactory. The size of the cutting is limited by convenience and economy in handling, by the amount of suitable growth available and by the percentage of reserve food in the tissue. A very large cutting made from tissue that grew rapidly might well have a smaller relative amount of available food stored within its tissues than a smaller cutting that grew more slowly, since the larger cutting may have used its food supply in making the vigorous growth. Various experiments have demonstrated that a 2-oz. "seed piece" of the Irish potato is a very satisfactory size. Such a seed piece will furnish an adequate supply of food for the new plant until it is amply able to grow independently of the food supply contained in the seed piece. Size of seed and vegetative portion of the plant used for propagation is significant largely because of its relationship to available food reserves. Numerous experiments with hardwood, green-wood and herbaceous cuttings have shown that the more mature cuttings, or those with a high carbohydrate-nitrogen relationship or a high starch reserve, form roots sooner and in greater quantity than do corresponding cuttings that are less mature or more succulent and that have a low carbohydrate-nitrogen relationship or a low starch reserve.

Age of Seed or Vegetative Part of Plant Used

There are stories of seeds that have retained their viability for centuries, but these stories are unsupported by facts. A few instances are known of seeds that have retained their viability for one hundred or two hundred years, but such seeds were dormant owing to impervious seed coats and had an exceedingly low rate of respiration. Under favorable storage conditions in the soil, seeds that are dormant because of impervious seed coats may remain viable up to fifty years. Non-dormant seeds would germinate and be destroyed under similar conditions. Some seeds deteriorate so rapidly that they are practically worthless the second year; many remain suitably viable for three to five years, after which they lose their viability very rapidly. Under favorable conditions some seeds will remain viable over much longer periods of time, but in the majority of cases the one-year-old seeds are preferred to the older seeds, and the older the seeds the less satisfactory they become. Some seeds keep longer if air dried and kept in tight containers at temperatures too low for germination. There are a few instances where, because of dormancy for one reason or another, the seed will germinate more satisfactorily the second year than it will the first. The chief causes of such action seem to be due to an immature embryo or the presence of certain enclosing structures about the embryo which must disintegrate at least partly before germination can take place.

In propagating by vegetative parts such as cuttings, the very young growing shoots root most satisfactorily in some cases, while in other cases more mature tissue or even a small portion of the two-year-old wood known as a "heel" produces the most satisfactory results. The advice is often given that herbaceous and green-wood cuttings should be taken when the tissue will snap and not crush at the position of the basal cut. In actual practice the basal cut is generally made at a considerable distance below the point where the stem will snap when bent.

Position of Cut

Considerable diversity of opinion exists relative to the position that the basal cut on a cutting should occupy in relation to the node. At one time it was considered essential for it to be made immediately below the node. This has been found unnecessary with some plants, for they will root readily irrespective of the position. In some that do not callus quickly and have a comparatively large pith area, the pith may decay up to the first node above the cut. The position of the cut is really determined by the manner in which the particular cutting forms roots. Roots from cuttings are usually of two types: morphological and wound. Morphological roots arise from root primordia, or root initials, that are laid down in the stem and have a definite relationship to the anatomical structure of the stem. These root initials are generally most abundant within an area extending about ½ in. below a node. Wound roots, on the other hand, have no apparent relationship to the original anatomical structure of the stem and usually arise just above, but rarely from, the callus. Many kinds of plants will develop both types of roots, and many will develop an adequate supply of wound roots. Since wound roots arise directly above the callus and the amount of such roots formed seems to be greater the smaller the amount of callus, cutting just below the node removes most of the root initials of those plants which possess them, encourages rooting at several nodes in some cases and may cause excessive callus formation in others. Some plants form better roots when cut at the node, others when cut above the node, but most root better when the cut is made about ¼ in. below the node. There is no one best place, however, for all the different kinds of plants, for the propagator is interested in the speed of root formation, the amount of roots formed and the position on the cutting from which the roots arise.

Leaf Area

The healing of the wound and the formation of roots in cuttings are influenced by food supply. In hardwood cuttings that do not possess leaves, all the food reserves are stored within the cutting; but greenwood, herbaceous and evergreen cuttings possess leaves which carry on photosynthesis and thus augment the carbohydrate supply of the cutting. The leaves also transpire water, and the greater the leaf area the more water transpired. A large leaf area is beneficial in one way and detrimental in another; consequently the propagator must balance these two factors to produce the greatest number of suitable plants on a given area. Foliage is removed for the convenience of the propagator, not for the benefit of the cutting. Under favorable conditions the larger the leaf area the greater the amount of carbohydrates synthesized. Consequently the most satisfactory results in speed of rooting and amount of roots produced are obtained from those cuttings which possess the maximum leaf area permissible under the conditions of temperature, moisture, humidity, light and prevalence of disease existing in the propagating bench.

Rooting Medium

When a plant propagator speaks of a rooting medium, he usually has in mind the medium in which cuttings are placed. In a broad sense, however, the stocks used in graftage might also be considered as rooting mediums. Cuttings may be rooted in air, water, soil, sand, peat, various fibers and combinations of any of the foregoing. The mediums most generally used in the United States are sand, peat and combinations of these in various proportions. Some cuttings produce better roots in one medium, and some in another. Furthermore, different mediums may give different responses at different seasons of the year. Peat holds more water than sand and may keep the rooting medium too moist and favor the development of disease during the shorter, darker days of winter but be admirably suited to the same kind of plants under a different set of growing conditions. There is no one best medium for all kinds of cuttings under all conditions.

Chemical Treatments

Within recent years various substances have been applied to grafts, bulbs, cuttings, seeds and the rooting mediums in which they were to be grown with the purpose of hastening or increasing growth. These substances are now quite numerous and are known by many and various names even for the same substance. They may all be grouped conveniently under the heading of "growth substances." Depending upon the concentration used and other factors, they both accelerate and retard growth of roots. Generally they retard the growth of buds. They are used in the form of liquids, dusts and pastes, being applied directly to the plant part or to the medium in which the plant is to be grown. As a whole they have been found to stimulate the growth and increase the amount of roots on cuttings and seedlings and to increase the rate of callus formation on grafts and cuttings. Some progress has been made in rooting a few plants that are difficult to root. Further tests will determine the commercial value of such substances, and uniform and standardized methods of procedure will be determined for the various plants under different environmental conditions. The propagator will find it profitable under favorable conditions to treat cuttings of such plants as carnation, poinsettia, gardenia, lilac (green wood) and some of the junipers with growth substances.

PROPAGATING REPRESENTATIVE TYPES OF HORTICULTURAL PLANTS

Propagation of horticultural plants may be understood more clearly if we study representative types of such plants propagated by seed, by herbaceous stem cuttings and by graftage. Practices for other plants propagated in similar fashions will vary in particular details, but the principles will be the same.

Tomato

The best varieties of tomatoes are quite homozygous for their major characteristics and will reproduce true to variety from seed if no other tomatoes have been grown close enough to allow cross-fertilization. In commercial seed production, the seed are obtained from fruits from a large isolated block of a single variety. In extracting the seed the ripe fruits are crushed and placed in wooden vats, or containers. The crushed pulp is allowed to ferment for two or three days, which removes the gelatinous covering that surrounds each seed. After adequate fermentation, water is added, and the pulp is stirred. The pulp and light seed will rise to the top where they can be floated or raked off, and the heavy plump seed will sink to the bottom of the container. These heavy seed are then washed thoroughly and dried by centrifuging by being placed in thin layers on a screen or by being exposed to a warm air current. The dried cake of seeds is then mascerated to separate the seeds, after which they are graded by fanning and stored in sacks.

Under favorable conditions the seeds are planted in pots, flats, seed beds or in the open, depending on local conditions. The plants are then transplanted several times or thinned until they are set in their permanent localities in which they grow and produce more tomatoes and more seed.

Geranium

The geranium, a herbaceous plant, will serve as a suitable example of a horticultural plant propagated by green-wood and herbaceous cuttings. Vigorous, healthy, well-grown but not rapidly growing plants will have a suitable reserve of food and give excellent results when used as cuttings. Usually only the terminals are used, but suitable stems can be cut into two or more pieces. Cuttings can be made any time during the year, provided suitable wood is obtained, is handled properly and is placed under favorable conditions. Make the cuttings about 3 in. long, clean and smooth, about ¼ in. below a node. Remove the leaves attached to the portion of the stem that will be below the surface of the propagating medium. Part of the remaining foliage may be removed. Foliage is removed to assist in the control of diseases, to economize on space and to decrease the loss of water by transpiration. Under suitably controlled conditions the cuttings with the most foliage will make the most growth, so one should leave as much foliage on the cutting as the local conditions permit. Prepare the cutting bed by filling with about 4 in. of clean, moderately fine sand, and pack with water or by tamping. Other rooting mediums are used satisfactorily. Make a trench about 1 to 1½ in. deep, and place the cuttings in it. Fill the trench, and pack with a tamper or with water. Protect the cuttings from excessive sunlight and strong air currents, which will dry them out before they form adequate roots. The amount of shading will vary with conditions. Roots will be visible in about two weeks. The rooted cutting should be potted in soil as soon as it develops a mass of roots about 1 in. long.

Apple

Many of the named varieties of horticultural fruits will not come true from seeds and consequently are propagated by asexual means, as cuttage and graftage. With graftage the plant is composed of parts of two—sometimes three—individuals. One plant forms the root, commonly known as "stock," and the other the top, commonly known as "scion." In the case of double-worked trees, three individuals are represented: the stock, the scion and an intermediate stock which is placed between the stock and scion.

Formerly the seed for the apple stocks were imported chiefly from France, but now much of it is obtained from commercial apple varieties in this country. Satisfactory stocks can be grown from seeds of Ben Davis, Winesap, Wealthy, McIntosh and Whitney. Apple seed is obtained in a fashion similar to that used to obtain tomato seed. The seed is usually obtained from the pomace at cider mills. The fresh pomace containing the seed is placed in tanks of water and agitated thoroughly. The pulp and light seed will float to the top, where they can be flooded or skimmed off, and the heavy seed will settle to the bottom of the container. After thorough washing these seed may be handled in one or two ways. They may be dried, then mixed with moist sand, placed in containers that allow for drainage and buried in the soil or placed in cold storage at a temperature ranging from 32 to 40°F. during the winter. Afterripening will take place in about eight weeks, and the viable seed will germinate when placed under suitable conditions. After cleaning, the seed may be dried in the air, placed in containers and stored dry. In this case, it will be necessary to soak the seed and stratify them in a moist medium, such as sand, at a temperature of about 37°F. for two to three months before germination will take place. Afterripened apple seed should not be allowed to dry out before being planted, as the embryo will be injured.

Apple seedlings are grown in the United States chiefly in parts of Kansas, Washington and Oregon. Deep friable soil is essential for long unbranched roots. After the soil is prepared, the afterripened but moist apple seed are planted in drill rows about 3½ ft. apart. The seedlings are cultivated similar to vegetable crops. If large enough, they are dug in the fall with a digger that cuts the roots 10 to 12 in. below the surface of the soil. Part of the top may be cut off; the seedlings are graded, tied in bundles of 100 and stored in shingle tow in a cool storage cellar until they are needed for grafting during the winter or for planting in nursery rows the following spring to be budded in the fall.

If grafting is to be practiced, scions are gathered from the desired kinds of trees shortly after the leaves drop in the fall. One-year-old twigs with mature plump buds should be taken. The base and tip portions of the scions are usually discarded. The scions are stored under conditions similar to those for the stocks. The grafting operation can be done any time during the winter or any time that the scions are dormant.

Many different types of grafts are used, but the most generally practiced nursery method is the whip, or tongue, graft. To lessen the amount of callus formed, a double-tongue graft may be used. Both "whole-root" and "piece-root" grafts are made. The only difference is in the lengths of the scion and stock. The completed graft is about 9 in. long. In the whole-root graft, the root or stock is 6 in. long and the scion is 3 in. long, whereas in the piece-root graft these measurements are reversed. Preparatory to grafting, the fibrous lateral roots are trimmed off close to the main root, which is from 3/16 to 1/4 in. in diameter at the crown. In making the tongue graft, an oblique cut about 1½ in. in length is made at the crown of the stock. This is followed by a second cut which forms the tongue. This cut is started about two-thirds up the face of the oblique surface of the stock and extends downward and slightly forward, which cuts across the grain of the wood. With this type of cut the tongue is thinner than the back of the stock, and a closer, smoother union can be obtained with the scion. This second cut extends for a distance of about 1 in. By means of a straight cut the root is made the desired length. The scion is prepared in a similar fashion. The stock and scion are then fitted smoothly together by overlapping the tongues, so that as large areas as possible of the cambial regions of the stock and scion are in contact. It is well to have the stock and scion nearly the same size; but in any case, the cambial regions of the two pieces should be in contact over as great an area as the sizes of the two pieces allow. After fitting snugly together, the graft is held firmly by wrapping with grafting rubber or adhesive grafting tape. As the graft grows, the rubber will stretch and finally decay, but the tape should be cut through with a knife at the time when the graft is made, which will allow the tape to separate as the graft enlarges. If the tape is not cut in this fashion, it will girdle the graft; for the tape does not decay until a considerable time later. After the grafts are wrapped, they are tied in bundles and stored in moist sand, moss or shingle tow at a temperature of about 40°F. Under such conditions callus growth will take place, continuous cambium will form across the union and the stock and scion will become as one plant. Early in the spring these grafts will be planted in the nursery row, leaving only the upper bud of the scion above the surface of the soil. They will be left until dug as one- or two-year-old grafted trees. The age of the tree is counted from the age of the top that arises from the scion of the graft.

With budded trees the operations are somewhat different. The stored seedling stocks are cut back both in top and root and planted 6 to 8 in. apart in nursery rows. Budding may be done at any time during the growing season that the bark "slips," but in the North the budding is done in August until the bark sets. Budding done at this time permits union to take place between the stock and bud, but the bud does not develop into a shoot until the following spring. Two or three days before the buds are to be inserted, the leaves are rubbed off the stock for a distance of about 8 in. above the surface of the soil. The stocks are about ½ in. in diameter. A cut shaped like a capital T is made on the shaded side of the stock from 2 to 4 in. above the soil.

The freshly cut bud is inserted in this cut and fastened with rubber bands or grafting tape. If tape is used, it must be cut as it was in grafting.

The buds are taken from "bud sticks." These are growing shoots taken from fruiting trees or from stock trees. Both the basal and the terminal end of the bud stick are discarded, as the plump axillary central buds are the most satisfactory. The leaves are cut off, leaving about ¼ in. of each petiole, which serves as a handle while working with the bud. The trimmed bud sticks are wrapped in moist burlap.

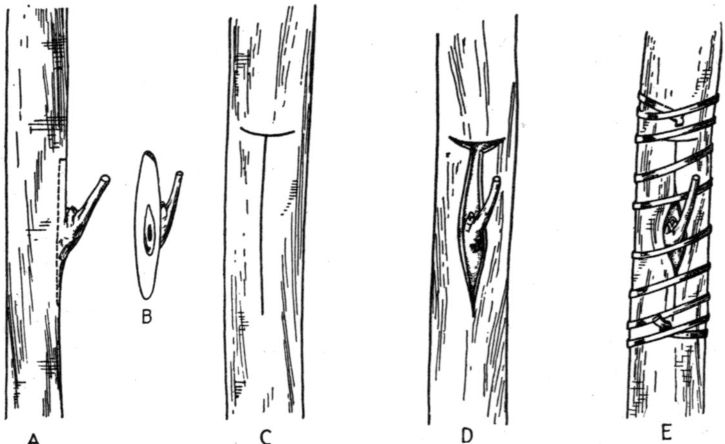

FIG. 99.—Diagrammatic sketches showing the steps in budding: (*A*) bud stick, showing taking of bud by cutting from below; (*B*) bud removed; (*C*) stock with T-shaped incision in bark; (*D*) stock with bud inserted and (*E*) bud tied.

A shield-shaped piece of bark, with or without some of the wood, is cut from the bud stick and inserted into the T cut made just previous to cutting the bud. The bud can be forced into the cut by slight pressure as the bark slips. The bud will knit or unite with the stock. The following spring the stock is cut off just above the inserted bud. This encourages growth from that particular bud.

Chiefly on account of winter injury, apples are being double-worked in certain sections of the United States. By this method the tree is composed of three rather than two individuals. The apple seedling is grafted or budded as before, but the scion is from a variety as Virginia Crab, Hibernal or some other variety that is resistant to winter injury at the collar and in the crotches and makes a favorable union with desirable varieties that are not so resistant to winter injury. The resistant variety is grown for one or two years in the nursery row and then planted in the orchard. A good method to follow in developing such double-worked trees is to train the resistant tree to form the main body of the tree and bud the desired variety into these framework branches. This will require three or four years or even more from the one-year-old tree, as the buds will be inserted in the primary scaffold branches as such branches are developed either every year or every other year. The budding is similar to that done in the nursery row, except that it is done on the underside of the primary lateral branches about 12 in. from the main trunk. Buds can be placed in either one-

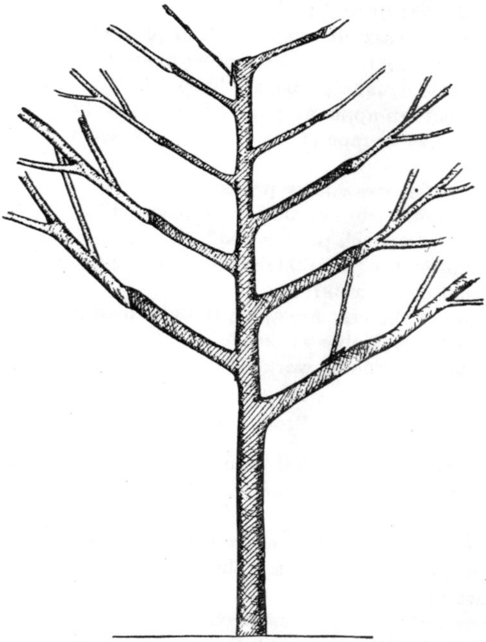

Fig. 100.—A diagrammatic representation of a double-worked tree; shaded areas indicate the original tree, light areas show growth from buds which were inserted.

or two-year-old branches. It is advisable to place two or three buds on the underside of each scaffold branch that is to be retained. The following spring the branch is cut off just beyond the one bud to be retained to form the permanent branch. Other primary scaffold branches can be worked the following years until the entire tree is worked and the necessary primary branches are all worked over to the less resistant variety. The tree now consists of an unknown seedling stock, an intermediate stock extending from the primary stock and forming the main trunk of the tree, the basal or crotch portions of all permanent scaffold branches and the less resistant scions which were budded into the branches of the intermediate stock.

In many cases the intermediate stock forms only a portion of the trunk, and the scion variety forms the entire top of the tree.

Review Questions

1. What is plant propagation?
2. Why are horticultural plants hybridized?
3. Are mutations of any importance in obtaining improved horticultural plants?
4. What is meant by sexual propagation of plants?
5. What kinds of horticultural plants can be propagated successfully by seeds?
6. What is meant by asexual propagation of plants?
7. What is the basic requirement for successful asexual propagation of plants?
8. Why is the edible banana propagated vegetatively?
9. Why is the Jonathan apple propagated vegetatively?
10. Why is the European grape (*Vitis vinifera*) grafted on roots of other species of grape?
11. How is the Premier strawberry propagated?
12. How is the Red Spectrum carnation propagated?
13. How is the Elberta peach propagated?
14. What are some of the variable factors influencing the successful asexual propagation of horticultural plants?
15. At what season of the year are apple trees budded in New York nurseries?
16. How are apple trees double-worked?
17. Why are apple trees double-worked?

Problems

1. You have been given a plant that is entirely unknown to you and have been asked to determine how it can be propagated satisfactorily. Explain your method of procedure.

2. Make a diagrammatic sketch of a propagating bench in a greenhouse suitable to use for propagating geraniums. Show all essential features including a recently inserted cutting.

3. Contrast the complete operations of obtaining apple trees by budding and grafting by completing the following table:

Time	Operation
Fall 1940	Obtain seed.
________	Stratifying seed.
________	Planting seed.
________	Seedlings topped, dug and stored.

Root Grafting	Budding
Winter—Year________	________
________	________
________	________

4. You noticed that one of the branches on a Rome Beauty apple tree produced apples much more colored than other Rome Beauty apples. Explain how you would proceed to obtain some trees that produced apples like the more highly colored ones on the Rome Beauty tree.

Suggested Collateral Readings

1. ADRIANCE, G. W., and F. R. BRISON, "Propagation of Horticultural Plants," McGraw-Hill Book Company, Inc., New York, 1939.

2. CRANE, M. B., and W. J. C. LAWRENCE, "The Genetics of Garden Plants," Macmillan & Company, Ltd., London, 1938.

3. KAINS, M. G., and L. M. MCQUESTEN, "Propagation of Plants," Orange Judd Company, New York, 1938.

4. LAURIE, A., and L. C. CHADWICK, "The Modern Nursery," The Macmillan Company, New York, 1931.

5. YERKES, GUY E., Propagation of Trees and Shrubs, *U.S. Dept. Agr. Farmers' Bul.* 1567: 1–51, 1932.

CHAPTER XII

SOIL MANAGEMENT OF HORTICULTURAL PLANTS

The final object sought in growing horticultural plants commercially is optimum growth and production at minimum costs. The amateur is interested primarily in obtaining satisfactory growth and production, and the cost is of secondary consideration. Soil moisture, nutrients and organic matter are important factors in obtaining the desired objectives, and these soil conditions are influenced greatly by the manner in which the soil is handled.

OBJECTS

The chief objects sought in soil management are (1) to provide a favorable moisture supply; (2) to supply sufficient nutrients for optimum growth and production; (3) to add enough organic matter to offset that lost by decomposition, by organisms and by erosion; (4) to prevent erosion and (5) to avoid injurious compacting of the soil. It is difficult to separate the foregoing objects, since with the addition of organic matter, the physical condition of the soil is improved which in turn would increase the rate of decomposition and liberation of plant nutrients and also improve the water-holding capacity of the soil.

The importance of moisture in relation to plant growth has been discussed previously. Here it is intended to call attention only to those practices associated with the maintenance of a satisfactory soil moisture. Cultivation to destroy weeds, addition of organic matter to retain moisture, irrigation to add needed water and drainage to remove excess water are all important factors in maintaining a favorable moisture supply in the soil.

The soil is constantly being depleted of its minerals through absorption by the plants and by percolation and erosion. The horticulturist conserves and replenishes these materials by plowing under plant refuse, checking too rapid percolation, preventing erosion, increasing the rate at which the unavailable elements in the soil are made available and the actual addition of elements used by the plant.

Organic matter in the soil is decomposed largely through biological activities. This decomposition is necessary in order to liberate the plant-food elements contained in the organic matter and make them available for plant use. The decomposition processes occur most rapidly with favorable conditions of temperature, moisture and aeration. To overcome this loss of organic matter, the horticulturist adds it by growing and plowing under cover crops and by the addition of plant and animal refuse.

The soil contains an aggregation of small connected cavities which are filled with water and air. Air is necessary in the soil for favorable biological activity and for root growth. Suitable aeration of the soil is provided by establishing and maintaining a favorable texture and structure and by suitable tillage practices with proper tools. A new process of soil aeration is being used for old shade trees in compact sods whereby compressed air is forced deeply into the soil.

SYSTEMS OF SOIL MANAGEMENT

Many variable factors, as kind of plant, age of plant, varying conditions of soil, weather and topography, must be considered. Certain soil practices might be satisfactory under one set of conditions and unsatisfactory under different conditions. Before determining upon a practice to be used in managing the soil it is advisable to consider the effects on the plant of that practice in comparison with other practices that might be used. By this method one arrives at a system of soil management that best meets his requirements. There are four main systems of soil management—sod, tillage, mulch and rotation—that may be used by the horticulturist. The management of the soil of any particular planting will probably be a combination of one or more of these four systems or a modified type of one or more systems.

Sod

Sod culture is that system of soil management wherein the plants are grown in permanent grass without tillage and without the addition of any litter. It is one of the oldest systems of soil management in the orchard and is also used universally for shade trees.

There are several recognized types or modifications of the sod system for tree crops. The most common are (1) grass allowed to grow without being cut; (2) grass cut one or more times during the season and let lie in place; (3) grass cut one or more times during the season and removed as hay; (4) grass pastured and (5) temporary sods, as clover, which is disked and reseeded every two or more years. The third and fourth types are decidedly objectionable and can hardly be termed systems of soil management.

Trees grown in sod are more shallow rooted than those grown in tilled soil. For this reason it is advisable to till the orchards for the first few years if conditions will permit such operations. The shallow growth of the roots seems to be associated with a decrease in the available nutrient supply—chiefly nitrates—in the soil. This decreased fertility is believed to be due to a reduction in the biological activities in the soil because of decreased aeration as a result of the presence of the sod.

FIG. 101.—A hillside orchard; steepness of slope would make clean cultivation in this orchard difficult and would lead to soil erosion. (*Gardner, Bradford and Hooker.*)

Some type of sod culture is used in commercial orchards in areas where the topography makes any system of tillage inadvisable or impracticable. Many home orchards are grown in sod; and sometimes the grass is removed for hay, or the orchard pastured. Although a small amount of hay might be removed for a few years or a few small animals or poultry pastured without injury to the orchard, too often both these practices lead to injury of the orchard.

The type of sod culture that appears to be the most satisfactory is the one in which the grass is cut and let lie. This type really approaches the mulch system which is discussed later. Another general modification of the sod system is the addition of fertilizers, usually a form of nitrogen. Practically all experimental data prove the advisability of this practice. Such fertilization is becoming a common and beneficial practice with established lawn trees where tillage is impossible because of the necessity of maintaining a turf.

TILLAGE

Tillage is that system of soil management in which part or all of the soil is cultivated in some way each year. A large number of modifications of the tillage system are used under particular conditions or for specific purposes. Clean cultivation is that type in which the soil is kept free of all vegetation other than the crop being produced. Clean cultivation and cover crop is that type in which the soil is kept free of vegetation until early summer when a crop is seeded that will make a good growth before being stopped by cold weather. Clean cultivation and intercropping is that type in which a cash crop is grown between the rows of trees while they are still young. Strip cultivation is that type in which a portion of the space between rows is cultivated, and the area occupied by the rows of plants is not cultivated. Row cultivation is merely the opposite of strip cultivation, for in row cultivation the area occupied by the row of plants is cultivated and a non-cultivated strip is left between rows. Various modifications and combinations of these types will suggest themselves to meet particular conditions. The tillage practices must be adapted to the crop and to the particular environmental conditions.

In recent years clean cultivation has lost favor in orchards, but it is still used extensively with small fruits and nursery plants and is the universally accepted practice for nearly all flowers and vegetable crops. It is conceded by all that clean cultivation destroys weeds and thus conserves for the other plants the moisture and plant nutrients utilized by the weeds. By tillage the aeration of the soil is improved, organic matter in the soil is decomposed more rapidly and plant food elements are liberated more rapidly and more abundantly. This same tillage practice, however, dries the soil more rapidly, "burns out" the organic matter in the soil too rapidly and produces a less desirable soil structure and may liberate the plant-food elements too rapidly for the use of the crops or at a season of the year when the crops cannot use them advantageously and consequently they are lost by leaching.

The production of many plants is increased by cultivating the soil and keeping it free from weeds. This increased production is more pronounced with some kinds of crops than with others. The beneficial effects are undoubtedly due to a number of interacting factors, the most important being elimination of competition for water and essential elements by the removal of the weeds and the actual increase in the amount of soil nutrients available for the plant because of the tillage operations. In comparing cultivation and the suppression of weed growth with no cultivation and the growth of weeds in corresponding planting of beets it was found that the cultivated, weedless area produced nearly four times as many beets that weighed nearly six times as much as was produced in the non-cultivated weedy area.

In some instances it appears that the elimination of the weeds and the resultant conservation of moisture and plant nutrients used by the weeds are the chief causes of increased yields. The tillage operation itself seems to be actually injurious in certain cases. This injurious

FIG. 102.—Clean cultivation of cauliflower, Long Island, New York.

effect appears to be due to one or both of two causes. The relative importance of the two causes appears to be determined largely by the kind of plant and the frequency and depth of tillage in relation to the frequency of rain and the amount of rainfall at each time. Cultivation should be such as to cause the least disturbance to the root system of the plant. Early tillage of annual crops can be deeper, but later tillage should be shallow and chiefly for the purpose of destroying weed growth. Under certain conditions cultivation may actually cause a loss rather than a gain in soil moisture. Weedless soil cultivated shortly after a light rain of ½ in. or less may increase the loss of water from the soil by exposing a larger soil area to the drying action of the air. The light rain may not have wetted the soil more than one-half the depth of the cultivated soil. The following table indicates the necessity of keeping the soil free of competing weed growth and at the same time shows that a large number or frequent cultivations throughout the season of weed-free soil is not necessary.

TABLE 29.—EFFECT OF CULTIVATION ON YIELD*

Kind of crop	Average yield of marketable portion of crop, pounds per plot			
	Cultivated all season	Cultivated half of season	Scraped	Weeds allowed to grow
Carrot	505.3	506.4	519.5	27.9
Beet	240.3	239.7	233.2	45.6
Cabbage	233.6	234.6	207.5	129.1
Onion	67.7	69.6	64.3	3.6
Tomato	164.0	166.6	166.8	23.3
Potato	148.3	150.4	158.8	52.7

* Adapted from THOMPSON, H. C., *et al.*, Cultivation Experiments with Certain Vegetable Crops on Long Island, *Cornell Univ. Agr. Expt. Sta. Bul.* 521, 1931.

Continuous cultivation resulted in a rapid depletion of organic matter, a loss of fertility and eventually an unsatisfactory soil texture. To remedy these difficulties a modified type of clean tillage known as "clean tillage with cover crops" was developed. In this type the soil is kept free of weeds and other vegetation except the desired crop by various tillage practices until early summer when cultivation is discontinued and volunteer plants or seeded plants are allowed to grow in the areas between the rows of the crop plants. The following spring the cover crop is turned under, and the tillage operations repeated. In locations where topographic, climatic and soil conditions are suitable, this type is quite popular, especially with tree fruits, bush fruits and grapes. The crop used as a cover crop will be determined by its adaptability to the location and soil to the specific requirements of the particular planting. Other than the volunteer weeds and grasses, the kinds of plants most generally used are winter vetch, oats, barley, rye, buckwheat, millet, rape, soybeans, cowpeas, clovers and various combinations of these plants. A number of advantages are attributed to cover crops. They check the growth of the perennial crop plants by competing with them for water and mineral elements and cause them to mature their wood earlier and more fully and thus enable them to withstand trying winter conditions more satisfactorily. By the use of cover crops the organic matter content of the soil is not only maintained but generally increased. The fertility of the soil is increased because of the more favorable conditions for biological activities, the temporary storage of elements in the tissues of the cover crops thus decreasing the loss by leaching and by the actual addition of nitrogen from the air when leguminous plants are grown. The ample organic content maintains a favorable physical condition in the soil with all its attenuating benefits of increasing the holding capacity of available moisture, checking loss by percolation and reducing soil erosion by decreasing the speed and amount of surface runoff. The presence of the organic matter on the surface of the soil during winter holds more moisture in the form of snow and in itself acts as a blanket which keeps the soil from changing temperature so rapidly and freezing to so great a depth. If the cover crop lives over winter, its growth and rapid transpiration of moisture in the early spring assists in drying out the top soil and facilitates earlier tillage of the soil or movement of machinery over the soil in the case of tree crops.

Intercrops are often used in young orchards, plantings of bush fruits, vineyards, certain nursery plantings and perennial vegetables. The length of time that such intercrops are used will vary from one to several years depending upon the kind of permanent crop. The best intercrops to use are cultivated crops, as beans, tomatoes, potatoes, cabbage and corn. It is better to avoid crops that make their growth early in the spring and compete with the permanent crop for moisture and nitrogen. The intercrop should not be planted so close to the permanent crop that it hinders the growth of that crop, and the use of companion crops should be discontinued just as soon as the permanent crop needs the area. Generally, corn as a companion crop in an apple orchard should not be planted closer than 5 ft. to the trees the first year, 7 ft. the second and third years and 10 ft. during the fourth and fifth seasons.

Strip cultivation consists in cultivating a portion of the space between the rows of plants and not cultivating the space occupied by the plants. Row cultivation is the reverse of strip cultivation in that the spaces between the rows are left uncultivated and the areas occupied by the plants are cultivated. These types are limited to orchard plantings and are often used in the home orchard.

Border cultivation is the next step in row cultivation in which a small area surrounding each plant is cultivated. This is particularly adapted to young shade and fruit trees and shrubs that are set in sod. The ground around these plants should be spaded well beyond the spread of the branches during the first years after transplanting and until the plant becomes well established. With many shrubs the tillage should be continued each spring to decrease the competition of grass.

Mulch

The mulch system of soil management is that system in which material such as hay, straw, special mulch paper or glass wool is placed on top of the soil. The term as applied to the orchard has had a wide interpretation so that anything from mowing sparse grass and allowing it to lie where it falls to spreading straw or other material to a depth of several inches or a foot beneath the trees has been designated as a mulch. From this it is clear that the line of demarcation is not

Fig. 103.—Orchard mulched, straw mulch extending beyond branches. (*Iowa Experiment Station.*)

clear between some of the types of the sod system and some of the types of the mulch system. The mulch system is really a progressive development from the sod system. The orchards were pastured; the grass was cut for hay; the grass was let grow; and, finally, the grass was cut and let lie. Next someone began raking the cut grass from between the rows and spreading it under the branches. It was then only a step to bring in additional mulch material and cover all the orchard area. It was a system that evolved because of the benefits that were noticed to develop from the better forms of the sod-mulch system. The mulch system, however, is distinctly different from the sod system in that no vegetation is permitted to grow other than the crop being produced. This eliminates the possibility of the formation of a compact sod, with its attendant difficulties.

Besides cost in areas where suitable mulching material is scárce, the objections to organic mulches are the increased possibility of injury by fire and increased possibilities of injury by mice. Strawberry plantings have always been mulched; raspberry and grape plantings have been mulched to a limited degree in some sections for many years; but only comparatively recently have commercial orchards been mulched. The advantages of mulch over sod are: Competition for water and nutrients between the grass and trees is absent; permeability to rainfall is increased and surface runoff practically eliminated; loss of water by transpiration from grass and evaporation from the soil is reduced to the minimum; no sod is present to become compact with its accompanying evils; decomposition of added organic matter adds a small amount of fertility to the soil; the presence of mulch makes it easier to move spraying and harvesting machinery throughout the orchard; and the fruit that drops is less injured and often salable.

Where conditions permit, the young orchard should be cultivated for a couple of years before being mulched, as this induces deeper rooting. A strawy mulch about 6 in. deep should be maintained. This will require 6 or 7 tons per acre the first year, with small additions every two or three years following.

The principal indexes used by fruit growers in determining the system of soil management to be practiced are operating costs; yield, color, size and price of marketable fruit and favorable tree growth. The data in the accompanying tables indicate that the use of a mulch had no detrimental effect on tree growth as indicated by the increase in the circumference of the tree, that a mulch prevented the deep

TABLE 30.—INFLUENCE OF SOIL MANAGEMENT ON GROWTH AND SIZE OF APPLE TREES. TREES PLANTED IN 1915*

Variety	Soil treatment	Circumference of trunk 12 in. from ground		
		1916	1929	1934
Stayman	Cover crops	3.50	30.41	36.6
Stayman	Mulch	3.53	31.56	38.1
Delicious	Cover crops	3.10	30.95	37.9
Delicious	Mulch	3.04	30.90	39.2

* ELLENWOOD, C. W., and J. H. GOURLEY, Cultural Systems for the Apple in Ohio, *Ohio Agr. Expt. Sta. Bul.* 580, 1937.

penetration of frost and kept the soil temperature more equable and that it was favorable for the retention of moisture in the soil. The tendency of roots under a heavy straw mulch to occupy the surface soil and grow in the loose decaying mulch has been advanced as an objection to the use of a mulch. In favorable soils, deep rooting is not inhibited by the development of the shallow roots just underneath the mulch.

TABLE 31.—DEPTH OF FROST PENETRATION UNDER DIFFERENT CONDITIONS OF SOIL MANAGEMENT, INCHES*

Soil treatment	Date of examination		
	Mar. 3, 1934	Feb. 3, 1936	Feb. 26, 1936
Heavy straw mulch	3.0	3.0	9.0
Sod, not mulched	9.5	7.0	18.0
Cover crop	10.5	11.0	30.0
Clean cultivation	14.3	18.0	

* ELLENWOOD, C. W., and J. H. GOURLEY, Cultural Systems for the Apple in Ohio, *Ohio Agr. Expt. Sta. Bul.* 580, 1937.

TABLE 32.—PERCENTAGE OF MOISTURE IN SOIL IN MULCHED, SOD AND CULTIVATED AREAS, AUGUST, 1930*

Date	Soil depth	Soil treatment			
		Mulch	Cover crop culture	Clean culture	Sod not mulched
Aug. 2	Upper 6 in.	8.88	5.11	8.58	6.04
	Second 6 in.	7.35	7.65	8.57	6.02
	Average for 12 in.	8.12	6.38	7.58	6.03
Aug. 4	Upper 6 in.	13.98	5.92	6.47	6.55
	Second 6 in.	10.88	6.49	7.17	6.67
	Average for 12 in.	12.43	6.20	6.82	6.61
Aug. 6	Upper 6 in.	8.49	6.28	7.39	5.90
	Second 6 in.	7.13	6.80	7.19	6.24
	Average for 12 in.	7.81	6.54	7.29	6.07

* ELLENWOOD, C. W., and J. H. GOURLEY, Cultural Systems for the Apple in Ohio, *Ohio Agr. Expt. Sta. Bul.* 580, 1937.

Growers of vegetable crops use straw mulch on some crops but use a special paper mulch in many cases. Pineapple growers use a similar paper. Within the last few years the vegetable growers and florists have been trying a new mulch made of glass wool. Under a straw mulch the soil temperature and available nitrates are usually lower than those of unmulched soil, and under a paper mulch they are higher. Moisture is conserved with either one. Straw is valuable as a mulch on crops, such as the potato, which grows better in low soil temperatures during the summer months. Increase in yield that could be attributed chiefly to more abundant water supply is more pronounced in dry seasons than in seasons of normal or heavy precipitation. Weeds will penetrate a straw mulch of less than 8 tons per acre. Available nitrates may be decreased under straw in the early part of the season but increase later. The depression of available nitrates under a straw mulch is thought to be due to the use of available nitrogen by the microorganisms engaged in the decomposition of the straw adjacent to the moist soil surface. This will have some influence on the time of adding the straw mulch.

Paper mulch was first used commercially in Hawaii on sugar cane to control weeds. Lightweight paper was laid over the rows of sugar cane; the young plants readily broke through the paper; all weed growth was suppressed. Increased growth on mulched sugar cane led to further trial on pineapples with good results. It was then tried in the United States on vegetable crops and found to increase yields on most warm-season crops. Because paper mulch increases the soil temperature, it is valuable in stimulating early growth of many crops but may be harmful, especially to cool-season crops such as the potato. Increased returns from vegetable crops in northern United States rarely pay for the cost of the paper and labor of applying it.

Coarse peat, free from weeds, is frequently used for mulching flowers out-of-doors on small areas. Peat readily conserves moisture, depresses weed growth when thick enough and lowers the soil temperature. It is usually too expensive for mulching crops grown commercially.

Rotation

The rotation system of soil management is that in which different kinds of crops are planted on the same area of land in a somewhat regular sequence. Systematic crop rotation is very important with certain nursery plants, with flowers and especially with annual vegetable crops. Rotation aids in the control of insects and diseases, in the equalization of the depletion of available minerals in the soil, in the improvement of the physical condition of the soil and in the avoidance of the deleterious effect of one crop on the succeeding crop.

Many serious diseases may be controlled if the host plant is not grown on the soil for three or four years. Clubroot of cabbage and related crops, as cauliflower, broccoli and Brussels sprouts, can be controlled if the soil is kept free of cruciferous plants for a two-year period, but not all diseases that are harbored in the soil can be controlled by rotation of the land. The organism causing cabbage yellows will survive in the soil for many years after the crop has been removed and will infect a susceptible variety of cabbage or related crop if planted on the infected soil even after many years. Some organisms attack only one host plant; others, such as nematodes, attack many kinds. Those which attack but one kind of plant are easily controlled by crop rotation, especially if the pest can live in the soil but a short time. Some few insects can be controlled through rotation if they feed on one crop only and do not move very far. If the host plant is not present when the insects or disease organism appears, the pest dies for lack of food. Pests that feed on many kinds of plants, such as the chinch bug which feeds on many species of Graminaceae, and insects that move long distances, as the corn-ear worm, cannot be controlled through rotation.

Crops that are heavy feeders, removing large quantities of all or certain nutrient elements, may be followed by a crop that is not so exacting in its requirements. Crops such as onions and tobacco which use large quantities of potash may be followed by such leafy crops as cabbage and lettuce, which are not heavy users of potash. Some crops may make better use than others of manures or commercial fertilizers applied to the previous crop, and some are better able than others to use the residue of the preceding crop.

Planting a green manure crop, such as rye, in late summer or early fall, after the production of a money crop, will aid in maintaining a good physical condition of the soil when the cover crop is turned under the following spring and will avoid losing the use of the land for one year. When manure or other plant residues are scarce and a nitrogenous fertilizer is necessary, a leguminous cover crop should be included in the rotation.

Some crops leave the soil more acid than others. This greater acidity may be reflected in lower yields of some of the following crops. Experiments have demonstrated that the yield of carrots differs very little following various crops, because the carrot is resistant to acid-soil conditions and is able to assimilate mineral nutrients from soil not fertile enough to maintain maximum yields of other crops. On acid soils the lettuce yields are greater after beets or potatoes than after peas or cabbage. Those crops which remove more of the basic elements from the soil should not be followed by acid-sensitive crops unless lime is applied.

USE OF FERTILIZERS

The soils in which plants grow are not alike in fertility; more important still, the plants themselves are not alike in their need for and use of nutrient elements in the soil. Hence fertilization is necessary and extremely profitable sometimes and distinctly unnecessary and unprofitable at other times.

A commercial horticulturist is interested in increasing his profits, whereas an amateur gardener is interested in improving the appearance and growth of his plants. The application of fertilizers may assist in one or more of four ways: (1) by making a larger plant with more buds in which blossoms may be formed, (2) by increasing the number of

Fig. 104.—Note the difference in size of between the two rows of seven-year-old Montmorency cherry trees. The row of trees on the left has been in alfalfa sod plus the annual addition of nitrogen-containing fertilizers, while the one on the right has been under a clean-culture, cover-crop system of soil management. The difference in size of these trees indicates that the cherry does not thrive with sod culture. (*Gardner, Bradford and Hooker.*)

buds that form flowers, (3) by increasing the percentage of flowers that set fruit and (4) by increasing the size of individual fruits.

Before discussing the use of chemical fertilizers in the production of horticultural crops, it will be well to review the functions of various elements in the soil that are necessary for plant growth.

Plant-food Elements

In 1840 Liebig considered the following elements essential to plant growth: carbon, hydrogen, oxygen, nitrogen, phosphorus, potassium, sulfur, calcium, magnesium and silicon. Later it was discovered that iron should be added to the list; and although silicon was present in many plants, it was not essential. Within the last few years other mineral elements, as boron, copper, manganese and zinc, have been found to be essential to plant growth. In most soils the grower is concerned with deficiencies of nitrogen, phosphorus or potassium and not with the so-called "minor," or rarer, elements, as boron, copper and zinc. These minor elements, however, are no less indispensable to the life of the plant than the so-called major elements nitrogen, phosphorus and potassium.

Nitrogen.—Nitrogen is the fertilizing element most generally deficient in soils and is likely to be the limiting element in soil fertility. It is a constituent of the protein substances that are a part of the protoplasm of the plant cells. It is available to the plants in the soil as potassium, sodium or calcium nitrate and sometimes as ammonia. Nitrogen influences the development of the vegetative parts of the plant and is responsible for the deep green color of the leaves. When nitrogen is deficient, the leaves are yellow-green in color and the growth of the plant is slow. Nitrogen can be supplied in mineral form such as nitrate of soda (15 to 16 per cent N), sulphate of ammonia (20 to 21 per cent N), calcium nitrate (15 per cent N), calcium cyanamide (20 to 21 per cent N) and urea (34 per cent N), which occur as natural deposits or are manufactured commercially. Plant products used as nitrogenous fertilizers to furnish nitrogen are cottonseed meal and linseed meal, and commonly used animal products are blood meal, tankage and fish scraps. The nitrogen is not so quickly available from the organic materials as from the mineral forms; consequently, the form used will be influenced by the speed desired.

Immense quantities of animal manures are used as fertilizers; and although the nitrogen content is only 1 per cent or less, these manures contain also small amounts of phosphorous and potassium. If a legume is used as green manure, it may add nitrogen from the air by bacterial action in the root nodules. The supply of nitrates in the soil varies with temperature and moisture. In cool, wet weather the nitrifying process by which nitrogen is made available to the plants through the action of microorganisms is slower than in warm dry weather. Applications of quickly available nitrogen are necessary to produce vigorous vegetative growth on poor soils. In peat and muck soils, which are high in nitrogen in warm weather, crops like onions and celery will respond to applications of a quickly available nitrogenous fertilizer in early spring, because the microorganisms may not be numerous or active enough for adequate nitrification. The same application made later in the spring or early summer may be of no benefit because sufficient nitrogen has been made available by increased biological activity at the higher temperature.

Phosphorus.—Considerable quantities of phosphorus may be present in the soil but may not be in a readily available form. A deficiency of phosphorus may cause small, poor root growth, late maturity and shrunken seeds. The chief sources of phosphorus in commercial fertilizers are phosphate rock and bones. Raw bone meal or ground rock phosphate is seldom used on horticultural crops because the tricalcium phosphate in them is not readily available to the plants. Superphosphate is produced by treating the rock phosphate with sulfuric acid and thereby changing the phosphorus to a more readily available form. Raw bone is steamed and treated with sulfuric acid to form readily available phosphorus. Superphosphate in more concentrated form, known as double or treble phosphate (40 to 48 per cent P_2O_5), is used in preparation of high-analysis fertilizers.

Potassium (Potash).—Potassium is supposed to be essential in the formation and translocation of carbohydrates and seems to increase resistence to certain diseases. Abnormal color of leaves, weak stems, stunted growth and poor development of roots and tubers are associated with a deficiency of this element. Large quantities may be present in the soil without harmful or toxic effect to most plants. All soils except mucks and peats contain relatively large amounts of potassium, but the amount available to plants in any soil may be low; especially is this true of sandy soils.

This element may be supplied in commercial fertilizers in several compounds of which muriate of potash (48 to 54 per cent) or sulfate of potash (48 to 52 per cent) are the forms most commonly used. New synthesized potassium salts, now in use to a limited extent, are potassium nitrate (44 per cent K_2O + 12.3 per cent N) and potassium ammonium nitrate (28 per cent K_2O + 16 per cent N). Wood ashes are commonly recommended as a potassium fertilizer but contain only 5 to 6 per cent of K_2O. Kainit contains 12 to 13 per cent K_2O.

Calcium.—Although calcium is one of the essential elements for the growth of plants, it is used mainly in the form of lime to neutralize acidity. Soils seldom lack calcium as a plant nutrient, but many crops require applications of lime in order to reduce the soil acidity. Certain toxic materials, such as soluble aluminum, may be rendered harmless by liming. The physical conditions of heavy soils may be improved, since liming causes flocculation, or grouping of fine particles into larger groups, making the soil more porous.

Limestone and hydrated lime are the forms most commonly used in adding lime to soil. Limestone may contain as high as 90 per cent $CaCO_3$, but many grades contain less. If magnesium carbonate is present in the limestone, it is equal to the calcium carbonate. Dolomitic limestones contain considerable magnesium carbonate and are very valuable on soils deficient in magnesium. Hydrated lime (calcium hydroxide) is quicker acting than limestone, and 74 lb. of the calcium hydroxide is equal to 100 lb. of calcium carbonate. Gypsum (calcium sulfate) should not be used to correct acidity, as it increases rather than decreases it.

Magnesium.—Magnesium is an essential element in chlorophyll formation, as it is a part of the chlorophyll molecule. Low magnesium is usually associated with low calcium in medium to strongly acid soils. In the Atlantic states from Maine to Florida, magnesium deficiency is common because there is little magnesium-bearing rock. Plants growing in soils deficient in magnesium develop chlorotic leaves, the lower leaves showing the symptons first. To correct this condition of the plant quickly, crude Epsom salts, containing 30 to 32 per cent MgO at the rate of 40 to 100 lb. per acre, may be applied. Dolomitic limestone containing magnesium carbonate may be used for correcting acidity where magnesium is deficient.

Iron.—Many plants suffer from a lack of iron when grown in calcareous soils, even though iron is present in sufficient amounts. The alkaline reaction of such soils, due to excess calcium salts in solution, produces a change in the available iron compounds and renders them unavailable to the plants. In Puerto Rico considerable difficulty is encountered in growing pineapple because of the lack of available iron in the calcareous soils of that section. In Florida, chlorosis of grapefruit trees has been found to be caused by too much limestone in the soil, and the more limestone present the greater the injury. The alkaline reaction changes the iron into a form unavailable to the plants. A solution of iron sulfate sprayed on the leaves, injected into the plant or added to the soil will remedy the condition.

Boron.—Boron is necessary, in minute quantities, for the development of many species of plants. A good illustration of this type of nutritional disturbance traced to boron deficiency is the disease known as "cracked stem" in celery which has been reported in Florida and New York. The appearance of water core or dark center in rutabagas and turnips has been a problem in the New England states. The application of 20 to 30 lb. of borax per acre remedies this condition. Cauliflower plants grown in Delaware County, New York, are deficient in boron and have hollow stems, frequently accompanied by browning of the stems and the heads, which renders the plants worthless for market purposes. Browning of cauliflower can be controlled by the application of borax at the rate of 2 to 5 lb. per acre. In the Lake Champlain Valley of New York, drought spot, cork, rosette and dieback of apple trees, all symptoms of the same disease, are due to boron deficiency. The injection of dry crystals of boric acid in holes in the trees will control the disease. Boron can also be very toxic to many plants if present in the soil in excessive quantities. In certain irrigation waters of southern California concentrations of boron are so great that injury to citrus and walnut trees occurs. The trees show the toxic effect of boron by yellowing around the margins of the older leaves and between the veins as well as dying back of tips and margins. Affected leaves of citrus and walnut trees fall off prematurely.

Zinc.—Malnutrition of plants which can be corrected by the application of zinc sulfate occurs in several places in the country, notably Florida and California. In Florida, zinc sulfate applied to acid soils will correct a form of chlorosis of the corn plant called "white bud." Tung trees growing in similar soils have shown symptoms of malnutrition which were corrected by the application of zinc sulfate. Mottle-leaf of citrus trees (chlorotic areas in the leaves) has been controlled on trees in California by spraying with 5 lb. of zinc sulfate plus 2½ lb. of hydrated lime per 100 gal. of water.

Copper.—In severe cases of copper deficiency, the plants may become yellow or chlorotic. Copper is more often deficient in peat and muck soils than in mineral soils. The cause of poorly colored onion scales produced on muck soils has been attributed to lack of copper in the soil. The addition of 100 lb. per acre of monohydrate copper sulfate has remedied the trouble. Copper salts in high concentrations may be toxic to the plants.

Manganese.—Plants growing in soils deficient in manganese become chlorotic and stunted. Manganese deficiency is more likely to occur in soils that are alkaline than in acid soils. Manganese salts in solution in concentrations as high as 5 parts per million are toxic to plants. Usually 1 to 2 parts per million is sufficient for the needs of the plant. The exact function of this element in plant nutrition is still obscure. Chlorosis in spinach has been corrected in Rhode Island by the application of 8 lb. of manganese sulfate ($MnSO_4$) per acre. Tomatoes and other truck crops grown on calcareous soil in Florida were benefited by light applications of $MnSO_4$.

Kinds of Fertilizers Applied

Animal manures were practically the only fertilizing materials applied to the soil in the early days of production of horticultural crops. With an increasing acreage of these crops, with more intensive cultivation and with a decreasing supply of manure the use of commercial fertilizers has become necessary in many cases for profitable production.

Animal Manures.—Manure is not valuable alone for the nitrogen, phosphorus and potassium that it carries because these materials can often be applied more cheaply in chemical fertilizers than in manure. Manure adds humus to the soil, thus improving the physical condition, and it contains favorable organisms which aid in decomposing the organic matter in the soil and liberating plant-food elements.

TABLE 33.—AVERAGE COMPOSITION OF FRESH MANURES

Animal	Water, per cent	Nitrogen, per cent	Phosphoric acid, per cent	Potash, per cent
Hen	55	1.00	0.80	0.40
Sheep	64	0.83	0.23	0.67
Horse	70	0.58	0.28	0.53
Pig	73	0.45	0.19	0.60
Cow	77	0.44	0.16	0.40

The amount and kind of bedding and the care of the manure before application to the soil also influence the value of the manure. Sometimes heavy applications of manure containing large quantities of straw or shavings may have a depressing effect on crop yields. The injury is due to a deficiency of nitrates because such nitrates are used as food by the soil organisms themselves which are breaking down the carbonaceous material. The plants must compete with the organisms for the nitrates in the soil. This condition can be prevented by plowing under the manure a considerable time in advance of planting the crop. This allows for partial decomposition before the crop is planted and before the crop needs nitrogen in large quantities. Rotted manure rarely has a depressing effect because decomposition of the carbonaceous material has proceeded far enough so that there are sufficient nitrates for both plants and organisms. Another remedy is the application of a quickly available mineral nitrogen compound to take care of the temporary deficiency.

Commercial Fertilizers.—A commercial fertilizer may contain but one compound and one fertilizing element, as nitrate of soda, acid phosphate or muriate of potash, or it may consist of several compounds containing the same element or different elements, as nitrate of soda and ammonium phosphate or tankage, cottonseed meal and bone meal. A complete fertilizer is one that contains the three elements nitrogen, phosphorus and potassium. The composition and analysis, which are always stated on the container of such fertilizers, determines their value. In the analysis the first figure represents the percentage of nitrogen; the second, the available phosphorus; and the third, the water-soluble potassium. It is becoming common to label fertilizers with three figures, though only one or two elements may be contained, as 0-9-27, in which phosphorus and potassium are present, or 0-16-0 in which phosphorus alone is present. High-analysis fertilizers are being offered for sale more than formerly. A high-analysis fertilizer, as 15-30-15, contains three times the quantity of each of the three nutrients as 5-10-5 fertilizer.

The effect of the material on the soil must be considered. Sulfate of ammonia may increase the acidity of the soil which may be detrimental to certain crops. On the other hand, in New Jersey experiments have demonstrated that apple trees made their greatest volume of growth at pH 7.5 and 8.5 when ammonium sulfate was used in comparison with nitrate of soda. Trees receiving nitrate of soda made their best growth at pH 3.5 to 6.5. Above pH 7.5 nitrate of soda was not equal to ammonium sulfate as a source of nitrogen.

Amounts of Fertilizers Applied

The amount and kind of fertilizer to apply to any crop depends on the requirements of the plant which is influenced by the kind and condition of the soil and on the previous crops grown. At present in the United States nitrogen is recognized as the only element that, in many cases, will pay for its addition to the orchard, and many orchards on fairly deep fertile soils in tillage or mulch will not respond profitably to an application of nitrogen. The same orchard if growing in sod would probably respond profitably to a nitrogenous fertilizer. Exceptions to this condition prevail in some of the older apple orchards of the eastern part of the United States. Each individual must decide from his own particular set of conditions and the growth of the plants whether or not the use of a fertilizer of any type is beneficial. Nitrogen deficiency is indicated in fruit plants by a lack of vigor and by pale green foliage. Mature apple trees making but little vegetative growth and bearing an immense crop of blossoms but failing to set fruit oftentimes indicate a lack of nitrogen. Small fruits indicate the lack of nitrogen by a decrease in the amount and vigor of vegetative growth and fruitfulness.

As previously mentioned, sodium nitrate, ammonium sulfate and calcium nitrate are the commercial fertilizers commonly used for quickly available nitrogen. In fertilizing trees the fertilizer should be broadcast on the surface of the soil from near the trunk of the tree to slightly beyond the ends of the branches. Spread the fertilizer over the area of soil occupied by the feeding-root system of the tree. The amount to apply will vary with the age of the tree and with the fertility of the soil. Various standards are used to estimate the approximate amounts required. One rule is to apply ¼ lb. of nitrate of soda for each year of the tree's age, and another is to apply ½ lb. for each inch of the diameter of the trunk of the tree 1 ft. above the surface of the soil. These indexes appear to be satisfactory for ornamental shade trees that show the need of nitrogen. Small fruits are treated in a fashion similar to tree fruits. The fertilizers vary according to the needs of the plants and are applied to the area of soil occupied by the feeding roots of the various plants.

Peat and muck soils commonly used for growing vegetables are notably deficient in phosphorus and potassium, and heavy applications of these fertilizers are usually necessary for most crops. Although peats and mucks are high in nitrogen, the application of a nitrogen fertilizer in the early spring may be beneficial because the microorganisms and chemical activities causing the liberation of nitrogen are less active at that time than later when the soil is warmer. Mineral soils in the Southern states are frequently deficient in nitrogen, and vegetable crops respond to high nitrogen-bearing fertilizers. Mineral soils in general throughout the Middlewestern states are frequently deficient in phosphorus and benefit from applications of this fertilizer. Sandy soils, as a rule, are deficient in the three elements and require heavy manuring and fertilizing for good results. For early tomatoes sufficient phosphorus must be available, or maturity is delayed; but overfertilizing with nitrogenous fertilizers may cause excess vegetative growth, so that fruiting may be delayed and yields seriously reduced.

Ornamental plants are subjected to a wide variation of soil conditions. Many people try to grow plants about their homes in subsoil dug from the basement and spread on top of the better surface soil. Only in rare cases is the excavated soil suitable for growing plants. The filled soil next to the foundation is likely to be alkaline because of the lime and plaster refuse that fell there. Proper soil preparation is essential in such cases, for fertilizers cannot overcome the ill effects of unsuitable soil and careless preparation. Fertilizers are usually beneficial and often necessary, but their kinds and amounts will be determined by the conditions of the various plants.

The amount and type of commercial fertilizer and the frequency of application on lawns vary for different parts of the country. Fertilizing the lawn will be influenced by the kind of grass grown, the soil reaction and the fertility of the soil. A long-time fertilization program should be established, and an effort made to maintain a satisfactorily fertile soil. A generally fairly satisfactory program is to broadcast a quickly available nitrogenous fertilizer in early spring at the rate of 3 to 5 lb. of 16 to 20 per cent nitrogen fertilizer per 1,000 sq. ft. Repeat this application in early June. In early September apply a top-dressing of peat or composted soil at the rate of 1 bu. per 100 sq. ft. Mix 5 lb. of a 4-10-4 fertilizer with each bushel of top dressing material. The phosphorus and potassium may not be necessary every year. An equal amount of a slowly available nitrogenous fertilizer could be added to the early spring application, and the June application omitted.

Time of Applying Fertilizers

The time of the year when fertilizers are applied depends on the plant. For spring applications of a quickly available nitrogenous fertilizer, a good general rule is to apply the fertilizer to fruit trees in early spring when the tips show green. If the fertilizer will have an opportunity to soak into the soil, it may be applied much earlier with results equal to or better than later applications. Organic and slowly available fertilizers should be applied earlier than the quickly available forms. If manure is used, it should be applied during late winter in order for decomposition to take place before growth starts. Applications of fertilizer to small fruits are generally made in the spring; but beneficial results have been obtained from early fall application to strawberries.

Fertilizers for crops planted annually are usually applied a few days prior to or at time of planting. Corn and potato planters have fertilizer attachments, and the fertilizer is applied as the seeds are planted. Commercial fertilizers may be more efficiently used when applied in a narrow band in the hill or row but not in direct contact with the seed, as injury to the germinating seeds will result. Heavy applications of broadcasted fertilizer should be well disked into the topsoil before sowing the seed.

IRRIGATION

The soil's water supply may be obtained as natural precipitation and by irrigation. From prehistoric times man has supplemented the natural precipitation with various types of irrigation.

Types

Irrigation in the arid and semiarid regions of the West is necessary for the growth of crop plants. In the humid regions of the Middle West and East irrigation is an insurance against drought. The irrigation water may be applied as surface irrigation, subirrigation or spray irrigation.

Surface irrigation is used commonly in the arid section of the West by means of flooding the surface of whole areas or by running the water in temporary furrows made between the rows of plants. For this type of irrigation the land must be nearly level to obtain even and satisfactory distribution of water. The water is allowed to flow through furrows between the rows of plants, or small areas are diked with soil and flooded with water. The cost of preparing the land for this type of irrigation and the cost of labor in distributing the water are high, but the cost of equipment is low.

Subirrigation is used in limited areas of Florida and in some peat beds of the Middle West. The application of water to growing crops successfully in Florida by this method depends on the presence of an impervious layer of subsoil, or hardpan, 3 to 5 ft. below the surface;

FIG. 105.—Furrow system of irrigation.

on a 1-ft. layer of coarse sand above the hardpan, which facilitates the even distribution of the water from the tile laid in this sand; and finally on a surface loam which conveys sufficient water by capillary action. Subirrigation has been used with limited success in small peat beds in the Middle West. The water is raised in the drainage ditch by means of dams and backed into the drain tile which is placed a few inches deeper than plow depth. Subirrigation is successful only under very special conditions.

Spray, or overhead, irrigation is most commonly used in humid sections east of the Rocky Mountains. This type does not necessitate that the land be leveled. To apply the water, parallel lines of pipe are placed about 50 ft. apart on posts. Each pipe contains small nozzles spaced 3 to 4 ft. apart which distribute the water at right angles to the pipe. The pipe lines can be rotated so that 25 ft. on each side of the line can be irrigated.

Recently a rotary portable spray system has come into use in many sections. The system consists of lightweight portable galvanized pipe 2 to 6 in. or more in diameter. Each section is 20 ft. long and the ends are equipped with quick-acting watertight couplings. A rotary nozzle can be attached near one end of each section of pipe. Each rotary sprinkler covers a circle 80 ft. or more in diameter. By spacing the sprinklers in the pipe line at 40-ft. intervals, a strip 80 ft. wide and as long as the pipe line can be watered at one time.

FIG. 106.—Overhead system of irrigation. (*Courtesy of Skinner Irrigation Co.*)

When the strip is sufficiently irrigated, the entire line is moved 80 ft., and another strip watered. About ½ in. of water per hour is applied by a line in which each sprinkler head discharges 16 to 17 gal. per minute. About 30 to 50 lb. pressure is necessary at the rotary nozzle to operate the line. The water is obtained from a stream or well by the use of a centrifugal pump. This type of system is much cheaper to install than the overhead spray system.

TIME

The time and frequency of applying water are dependent on the kind of crop and the climate of the area. Where irrigation is used to supplement natural rainfall, water is applied only during times of drought, but in the arid West water is applied at frequent intervals during the entire growing season. In all irrigation practices the aim should be to maintain an adequate water supply for the satisfactory growth of the plant. Temporary wilting due to excessive hot weather on a bright day, however, should not be taken as an indication of a serious shortage of moisture. Sampling the soil for moisture at frequent intervals and at various depths is a good practice. If the soil sticks together in a lump when pressed in the palm, it is likely that ample moisture is present.

The layman has discovered that the most efficient and effective time for watering the home lawn during the summer is late afternoon and evening. Where watering is at all possible, this practice can be used to good advantage during continued dry weather. Heavy weekly applications are much more satisfactory than light daily applications, because the former method promotes the development of a deeper root system. The watering of lawns, after the application of commercial fertilizer, is also important. If the fertilizer is not washed off the blades of grass and into the soil, it will cause burning. The chemical elements of a fertilizer must be in solution to be absorbed by the plant; hence, watering after applying a fertilizer is a good practice.

Amount

Obviously, the amount of water to apply will vary with the water requirements of the specific kind of plant, with the type of soil and with the climate of the particular area.

It has been estimated that 1 in. of rainfall a week would be abundant for most crop plants if it was evenly distributed. One inch of water over an acre of land, known as an "acre-inch," requires 27,152 gal. of water. Rarely does nature apply this much water at one time evenly distributed to a growing crop, but with irrigation equipment and a good supply of water it is easy to do so.

The application of water does not always solve the problem where irrigation is practiced. Irrigation has led to drainage difficulties and salt accumulations in some areas of the West.

In general, however, regardless of the method of irrigation, sufficient moisture should be applied to moisten the soil thoroughly. A single application on vegetable crops in the arid sections of the West is equivalent to a solid sheet of water 3 to 4 in. deep.

Review Questions

1. What are the principal objects in soil management for horticultural plants?
2. What are the principal systems of soil management for horticultural plants?
3. What are the various types of sod culture?
4. What are the advantages of sod culture?

5. What are the disadvantages of sod culture?

6. What are the various types of clean tillage?

7. What are the advantages of clean tillage?

8. What are the disadvantages of clean tillage?

9. What are the principal materials used in the mulch system of soil management?

10. What are the advantages of the mulch system?

11. What are the disadvantages of the mulch system?

12. What is meant by the rotation system of soil management?

13. What are the advantages of the rotation system?

14. What are the disadvantages of the rotation system?

15. How may the application of a fertilizer improve horticultural crops?

16. What three elements, essential for plant growth, are most likely to be deficient in the soil?

17. Name a commercial inorganic, a commercial organic and a non-commercial organic nitrogenous fertilizer.

18. Name a commercial fertilizer containing phosphorus.

19. Name a commercial fertilizer containing potassium.

20. What is meant by 6-10-4 when on a fertilizer tag?

21. What should serve as an index in determining the need of a fertilizer?

22. What should you use as an index to determine the amount of fertilizer to apply?

23. What time of the year should fertilizers be applied?

24. What are the two main types of irrigation?

25. When should horticultural plants be irrigated?

Problems

1. Outline a system of soil management for a home orchard.

2. Outline a system of soil management for the home vegetable garden.

3. Outline a system of soil management for maintaining a suitable lawn.

Suggested Collateral Readings

1. AUCHTER, E. C., and H. B. KNAPP, "Orchard and Small Fruit Culture," pp. 407–448, John Wiley & Sons, Inc., New York, 1937.

2. LAURIE, A., and G. H. POESCH, "Commercial Flower Forcing," pp. 93–167, P. Blakiston's Son & Company, Philadelphia, 1939.

3. THOMPSON, H. C., "Vegetable Crops," pp. 36–51, 52–73, 117–143, McGraw-Hill Book Company, Inc., New York, 1939.

4. WATTS, R. L., and G. S. WATTS, "The Vegetable Growing Business," pp. 133–158, Orange Judd Company, New York, 1939.

5. WHITE, E. A., "The Florist Business," pp. 139–158, The Macmillan Company, New York, 1933.

CHAPTER XIII

TRAINING HORTICULTURAL PLANTS

One needs but to observe a few of the many horticultural plants to note that each kind has certain rather definite growth characteristics. For example, the stems of a Tartarian honeysuckle shrub grow upright, whereas the stems of a shrub like forsythia are spreading and drooping. The characteristic growth habit of a Lombardy poplar is tall and erect; that of the weeping willow, drooping. Similarly, characteristic growth habits of herbaceous flowers and vegetables vary; some grow upright; some are spreading; and some sprawl over the ground.

Man has discovered that he cannot alter the characteristic growth habit of a plant materially but that he can induce the stems to grow in directions other than those in which they grow themselves. This process of controlling the directions in which stems grow is termed "training." Training determines the framework of the plant and the general, if not the detailed, outline of the plant. In general, man accomplishes this task by cutting the stems, by tying them to stakes or trellises or by coordinating the cutting practice and the tying practice. When the horticulturist cuts off part of a plant, he terms the practice "pruning," and this is the most important means used in training plants.

OBJECTS OF TRAINING

The objects sought in training horticultural plants are to develop a satisfactory framework and to promote the earliest desirable production. Training plants is a constructive operation. The pruner is a builder, and he should perform his work with the realization that he is building a structure that is to last many years.

Satisfactory Framework

Obtaining mechanical strength is primarily a problem associated with training. Potentially, trees are long-lived, but the length of life of a tree depends largely on a mechanically strong framework. The mature bearing fruit tree must have adequate mechanical strength to support heavy loads of fruit and withstand the added strain of wind and snow and sleet which frequently place extremely heavy stresses on both fruit and ornamental plants. Breakage of branches from the trunk frequently results when the framework is weak. The loss of a large branch from a mature bearing fruit tree results in financial loss to the grower, and the breakage of branches from a shade tree that has grown for fifty years or more destroys its beauty, lessens its value and may lead to its untimely death.

A satisfactory framework in fruiting plants must be mechanically strong, must facilitate cultural operations and must distribute the wood

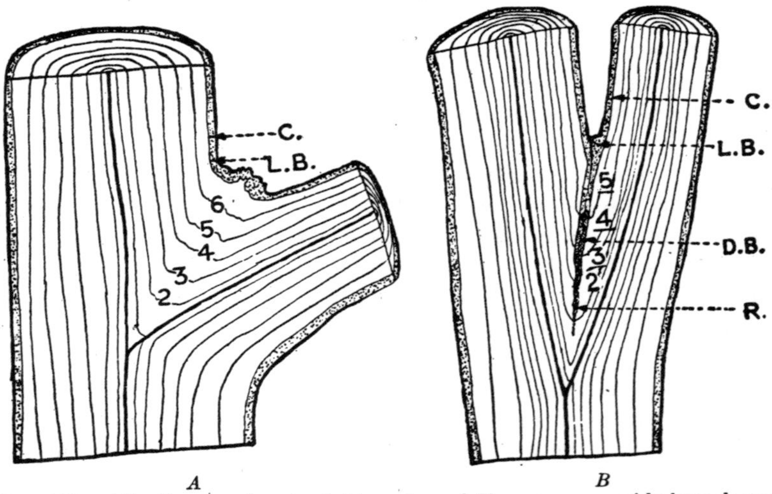

FIG. 107.—(*A*), diagram showing bridge of wood fibers across a wide branch angle. Such a crotch is strong; such branches will carry a heavy load of fruit without breakage; (*B*), diagram of a narrow branch angle showing no bridge of wood fibers across from branch to trunk. Such crotches are weak. (*C*), cambium; (*LB*), live bark; (*DB*), dead bark; (*R*), decaying bark. Figures represent years' growth. (*Cornell University Agriculture Experiment Station.*)

to the best possible advantages for fruit production. A similar framework is essential to many ornamental plants, but in some temporary or short-lived ornamental plants one may be justified in sacrificing strength and permanency to appearance.

An important means of developing and maintaining a mechanically strong framework is to keep the various branches properly subordinated to one another. Two main branches in a tree should not be allowed to develop. The primary lateral branches should be smaller than the main trunk, and the secondary branches should be subordinated to the primary branches. This system makes for strength.

Another means of obtaining a strong framework is to retain only those branches which arise at a wide angle. If the angle is relatively wide (45 deg.), successive layers of wood tissue will be laid down in the crotch, and it will be filled with strong tissues. When the angle is narrow, new wood is added to the branches, and the dead bark on the inside of the two branches forming the narrow crotch is pressed together before the crotch angle can be filled with new woody tissue formed by the cambium. The enlarging branches press away from one another, for their new growth in diameter acts as wedges prying them apart at the narrow angle. The fibers from the two branches run in a parallel course rather than tying together laterally, and the dead bark becomes embedded in the crotch. Eventually decay takes place, and one of the branches splits off.

In addition to possessing mechanical strength a mechanically satisfactory framework is one that facilitates the various cultural operations. The branches must not only be properly subordinated to one another and have wide strong crotches, but they must be so spaced that the operations of cultivation, spraying and harvesting can be carried on efficiently and economically. Soil management operations are difficult in orchards where trees are headed too low. Efficient spraying requires comparatively low trees with the branches spaced far enough apart to allow adequate penetration and coverage by sprays. Harvesting operations are done more efficiently and cheaply on lower trees with well-spaced branches on which the fruit is readily accessible.

The various branches on the tree should be distributed in such a relationship to one another as to promote the optimum growth, flowering and fruiting. Favorable conditions for these functions exist on those branches which are well exposed to sunlight and air. Very little fruit is produced in the densely shaded interior of large apple trees; a grapevine sprawling on the ground does not produce a satisfactory crop; and tomatoes trained to stakes in certain Southern sections are more free of diseases.

Often the object sought in training a plant is to make its appearance more pleasing to the eye. By careful training, such as is done in topiary work, plants can be made to grow into definite shapes to resemble a peacock, a squirrel or some other non-plantlike object. In general, however, topiary work results in objects of curiosity rather than objects of beauty. The training of certain types of evergreens, such as junipers and mugho pines, is done principally to obtain a more pleasing effect by symmetrically formed plants.

Earliest Desirable Bearing

Pruning decreases the total growth made by a plant but increases the vigor of the growth made by the remaining parts. The unpruned plant might produce 100 lateral shoots averaging 6 in. long, and the pruned plant would produce 25 shoots each 18 in. long. Moderately heavy pruning which is necessary in training the young plant maintains the plant in the C/N+ condition so there is little opportunity for flower-bud formation. For this reason, therefore, one should not prolong the training period in flowering and fruiting plants beyond that time when a satisfactory framework can be developed. The longer the training period the later the beginning of the productive period.

MANNER OF TRAINING

The manner of training horticultural plants depends upon the particular kind of plant and upon the objects that one has in mind. In general, the horticulturist trains plants by cutting or by placing branches in particular positions to guide the direction of growth or by a combination of cutting and placing. The principal method of training trees is by cutting; sweet peas, by placing the stems on a trellis constructed of string or wire; and grapes, by both cutting and tying to a trellis.

Cutting

The judicious training of horticultural plants by cutting requires a knowledge of a few fundamental principles associated with plant growth.

Heading Back.—Heading back refers to the removal of a terminal portion of a shoot or a twig. It is employed to stimulate lateral branching near the cut and to suppress growth. The development of the lateral branches is based on the principle that the greatest effect of a pruning cut is localized and the response of the plant to a pruning cut takes place mostly near the vicinity of the cut. When a particular twig is headed back, branching will usually occur within a space of 2 to 12 in. below the cut. In forcing hedge plants to branch close to the ground and thus form a dense compact growth it is necessary to head back the newly set plants to about 4 or 6 in. from the surface of the soil. When branching is desired in a one-year-old apple-tree whip at a height of 24 in. from the ground, the whip is cut at a height of 34 in. from the ground, and most of the lateral branches will arise from the upper 10 in. of the trunk. When branching is desired within 6 or 8 in. of the trunk, the twigs are cut at a distance of 16 to 18 in. from it. The principle of heading back is followed in the shearing of evergreens, such as juniper and arborvitae, when it is desired to make them compact or to keep them from becoming too large. Pinching back shoots of black raspberries and of dahlias is also a type of heading back employed to force branching at desired positions. In heading back to stimulate lateral branching the end of the shoot or twig is cut off a short distance beyond the area from which one desires the lateral branches to develop. When one of two similar twigs are to be subordinated to the other, the one to be subordinated should be cut back more heavily than the one that is to be kept larger. This heavier

FIG. 108.—Prevention of weak crotches by pruning; branch on left has been subordinated by cutting back to a lateral.

cutting of the one twig reduces its growth by removing more potential leaves, and this reduces the carbohydrate supply available for the growth of that particular twig. The unpruned or more lightly pruned twig has more leaves in proportion and consequently makes more carbohydrates which are available for use by that twig. This principle of suppressing growth in a branch is also important in thinning.

Thinning Out.—Thinning out refers to the removal of an entire twig or branch at its point of origin or at a point where a lateral branch arises from the main branch. When a branch more than one year old is removed back to a lateral branch or when a twig is removed completely, it is considered as thinning out.

Twigs and branches are thinned to space the remaining parts of the plant to better advantage, to suppress the growth of one branch in relation to another and to keep the plant within a prescribed area. Thinning out is less devitalizing to a plant than an equal amount of heading back. If all branches on a plant are left to grow, they will be entirely too numerous and too close together for the most satisfactory development of flowers and fruits. For this reason certain of the less desirable branches are thinned out, thus giving better spacing to the remaining branches. Just as one shoot or twig can be subordinated to another by more severe pruning, the total growth of an older branch can be decreased in proportion to the total growth of another branch by heavier pruning. Two branches of equal size were selected on the same three-year-old apple tree, and one of them was pruned heavily, whereas the other was left unpruned. At the end of the growing season the unpruned branch had increased in diameter approximately 150 per cent, but the heavily pruned branch had increased only 20 per cent. This practice of cutting one branch more heavily than another is used to prevent the formation of weak crotches.

Older trees eventually become so large that they extend beyond the area in which they should be confined. Certain branches on young trees may make a too rapid growth in relation to the other branches on the tree. Consequently such trees or such branches are kept within their prescribed areas by cutting off the offending branches back to their points of origin or to a suitable lateral. This type of thinning is really an advanced stage of heading back and an early stage of dehorning, which is sometimes practiced in old trees.

Another important principle of plant growth is that of two equal branches: the one more closely approaching the perpendicular makes the more growth. The vertical branch apparently has an advantage in securing nutrients, water and sunlight and other factors that contribute to the amount and speed of growth. This principle is important when selecting leaders and laterals. Vertical branches of nearly the same size as the primary branches from which they arise are removed when young, as they dwarf the branch from which they arise, interfere with the branches directly above them and often compete with the leader for supremacy. Branches arising from the trunk at moderate angles grow sufficiently rapidly to develop into excellent primary laterals and also form strong unions with the trunk.

Parallel branches should not be allowed to develop close together. Such branches arise directly above one another from the main trunk of the tree or from the underside of branches one or more years older than themselves. If those arising from the main trunk are too close together, the lower one will be shaded, and one of them should be removed. If a branch that arises from the underside of another branch develops to any appreciable extent, it will grow immediately below and parallel to its parent branch. The two branches will occupy or can easily be trained to occupy practically the same space in the plant, and consequently one of them should be removed at an early stage in its devlopment.

Placing

Plants are often trained by placing the stems in definite positions and causing growth to incline itself in the desired directions. This method of training is employed principally with vines when they are grown on trellises, walls and arbors. It is used also with the espalier training of trees.

SYSTEMS OF TRAINING

There are many systems of training the various kinds of horticultural plants. Furthermore, several systems are well adapted to the same kind of horticultural plant. Choice of system will be influenced by the growth habit of the plant and by the purpose for which it is being grown.

Trees

At the present time trees are trained in five general types or forms: the natural, the central leader, the open center, the modified leader and the espalier.

Trees have a deliquescent, or open, branching form in which the main trunk, or leader, is lost by dividing into several subordinate branches; and an excurrent form in which the main trunk extends from the ground to the top of the tree with subordinate lateral branches arising from the main trunk. Mechanically strong trees are produced in nature by both types of growth, but the deliquescent form, which is the common form of most deciduous trees, has many weaknesses from the standpoints of suitable framework and early production that can be eliminated or improved by intelligent training. Trees grown without any training develop too many weak crotches, have the branches too low to the ground, are too dense and too tall for efficient spraying and economical harvesting of fruit. Early bearing may decrease the total amount of crop produced and the duration of profitable production. To overcome these difficulties fruit trees and some ornamental trees are trained in various systems.

Natural.—In the early fruit orchards of the United States the trees were allowed to grow naturally, and very little attention was given to training the trees to any particular form. The disadvantage of this type of tree for fruit production was the development of numerous small branches which made the trees compact and the foliage dense. Much of the fruit was poorly colored, and spraying, thinning and

FIG. 109.—European linden used as an avenue tree, developing a central leader but with lower scaffolds entirely too low.

picking became difficult. These natural-growing fruit trees produced sufficient fruit for cider, dried fruit and cooking. No particular preference existed for well-colored fruit, and this type of tree served the purpose for which it was intended.

Central Leader.—The central-leader tree is one in which a main trunk extends from the surface of the soil to the top of the tree. Primary lateral branches arise from this main trunk at various intervals throughout its entire length. It differs little from the natural excurrent type of many of the conifers. Attention is directed chiefly to selecting, spacing and arranging the primary branches on the main trunk and maintaining the proper degrees of subordination among these primary laterals. Most coniferous ornamental trees and many nut trees are trained to the central leader type. The chief disadvantage of this type of training for fruit trees is that some kinds grow too tall, and such operations as spraying, pruning and harvesting are made difficult and expensive.

The Open Center.—The open-center tree is one in which the main trunk is cut out to a lateral branch and three to five nearly equally developed primary lateral branches arise relatively close together

Fig. 110.—Result of training to an open-center type and permitting all scaffold limbs to originate at practically one point on trunk.

from the short trunk. Eventually as these branches develop and increase in diameter, their bases crowd one another on the trunk. This arrangement and the equality in the size of the primary laterals often produce a mechanically weak structure which requires artificial bracing or the use of props to prevent serious breaking or splitting of limbs. The open-center tree exposes the maximum amount of the tree to the sun, and this results in a more uniform distribution of fruit on the tree and in more highly colored fruit. The open-center tree is closer to the ground than trees trained by other systems, and this facilitates spraying and harvesting operations. This system of training is most widely used in training peach trees and to a limited extent in some sections for the apple, both sour and sweet cherries and the American type of plum.

The Modified Leader.—The modified-leader tree is one in which the central leader is retained for a distance of several feet and then cut back to a lateral, as was done in training the open-center tree. This tree is a combination of the central-leader and the open-center type. It is an attempt to eliminate the major defects and retain the better features of both systems. The fact that it is a very popular system and used universally proves that it is quite satisfactory in

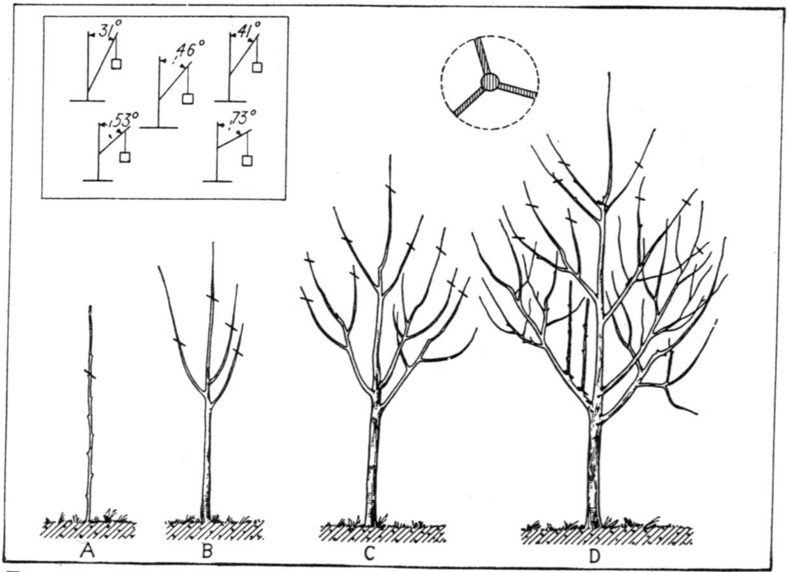

FIG. 111.—Diagrammatic sketches showing the training of an apple tree to the modified central-leader type; (*A*), one-year-old tree; (*B*), two-year-old tree; (*C*), three-year-old tree; (*D*), four-year-old tree.

attaining its aims. This system of training develops the central leader until it reaches a height of 6 to 8 ft. Beginning at a distance of 2 ft. or more from the ground, primary lateral branches are developed from the main trunk. During the four or more years of training, these laterals are selected to ascend somewhat spirally about this main stem and 6 in. or more apart up and down the trunk. If the leader has not lost itself naturally by the time the main trunk has attained the desired height, it is removed by cutting it out to a lateral branch at that height.

The skeletal structure of the tree at four to six years of age consists of a central trunk which extends from the ground to a height of 6 to 8 ft. Beginning at about 2 ft. from the ground, 7 to 12 primary lateral branches arise in a spirally ascending order from this central trunk, at

6- to 10-in. intervals. Since the bottom primary branches are older and larger, the tree appears as a truncated cone supported on a short stem. These primary scaffold limbs should be attached strongly to the trunk. The strength of their attachments depends largely upon the width of the angle between the trunk and the branch, the relative size of trunk and branch and the distance on the main trunk between the place of origin of adjacent laterals. The most strongly attached laterals have wide angles, are noticeably smaller in diameter than the main trunk from which they arise and are spaced far enough apart on the trunk so that there is no choking or crowding of branches. Low-headed fruit trees are preferable to high-headed trees, because they require less pruning, grow more rapidly and bear at an earlier age. In comparison with high-headed trees, low-headed trees are more easily pruned, more efficiently sprayed and more economically harvested. The crop and branches on these low-headed trees are less subject to injury by the wind, and the trunk is less liable to injury by sunscald or winter injury.

The principal methods of obtaining the modified-leader tree are by heading back and by disbudding. The heading-back method is perhaps the more satisfactory for the average grower. The discussion that follows applies particularly to apple trees but may be taken in a general way to apply to the training of other fruit trees. The time to start training a tree is the day it is set in the orchard. Most of the major framework should be developed during the first four to six years of growth.

Apple trees as they are obtained from the nursery are either one-year-old unbranched whips or two-year-old branched trees. After planting, the one-year-old whip is cut back to cause the development of the laterals at the desired height for the head. This is usually about 34 to 40 in. from the surface of the soil. The lateral that develops from the topmost bud immediately below the cut and on the side of the trunk toward the prevailing summer winds will usually make the most growth, grow directly upward and assume the role of the leader. Other lateral branches will develop from the region of the trunk about 10 in. below the cut. At the close of the growing season the tree will have an appearance similar to a two-year-old nursery tree. The two-year-old tree, whether grown in a nursery or in an orchard, has from 3 to 10 primary branches. Some have one central branch, others two or even three branches competing for the leadership. The first step in training this tree is to choose one central branch as the leader and remove any others that might compete with it. Next select a well-developed, strongly attached branch 20 to 30 in. from the surface of the soil and preferably on the side of the tree of the prevailing summer winds. Eight or more inches above this first scaffold branch, select a similar one located in a spiral position about one-third of the way around the trunk of the tree from the first scaffold. If spacing allows, select a third scaffold about 8 in. above the second one and about one-third of the distance around the main trunk.

Remove all other branches, and make the cuts close to the trunk. The two-year-old tree now has a central leader which may be designated as *D* and three primary scaffold branches designated as *A*, *B* and *C*. The lowest lateral *A* is about 2 ft. from the ground, and the other laterals *B* and *C* are about 8 in. apart and spaced spirally about the tree in such a fashion that the area of the circle is evenly divided among them. Each scaffold branch should be cut back to 16 in. or more in length, making the cut in such a fashion that the last bud is on the underside of the branch. The leader should be cut back, leaving it 8 in. longer than the upper lateral branch and with the last bud in the direction of the prevailing summer winds. During the summer, secondary laterals will develop on each of these one-year-old twigs, and the shoots nearest the end will become the leader of that branch, continuing the growth outward from the laterals and upward from the central leader.

It is an unusual tree that has its branches so spaced that more than three scaffolds can be left 8 to 10 in. apart and symmetrically arranged around the trunk. Often only two primary laterals can be obtained in one year, and their selection must be modified accordingly.

At the time of pruning the third year *A*, *B*, *C* and *D* will each appear similar to two-year-old trees attached to a common trunk. The leader *D* will be pruned much like the original two-year-old tree, but the lateral branches *A*, *B* and *C* will be trained to occupy a flattened or oval area rather than a cylindrical one. Pruning the third year, therefore, will consist of prolonging the leader from the most favorable growth on *D*, selecting additional primary laterals on *D* and continuing the leader and selecting secondary laterals on each of the primary laterals *A*, *B* and *C*. The prolonged terminal from *D* is left about 16 in. long, being cut in such a fashion that the top bud points in the direction of the prevailing summer winds. Additional primary laterals are selected at suitable places on *D*. To keep these laterals from overgrowing those lower down it is necessary to prune them more heavily than laterals *A*, *B* and *C* were pruned. These may be left about 10 in. long and cut to down buds. Laterals *A*, *B* and *C* are pruned alike. Take off any twigs that arise directly from the top or bottom of these primary laterals, and save two well-grown secondary laterals and the leader. The leader will continue the elongation of the primary lateral, and one of the two secondaries will usually be developed from one side of the primary and one from the other. The leader should be left longer than the laterals, and they should not be much over 16 in. long. Neither secondary should be closer than 10 in. to the main trunk. Remove all other twig growth except the very short ones that give evidence of early spur formation. The training process has been continued, and the degree of pruning has been lightened slightly; but the total amount of growth is still being sacrificed for arrangement and vigor of selected branches. The pruned tree now possesses a central leader 6 or 7 ft. long from which arise six or more primary laterals, the lowest three of which possess secondary laterals.

Pruning the fourth year and thereafter until the training process is completed will merely be modifications of the previous training practices. When a sufficient number of primary laterals has been developed, the leader should be removed, as it was in training the open-center tree. Care should be exercised to prevent the top of the tree from overgrowing the bottom scaffolds. It will be necessary to encourage some branches by little or no cutting and to suppress others by heavy pruning. Remove upright growing branches, parallel competing branches, large rubbing branches and branches that form narrow weak crotches. Decrease the degree of pruning as rapidly as is consistent with good training. This increases the total growth and promotes fruiting.

Training of many varieties of apples is influenced by their particular growing habits. The Delicious tree while young produces many water sprouts which must be removed, develops many narrow crotches which necessitates the annual removal of one of the branches forming the angle and develops many vigorous upright growths on the trunk which make narrow, weak crotches so that special care is required in selecting primary scaffold branches that form wide strong angles with the leader. The Jonathan is naturally more spreading and less vigorous in its habit of growth than the Delicious. It develops many laterals with wide, strong crotches, thus making easy the selection of satisfactory scaffold branches. A four- to six-year-old Jonathan requires the removal of a large number of upright, vigorously growing twigs which arise from the tops of the various branches. Minor details of pruning must be adjusted to fit the requirements of the particular kind and variety of plant that is being trained.

The directions given thus far have referred to trees that have received correct pruning and training each season from the time they were set in the orchard. There are many young trees up to ten years of age, however, that never have been pruned or have been pruned incorrectly. A tree whose pruning has been neglected for the first five or six years cannot be shaped to the modified-leader type without an excessive removal of branches. Such pruning would do the tree more harm than good, and the shape of a tree should not be changed materially after it has attained bearing age. Neglected trees younger than five years should, however, be given a corrective pruning. In such cases the pruning will have to be quite heavy to obtain the desired results. Growth will be retarded, but a longer lived, more productive tree with less future breakage of the limbs will more than compensate for this temporary setback.

Unpruned young trees contain entirely too many scaffold branches and often have either no leader or more than one leader. Very heavy pruning to produce a particular type or form is rarely advisable in well-grown trees seven or eight years of age. Corrective pruning should be done to remedy obvious faults and to eliminate greater difficulties which apparently would develop without such correction. In pruning such trees one leader should be developed if its development can be accomplished without removing too many branches. Careful attention should be given to thinning out the primary scaffolds and to spacing those left to the best possible advantage for their future development. If several main laterals grow from the trunk near one another, only a few should be removed in any one year, as this decreases the ringing effect of such a group of wounds. It is better to spread corrective pruning over several seasons.

The heading-back method previously described for the training of apple trees to a modified central leader will apply in a general way to other deciduous types of tree fruits. One exception, however, exists in the case of the cherry tree when grown in certain Southern and Midwestern sections. In these sections the branches should not be headed back, as the tree often dies or makes poor growth as the result of the injuries. In many sections the tree apparently makes a better start if the terminal bud of the branches is not removed. The young trees may be trained by the entire removal of the branches not required to form the main scaffolds. Proper selection and spacing of branches, therefore, can be accomplished with no heading back. In Wisconsin and Northeastern sections, apparently, heading back causes no injury to cherry trees.

The disbudding method of training is essentially as follows. A one-year-old apple tree 4 to 6 ft. high is planted and not headed back. Just as growth starts in the spring, groups of four buds are left at 8-in. intervals along the whip where the framework branches are desired, and all other buds are cut off. Lateral shoots develop from these buds and are allowed to grow for one year. Before growth starts the second season, a selection of laterals is made at each point where buds were left the season before. One lateral is chosen for the framework branch at each location, and all others are removed except the slow-growing horizontal laterals which do not compete with the selected branches.

FIG. 112.—Fruit tree trained to an espalier. (*Courtesy of Henry Leuthardt.*)

The slow-growing laterals are left because their leaf areas will add food to the tree. In some cases the laterals to be retained are selected shortly after growth starts the first year, and the other shoots are removed at once. It is suggested that by this method one may obtain three to five well-spaced and properly distributed framework branches during the first year after planting. Subsequent training is continued in a similar fashion until the main framework of the tree is formed.

Espalier.—*Espalier* is a French word meaning "trellis" and as used here refers to the training of fruit trees to walls, trellises and fences. This system of training is practiced commercially in certain sections of Europe, where lack of heat often prevents the proper ripening of fruits. The trees are grown against walls where they receive reflected heat. The system is used to some extent in the United States where economy of space is a factor.

There are many forms of training. The plant may be trained to stems lying in opposite directions and mostly horizontal, in which case it is called a "cordon." When the top is spread fan-shaped on

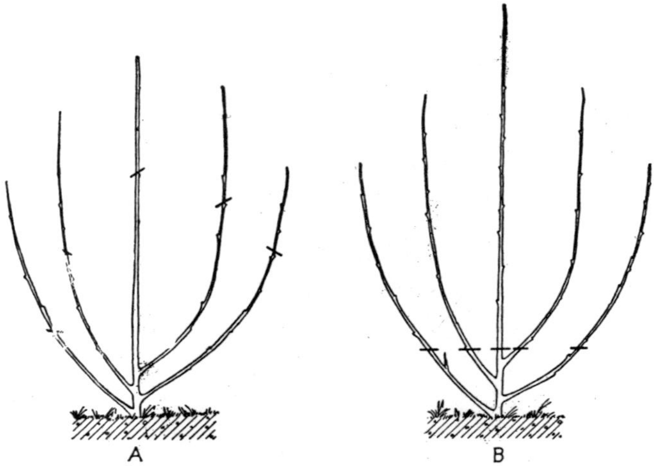

FIG. 113.—Diagrammatic representation of cutting a shrub by training to (*A*) informal shape, (*B*) formal shape.

the trellis or when any of the branches are trained perpendicularly, the system is referred to as the "espalier."

SHRUBS

Shrubs are generally trained to either informal or formal shapes.

Informal.—The natural habit of a shrub is generally its most attractive form. Consequently, in training shrubs that are growing informally, little cutting is necessary. One should not attempt to convert a tall, upright-growing shrub into one with a flat, spreading habit or a shrub with drooping, graceful branches into one that grows upright.

Formal.—Occasions arise where shrubs should be trained in formal shapes. Evergreen plants like junipers, arborvitae and mugho pine are often used in the foundation planting about a building, and it is often necessary to train these plants in formal shapes to keep them in proportion to the house; often a formal shape meets the requirement much more successfully than a natural shape does. Shrubs like the yew and boxwood are often trained to formal shapes in order to conform to the design of a particular garden. Shrubs like privet and honeysuckle are often trained as formal hedges where the growth is kept within certain well-defined limits of heights and width. Formal hedges do best when kept in a flattened ovoid shape with the widest part at the base. This shape facilitates the penetration of light to the leaf surface at the lower parts of the plants.

Vines

Vines generally require some type of support and consequently are trained on walls, arbors and trellises. The type of support provided for a vine requires some knowledge of the method by which the particular vine climbs or clings. The English ivy climbs by means of aerial roots and requires an irregular surface like a rough brick or stone wall for support. Virginia creeper and Boston ivy climb by means of adhesive disks which cling to wood, brick or stone. The aforementioned types are representative of vines that are trained against walls. Vines like clematis climb with the aid of leaf petioles which twine around any suitable support; and bittersweet climbs by twining its stem about the support. Various annual vines like morning glory and sweet pea seem to do better on vertical supports such as strings or slender poles.

The most widely grown vine for the production of fruit is the grape. Although many systems are used in training grapes, it is well to note the most widely used one, known as the "four-cane, single-stem Kniffin." The first problem in training a vine to this Kniffin system is to get the vine established on a trellis as quickly as possible. At planting time the one-year-old vine is cut back to a stub possessing two buds. This reduction in the number of buds results in increasing the vigor of the shoots that arise from them.

During the first growing season after planting, shoots will arise from one or both of the buds. These individual shoot growths may vary from a few inches to several feet in length, and the amount of growth made will determine the pruning the next year. Because of the variability in the amount of growth made by the shoots, the trellis may not be required until the third year. Since it may be required at the time of pruning the second year, it would be well to construct it before that time. It should be substantially made, using heavy end posts set about 3 ft. deep and securely braced. Smaller posts may be used for line posts which should be set 20 ft. apart. This will allow two vines between posts, and a vine will be but 5 ft. from a post. Stretch

No. 9 galvanized wire 36 and 60 in. above the surface of the soil. The staples are not driven entirely "home" on the line posts but should be left so that the wire can slip.

If neither of the two cane growths is sufficiently long to reach easily to the top wire of the trellis, the vine should be pruned in a fashion

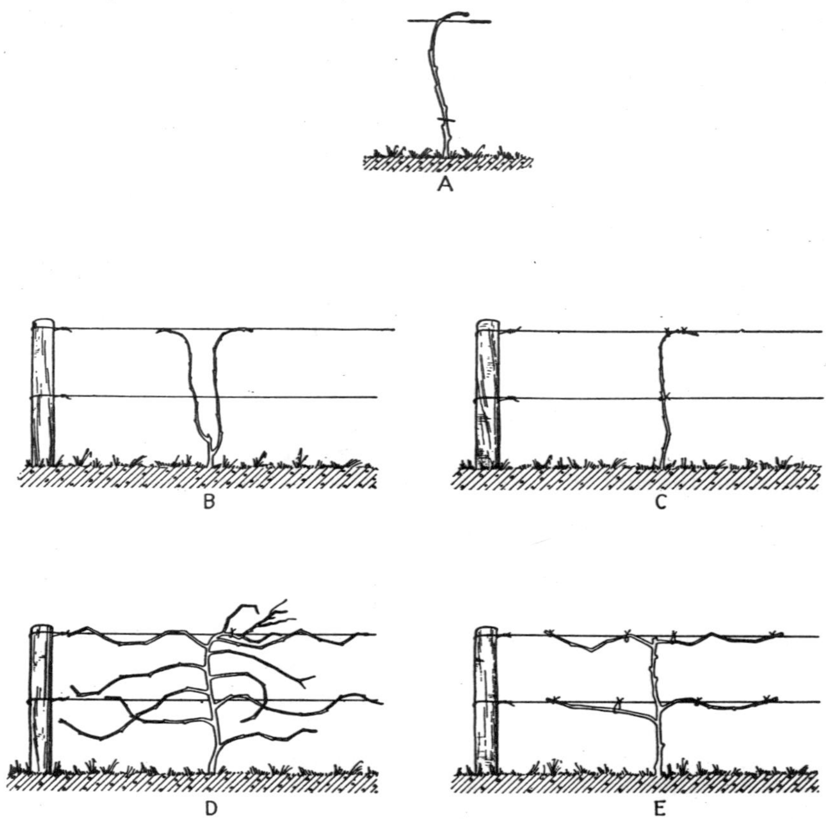

FIG. 114.—Diagrammatic sketches showing the training of a grape vine to the single-stem four-cane Kniffin system: (*A*), one-year-old vine cut back to two buds at place indicated by line; (*B*), two-year-old vine before pruning; (*C*), two-year-old vine after pruning; (*D*), three-year-old vine before pruning; (*E*), three-year-old vine after pruning.

similar to that of the previous year, leaving but one of the new canes and cutting it back to two buds. If, however, one of the canes is of suitable length, it should be pulled straight to the top wire, tied tightly, carried along the wire a short distance, tied securely again and the remainder cut off. The straight stem should be tied loosely to the bottom wire. All other canes should be cut from the vine at their points of origin. After pruning and training, one will have but a single cane extending directly to and a short distance along the top wire. During the second growing season lateral shoots will develop from the buds on this upright stem.

Pruning and training the third season will consist in selecting four suitable laterals on this stem. Two should be chosen that arise just below the top wire, and two that developed near, preferably just below, the bottom wire. All other canes should be cut away, and the main stem should be cut off a short distance above the highest of the four selected laterals. These four laterals should be cut back so that each possesses five or six buds, or eyes, and tied in two places to the right and left on their respective wires. Tie the end tightly to the wire, but the main stem and the basal part of the canes should be tied loosely to allow for growth without girdling. The pruned vine now has a central stem reaching to or nearly to the top wire and four lateral canes extending to the left and right on the two wires. The vine thus resembles one capital letter T arising on top of another. The plant now possesses its basic framework, and subsequent training will be directed toward increasing the vine to its allotted size and maintaining this general form.

Vegetables

Vegetables as a group receive practically no training. Like some of the herbaceous flower plants, however, such plants as tomatoes and cucumbers may receive a certain amount.

Tomatoes and cucumbers grown under glass are pruned and trained. The tomato is trained usually to a single stem or occasionally to two stems. When pruned to a single stem, shoots that arise on the main stem in the axils of the leaves are removed. Plants so pruned are planted closer together than when untrained. They are supported by upright stakes and every few inches are tied to the stake with jute twine.

Pruning and training tomatoes grown out-of-doors is commonly followed in the South but only to a limited extent in the North. Training to a single stem and tying to a stake, as in greenhouse culture, is practiced. Training to a single stem, however, in many sections of the Middle West may expose the fruits too much to the hot sun, so that they sunscald and become worthless. Pruning and training do not increase the yields from individual plants or hasten the ripening; but since more plants per acre can be grown when pruned and trained, the yield per acre may be greater and a larger early yield obtained. In the South where foliage diseases are severe, pruning and training allow better circulation of air around the plants, thus holding the diseases in check.

Cucumbers grown under glass are also pruned and trained. Lateral branches that arise in the axils of the leaves from the main stem are nipped off beyond the first female flower. The main stem is nipped off when it reaches the top of the trellis. The vines are tied to a suitable upright or sloping support.

FIG. 115.—Tomato plant trained to a stake. (*U.S. Department of Agriculture.*)

FLOWERS

Herbaceous flowering plants generally receive little training but are allowed to grow at will. Training to a single stem is often followed, however, with such plants as the chrysanthemum and the dahlia. Chrysanthemums grown in the greenhouse for specimen blooms are trained to a single stem for the production of one large flower or to several stems for the production of a flower on the end of each strong lateral branch. When trained to a single stem, the side shoots arising from the main stem are removed, and but one flower bud on the main stem is retained. When four or five large flowers are desired on a single plant, the terminal bud of the young plant is pinched out to induce lateral b r a n c h i n g. As the several laterals develop, the terminal bud of each lateral branch is reserved, and all other flower buds are removed. Pinching to force lateral branching is usually done when the stem tissues have hardened to some extent.

Dahlias grown out-of-doors are trained in a somewhat similar fashion. If one wishes to obtain 18 flowers from a single plant, one could proceed somewhat as follows: After six primary laterals had developed, the main stem should be cut back to the origin of the top lateral. Pinching out the terminal will promote the formation of laterals. Two secondary laterals should be selected on each of the six primary laterals, and all other secondary laterals rubbed off as they appear. No tertiary laterals should be allowed to form on the 12 secondaries. The plant consists of a short main stem from which six primary laterals arise. Each primary lateral is terminated by a flower bud and possesses two secondary laterals, each of which in turn is terminated by its flower bud. The plant will thus produce 6 large flowers from the terminals on the 6 primary laterals and 12 smaller flowers from the terminals on the 12 secondary laterals. The number of flowers can be regulated by the training and degree of pruning.

Review Questions

1. What is meant by training horticultural plants?
2. What are the principal objects sought in training horticultural plants?
3. What are the qualities of a satisfactory framework?
4. What is the influence of training on the time of coming into bearing?
5. What is meant by heading back?
6. What are two purposes of heading back?
7. What is meant by thinning out?
8. What particular branches should be thinned out?
9. What are the principal systems used for training trees?
10. What are the advantages of the central-leader type?
11. What are the disadvantages of the central-leader type?
12. What are the advantages of the open-center type?
13. What are the disadvantages of the open-center type?
14. What are the advantages of the modified-leader type?
15. What are the disadvantages of the modified-leader type?
16. What are the two principal forms to which shrubs are trained?
17. Is training important with vegetable crops?
18. Is training important with herbaceous flowering plants?

Problems

1. Make diagrammatic sketches of the one-, two-, three- and four-year-old apple tree before and after pruning when trained to the modified-leader system. Explain.

2. Make diagrammatic sketches of the one-, two-, three- and four-year-old American grapevine before and after pruning when trained to the single-stem, four-cane Kniffin system. Explain.

3. Make diagrammatic sketches showing the pruning of a formal privet hedge for each of the first three years after planting. Explain.

4. Make diagrammatic sketches showing the training of a Jane Cowl dahlia that will produce 15 flowers. Explain.

Suggested Collateral Readings

1. Auchter, E. C., and H. B. Knapp, "Orchard and Small Fruit Culture," pp. 214–225, 228–230, 244–246, 546–549, 578–579, John Wiley & Sons, Inc., New York, 1937.

2. Chandler, W. H., "North American Orchards," pp. 410–412, 416–418, 425–428, Lea & Febiger, Philadelphia, 1928.

3. Marshall, R. E., *et al.*, Pruning Young Fruit Trees, *Mich. Ext. Bul.* 148: 1–32, 1935.

4. Talbert, T. J., and A. E. Murneek, "Fruit Crops," pp. 127–147, Lea & Febiger, Philadelphia, 1939.

5. White, E. A., "The Florist Business," pp. 269–270, The Macmillan Company, New York, 1933.

CHAPTER XIV

PRUNING HORTICULTURAL PLANTS

More than two hundred years ago a great French gardener said, "Everybody cuts, but few prune." Since that time more facts have been obtained relative to the responses of plants to pruning, but unfortunately the statement of the French gardener is still pertinent. There is no topic in fruit growing upon which there is such a diversity of opinion as that of pruning. This great diversity of opinion relative to the value of pruning practices is not surprising when one considers the great diversities in growing and fruiting habits of even some of the most stable horticultural plants and the fact that the development of these characters is being influenced by all phases of the environment. As a result of the variation in plant growth and of the different concepts of what one may expect to accomplish by pruning, great differences exist in the execution of the practices even on a single kind of plant from year to year and in different parts of the country.

Pruning in its final analysis may be considered as the act of severing or removing a portion of the plant. It is done either for its direct effect on the growth of the plant or, as discussed in the last chapter, to mold the plant into a form desired by the pruner. Since training of plants by pruning was discussed in Chapter XIII, the discussion in this chapter is confined, with a few exceptions, to those plants which are past the training period. The three chief factors considered in pruning mature plants are the manner, the degree and the time of pruning.

OBJECTS

In a broad sense the only justification that a horticulturist has to prune a plant is to make money or to add value. The specific means used in attaining this object will vary with the kind and with the age of the plant and the conditions under which it is growing. In pruning plants after the training period, one strives to retain the structural qualities developed during the training period and to provide the optimum conditions for growth and fruitfulness. Specifically the objects of pruning mature plants are (1) to maintain the vigor of plant growth, (2) to improve the product of the plant and (3) to facilitate horticultural operations.

MAINTAIN VIGOR

Horticultural plants lack vigor when the shoot and leaf growth, flowering and fruiting are less than those of similar plants growing under favorable conditions. For example, if, in a bearing apple tree, shoots are thin and make less than 2 in. of growth and have small yellowish-green leaves; if the yearly growth on the spurs is less than ¼ in. and the leaves are only three or four in number, small in size and yellowish green in color; if the flowers either do not form, abscise or fall prematurely or are few in number and small in size and if the fruit is small in total quantity and in size of individual fruits the plant is below the desired vigor. The vigor can be restored to some extent if not fully by various practices of which pruning is an important one.

IMPROVE PRODUCT

Pruning improves the size, color and quality of the product that the pruned plant produces. This is true whether one is growing apples or roses. Pruning removes some of the flowers and fruits on the plant. If the root system has not been pruned, the reduced number of flowers or fruits receive a proportionately larger supply of water and mineral nutrients for a period of time. If the foliage area per fruit or flower has not been reduced below that required for satisfactory growth, the flowers or fruits increase in size and become larger individuals than would have been possible with a larger number of flowers or fruits on the plant. Competition between individual fruits for a limited supply of food and water on which to grow has been lessened by decreasing the number of competing individuals. The individuals that remained became larger than would have been possible under the restricted food supply that would have prevailed without pruning. Pruning reduces the total number of flowers or fruits produced per plant; it may decrease the total volume of the product; but it probably will result in an increase in the individual size of the remaining flowers or fruits. Often a plant such as the raspberry or blackberry sets more fruit than it can mature properly, and pruning is one means of thinning the crop. In the production of certain floral crops such as chrysanthemums, dahlias and roses, it is often desirable to develop one or several large flowers on a plant rather than many smaller ones, and this can be done readily by intelligent pruning. The thinning of flowers and young fruits is really a type of pruning.

The development of red color in fruits, like the apple and peach, as they mature depends mainly on sunlight. Pruning thins branches and foliage, and the fruit on properly pruned trees is better exposed to light and consequently develops more color and a more attractive appearance. A higher quality of fruit is associated with the better color.

Facilitate Horticultural Operations

Pruning the mature plant facilitates certain horticultural operations. In pruning, one should remove the weakly growing, nonproductive branches found chiefly in the interior of the tree and also in too densely branched areas. Remove also the badly rubbing and improperly placed branches as they develop throughout the tree. Such pruning permits better penetration of spray materials, makes harvesting of the fruit much easier and lessens the chances of some diseases by removing the rubbed bark areas where a disease might gain entrance.

MANNER OF PRUNING

A major consideration in pruning is the manner in which the work is done. Heading back and thinning out are the two methods employed in removing branches from plants. It will be recalled that these are the two methods discussed in Chap. XIII. Heading back is used extensively with mature plants where continued training is necessary. Continued severe heading back of the roots dwarfs the plant and induces flowering in vigorously vegetative plants, since the reduced root system absorbs less water and mineral elements and the relative amount of the carbohydrates in the top increases, thus changing the $C/N+$ condition to C/N. The roots of the shrub border often encroach on the lawn area, and they are kept within a restricted area by being cut back with a spade forced vertically into the soil a suitable distance from the shrub border.

Thinning refers to the removal of an entire twig or branch at its point of origin or at a point where a lateral branch arises from the main branch. Thinning is applied also to the removal of individual flowers, fruits and entire plants. Individual flowers and fruits are thinned to increase the size of those which remain. Often seedlings or other plants that reproduce readily become too crowded, and a reduction in the number of plants is advisable to decrease the competition for light, water and mineral nutrients. The removal of filler trees in an orchard is based on the same principle.

All pruning wounds should be smooth, clean and made close to the remaining branch or bud so that no stub is left as a place in which rot may start. In removing large limbs start about a foot from the limb to be left, and first saw from the underside until the saw binds, then starting a little farther out saw down. The limb will break off leaving the stub but will not strip the bark on the remaining limb or trunk. The stub should then be sawed off smoothly with the remaining branch. Except in the presence of diseases such as Illinois apple-tree canker, wounds 2 in. or less in diameter need not be protected.

DEGREE OF PRUNING

It is comparatively easy to learn the proper time and a suitable manner of pruning, as they are fixed rather arbitrarily. The degree of pruning or the amount of tissue to remove necessitates judging the ability of the plant to produce a satisfactory crop. The least productive and weakly vegetative wood is then removed. The severity of pruning the mature plant will be based on the fruiting habit of the plant and the apparent vegetative vigor of the plant or its ability to produce a crop. The least possible amount of pruning that will produce the desired results is the best amount to be given.

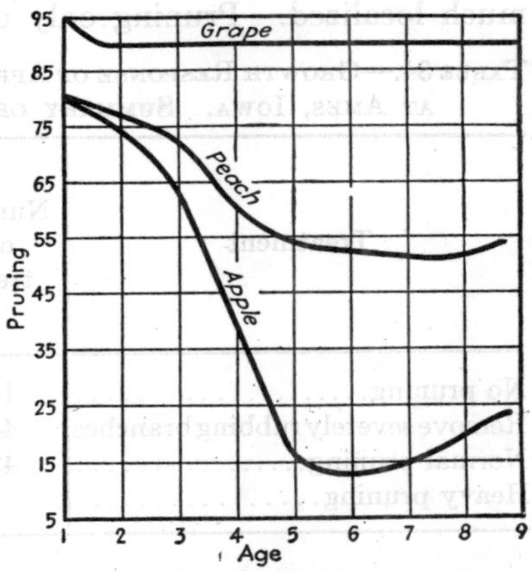

FIG. 116.—Graphs showing relative amounts of pruning required for the grape, peach and apple at different ages.

INFLUENCE ON AMOUNT OF GROWTH

Any amount of pruning decreases the size of the plant by the amount of the tissue removed in pruning and decreases the total amount of growth made by the pruned plant when compared with the same plant unpruned. Pruning reduces the amount of leaf area that was capable of manufacturing carbohydrates for the plant. Any practice that reduces the leaf surface lowers the potential food supply Other things being equal, the less the amount of foliage the less the amount of food for the plant; and the less the amount of food the less the total growth made by the plant.

INFLUENCE ON VIGOR OF GROWTH

Pruning invigorates the growth of shoots. Shoots on pruned branches are fewer in number, smaller in total growth but larger individuals than shoots on corresponding unpruned or more lightly pruned branches. Pruning involves the removal of wood and of buds; consequently, it removes stored food reserves, potential leaves and often potential flowers and fruits. This reduction in number of buds diminishes the number of growing points where water, nitrogen and other minerals are required without simultaneously reducing the absorbing root system correspondingly, and consequently, the supply of these materials for each of the remaining buds is temporarily increased. This results in each growing point's making more growth than it could have made without the increase in food materials. The individual laterals on a pruned branch, therefore, are longer than on a similar unpruned branch. The invigorating effect of pruning is very much localized. Pruning only on one side of a large tree has little or no effect in increasing the vigor of growth of the branches on the other side. Often old, but healthy, horticultural plants are rejuvenated by very severe top pruning. Fifty per cent or more of the top may be removed in one year, but the rejuvenation operations, which may result in a practically new top, are generally distributed over a period of two or three years. The gain in leaf surface on individual shoots due to increased vigor is insufficient to make up for the loss of foliage that would have developed on the shoots arising from the buds removed by pruning and for the loss of food reserves that were in the branches removed by the cutting. The net result of pruning for the plant as a whole is a decrease in size.

TABLE 34.—GROWTH RESPONSE OF APPLE TREE TO VARIOUS DEGREES OF PRUNING AT AMES, IOWA. SUMMARY OF 12 GROWING SEASONS, 1924–1935

Treatment	Number of trees	Average circumference, inches		Total inches gained	Total percentage
		Spring, 1924	Fall, 1935		
No pruning	15	2.12	20.85	18.73	884
Remove severely rubbing branches	45	2.17	20.17	18.00	830
Normal pruning	42	2.04	17.94	15.90	780
Heavy pruning	8	2.15	17.84	15.69	730

FRUITING HABIT

The age and position of the wood upon which the fruit is produced influences the severity of pruning given horticultural plants. Some plants, as the grape, produce fruit only on current season's wood; some, as the peach, produce fruit on one-year-old wood; others, as the apple, produce fruit mostly on the new wood of short spurs that may be several years old. In the case of the grape, pruning is really an annual rejuvenating process. Pruning removes all except enough one-year-old wood to produce a sufficient number of fruiting shoots and the minimum amount of old wood that will make a satisfactory framework for the vine. Each year about 90 per cent of the vine is pruned away. The peach bears fruit laterally on comparatively long, vigorous one-year-old twigs. In this case one is interested primarily in the fruiting of one-year-old wood and in the older wood only in its relationship to the distribution and production of satisfactory fruiting wood. Regular production depends upon the annual provision of an adequate supply of suitable shoots. Moderately vigorous twigs are necessary for fruiting, and to obtain such twigs a comparatively severe

FIG. 117.—Old peach trees cut back to two- and three-year-old wood after their fruit buds were killed by late freezes. The vigorous growth following lessens chances for a crop next year. (*Missouri Agriculture Experiment Station.*)

thinning and heading back is necessary. Such pruning also serves as a means of fruit thinning. Often one-third or more of the peach tree is pruned out each year. Since the apple bears its fruit chiefly on old spurs distributed throughout all parts of the tree, the object of the pruner is to maintain an adequate supply of sufficiently vigorous spurs. This is accomplished by a comparatively light thinning throughout the tree. A heavy pruning, such as that given the peach, would force the spurs into long vegetative growth, and fruiting would be decreased. The amount of pruning will be about one-third or less that given a peach tree.

TIME OF PRUNING

Plants may be pruned during either the dormant or the growing season. The time of pruning is influenced by the kind of plant and the purpose of the pruning. The time of pruning most plants out-of-doors is determined by the influence of the environment on the plant.

Dormant Season

It will be recalled that the dormant season is that period of the year during which the plants are not making any apparent growth owing to the fact that the temperature is too low for such activity. There is no apparent difference in the amount and character of the subsequent growth made by plants whether pruned early or late in the dormant season. There are, however, other factors that one should consider. If pruning is done early in the dormant season, wounds that are left do not heal so rapidly as similar pruning wounds made toward the end of the dormant season. In sections where the plants are subject to winter injury, considerable death of tissues often occurs adjacent to the wounds made early in the dormant season. Adverse winter conditions may cause the death of many buds, and the crop will be considerably reduced. If pruning is delayed until the extent of the winter injury is known, the severity of pruning can be adjusted accordingly and a more nearly normal crop obtained. Pruning late in the dormant season is often attended by bleeding. This bleeding, with the exception of the English walnut, in which case the moist exposed surfaces offer good opportunity for infection, does no harm. If pruning is delayed until the buds swell, many buds will be rubbed off as the prunings are pulled from the plant.

Growing Season

Some types of pruning are necessary during the growing season with certain plants and advisable with others. All pruning decreases the total growth made by the plant. Pruning during the growing season is more dwarfing than the same amount of pruning during the dormant season. The early shoot growth is made largely at the expense of reserve foods stored within the plant. The new leaves begin photosynthesis and are soon synthesizing sufficient carbohydrates for their own needs and for the requirements of the shoot to which they are attached. A short time later the leaves on the shoot are not only supplying the needs of that shoot but manufacturing a surplus of carbohydrates which is used by the developing flowers or fruit, by the expanding roots, by the growing parts of branches at some distance from the leaves and as stored reserves for future use. The younger the leaf or portion of a shoot the more it is dependent on the reserves of the parent plant. The amount of dwarfing will be determined by the amount and time of summer pruning. The greatest dwarfing would be caused by removing a large quantity of the shoots or parts of shoots just at that time when they had removed all the stored food from the parent plant that they required and just began to be self-supporting. Very late summer pruning might force the growth of succulent shoots that would not have an opportunity to mature properly and would be killed during the following dormant season.

Clipped hedges and sheared evergreens must be pruned several times during the growing season. Undesirable water sprouts can be rubbed off easily then. The terminals are removed from black raspberry shoots to force the development of desirable laterals, and similar summer pruning is used to shape other kinds of plants. Pinching out the terminals and disbudding certain ornamentals, as peonies, dahlias, snapdragons and chrysanthemums, are common and necessary practices which extend over several weeks for the same plant or kind of plant. Thinning fruit is merely a more advanced stage of disbudding. It is usually done shortly after the "set" of fruit is assured.

PRUNING REPRESENTATIVE PLANTS

In the preceding discussion some of the important general effects of pruning were discussed. The adaptation of pruning practices to a few representative horticultural plants is suggested in the succeeding part of this chapter.

Apple

It is well to review the growing and fruiting habit of the apple before discussing the pruning treatment, because the fruiting habit of a plant is one of the important factors determining the kind and severity of pruning. Although exceptions such as annual bearing spurs and the production of flowers laterally on one-year-old twigs occur, one may consider, for all practical purposes, that the fruiting habit of the apple is essentially as follows: The flowers of the apple are borne terminally on short growths or spurs, and these spurs commonly bear fruit every other year. They occur on wood that is two years old or older, and they may live for many years. If they are strong, vigorous and productive, they should be allowed to remain on the tree. Maximum fruit-spur formation is encouraged by leaving a tree unpruned or pruning it lightly during the fruit-bearing period in its life.

The kind and severity of pruning treatment suggested for apple trees varies principally with the age and vigor of the trees. The life of an apple tree may be divided into two periods: the non-bearing, or formative period, when the framework is being developed, and the bearing period. The bearing period may be further divided into the early bearing period, the period of maximum production and the period of decline. The pruning treatment varies with the stage of development of the tree.

When the growing and fruiting habit of the apple and its response to different kinds of pruning are considered, one may briefly summarize the procedure as follows: During the first few years in the orchard the tree is in its formative period and is pruned by both thinning out and heading back. This heavy pruning is for the purpose of developing a suitable framework. As the tree becomes older, pruning is decreased gradually in severity until at six to eight years of age the tree comes into the early bearing period. During this period, which may cover a period of ten or more years, very little pruning is done, but some thinning out and some corrective pruning may be necessary. After the tree reaches the period of maximum production, which may last for twenty to fifty years or more, some pruning will be advisable. The severity of the thinning will depend largely on the vigor of the tree; but in general, it will consist of the removal of small weak branches chiefly in the lower and interior parts of the tree. After the apple tree passes its period of maximum production and enters a period of decline, if it is healthy and fairly vigorous, it is often renovated or rejuvenated by a severe pruning which consists of the removal of large branches and a heavy thinning of those which remain. This practice often lowers the tops of the trees as much as 10 ft. or more by cutting to suitable lateral branches.

Peach

It will be recalled that the peach bears its fruit laterally on the one-year-old twigs. New shoot growth develops from the terminal bud and from some of the lateral shoot buds. In some instances, especially following the loss of crop by frost or excessively heavy pruning, as in rejuvenation, shoots develop from latent buds lower in the tree. The peach tree, therefore, carries its fruiting wood a foot or two farther out and up each year. Since satisfactorily bearing peach trees usually differentiate more than enough fruit buds for a good crop, and since the fruit is produced on the one-year-old wood, there is little danger of rendering a peach tree unproductive by removing all its fruiting wood. The problem of pruning the bearing peach tree is largely one of keeping the fruiting wood fairly close to the trunk and of thinning the crop.

A brief summary of the pruning of a peach tree beginning with the one-year-old tree follows. A one-year-old branched peach tree of average vigor is generally headed at a height of 16 to 24 in. from the ground. Three to five primary scaffold branches are selected which should be spaced spirally on the short trunk and as far apart as possible. All other branches should be removed, and the scaffold branches headed back to 6 or 8 in. in length. Until the tree reaches bearing age, some corrective pruning will be necessary in order to develop a suitable framework. This pruning will consist of thinning out unsuitably placed branches and heading back too vigorously growing branches. It should be light. The center of the tree is kept open, allowing only a few branches to grow inward and upward. The pruning treatment should keep the fruiting wood close to the tree and near the ground. This develops a stronger tree and facilitates the pruning, spraying and picking operations.

After the peach tree comes into bearing, the severity of both thinning and heading will depend largely upon the prospects of a crop. If a heavy crop is expected, the tree should be thinned and headed back quite severely, removing anywhere from one-half to two-thirds of the previous year's growth. If one had only a few trees, he might delay pruning until it was possible to determine the extent of winter injury and regulate the degree of pruning accordingly. When most of the flower buds have been killed by low winter temperatures or when the blossoms are destroyed by a late frost, one may take advantage of the situation and give a rejuvenating pruning. Cut branches back severely in order to lower the top and develop suitably located laterals. Too severe cutting back may stimulate shoot growth to the extent that no fruit buds will form for next year. Consequently, one should adjust the degree of cutting to the vigor of the tree and the results desired.

Shrubs

The kinds of pruning employed with shrubs consist of both heading back and thinning out. In general, when shrubs are trained to their natural shapes they can be kept in vigorous growing condition, in good form and in the maximum flowering condition by a yearly systematic thinning of old and weak wood. Heading back new shoots, which arise from the base of the plant, is often advisable in order to force lower branching. A general heading back or shearing of the tops will produce an unnatural shape and should be used only for hedges. The method followed in maintaining a formal compact hedge consists of heading back the tips of the new shoots at various intervals during the growing season.

The severity of pruning, as with other plants, depends upon the flowering habit of the particular shrub, the age of the plant and the vigor of growth. Many shrubs like spiraea Van Houttei, forsythia and lilac produce flowers from the buds that were formed on the past season's shoots. The degree of pruning given will depend on the number and distribution of these shoots. In general, if a systematic thinning has been practiced, there is usually a sufficient amount of this type of wood to produce optimum flowering. If the shrub has been neglected for a number of years, it may be necessary to remove a

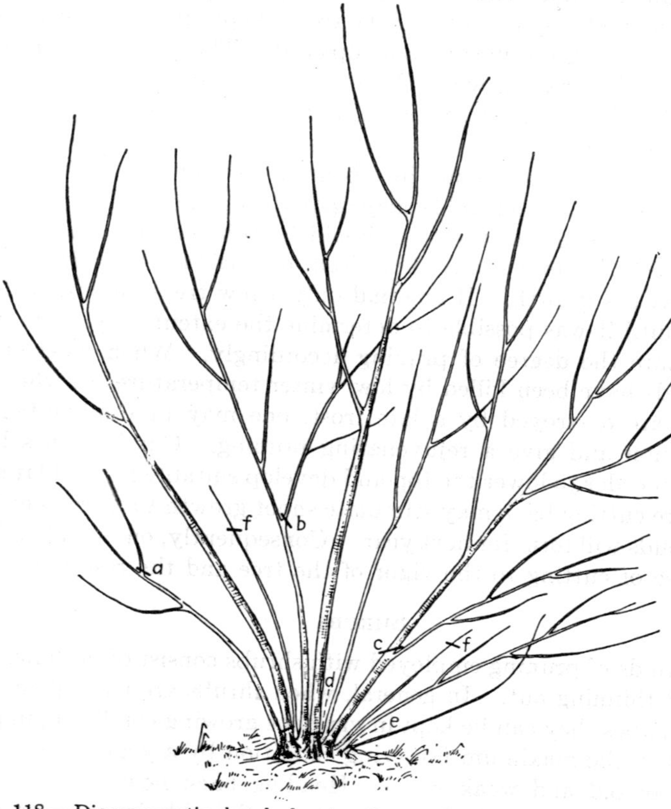

FIG. 118.—Diagrammatic sketch showing the pruning of a shrub; (*a*) thinning by the removal of one-year-old laterals; (*b*) thinning by the removal of two-year-old laterals; (*c*) thinning by the removal of old laterals; (*d*) thinning by the removal of an entire old branch; (*e*) thinning by the removal of a one-year-old stem from the crown; and (*f*) forcing the growth of laterals by heading one-year-old stems from the crown.

large quantity of older wood by thinning branches by cutting out at the ground line. One may have to sacrifice the production of flowers for a season while the shrub is being rejuvenated or reshaped by severe pruning.

Other shrubs, like some roses, hydrangeas and spiarea Anthony Waterer, produce flowers from the buds formed on current season's

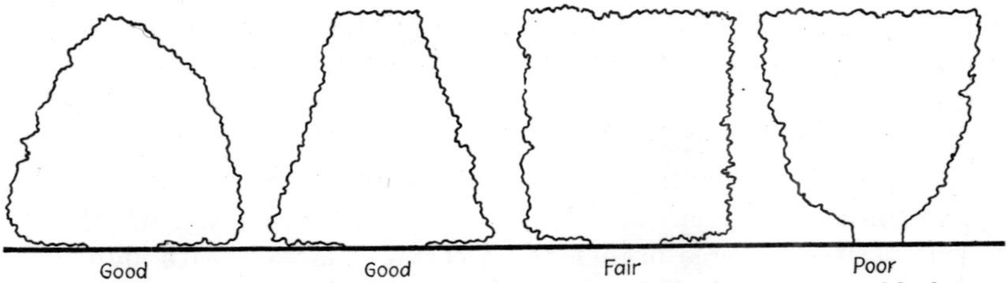

FIG. 119.—Diagrammatic sketches showing method of a shaping a formal hedge.

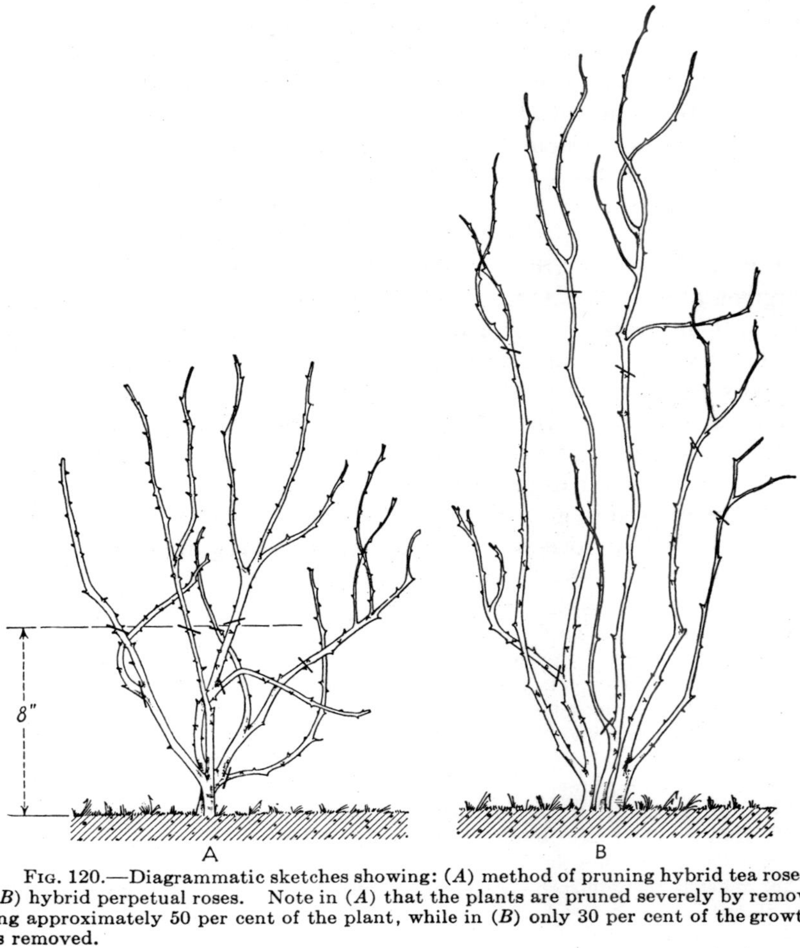

FIG. 120.—Diagrammatic sketches showing: (*A*) method of pruning hybrid tea roses; (*B*) hybrid perpetual roses. Note in (*A*) that the plants are pruned severely by removing approximately 50 per cent of the plant, while in (*B*) only 30 per cent of the growth is removed.

growth. This type of shrub may be headed back severely in order to produce vigorous new shoots. For example, hydrangea is often cut back to 6 to 8 in. from the ground, and hybrid tea roses are cut back to three or four buds.

The degree of pruning given a shrub immediately after it is set will depend somewhat upon the extent of the root system and whether or not one desires to force branching at certain places. Although some shrubs grow well and give a quicker effect without any cutting back, it is advisable to remove a portion of the top. The reduced root system caused by digging the plant is not capable of supplying adequate water to the original top. During the first three or four years after planting, the shrub will require only a corrective pruning consisting chiefly of the removal of irregular and weak branches. After this formative period a systematic light thinning out should be all the pruning that is necessary.

Shade Trees

Pruning of shade trees differs somewhat between deciduous and evergreen trees. The methods of training are similar to those used in training fruit trees. After a deciduous tree has passed the training period, very little pruning is necessary. The kind of pruning given is generally confined to the removal of dead, diseased, injured and badly conflicting branches. Occasionally it may be necessary to remove or brace branches that have developed extremely weak crotches. Evergreens that are grown as shade trees usually require very little pruning if they are allowed to grow naturally. Some evergreens, like junipers, are often trained to formal types and in this case require shearing once or twice a year in order to maintain a compact growth.

Review Questions

1. What is meant by pruning?

2. What are the objects of pruning mature horticultural plants?

3. How does pruning maintain vigor?

4. How does pruning improve the product of the plant?

5. How does pruning facilitate horticultural operations?

6. What are the three chief variable factors in pruning?

7. Contrast thinning out with heading back.

8. How should a large branch be removed from a tree?

9. What determines the amount of pruning to give a mature plant?

10. What is the influence of the amount of pruning on the total growth made by a plant?

11. What is the influence of the amount of pruning on the vigor of growth made by the plant?

12. Which is more preferable on an eighteen-year-old apple tree: 10 cuts removing branches about 2 in. in diameter or 100 cuts about the size of a lead pencil? The total growth removed in the two cases is the same in amount.

13. What is meant by the fruiting habit of a plant?

14. What determines the time of pruning?

15. What part of the dormant season is the most satisfactory for pruning?

16. Why is pruning during the growing season more dwarfing on the plant than an equal pruning during the dormant season?

17. Compare the pruning of a privet plant when grown as part of a formal hedge with the pruning of the same kind of plant when grown as an individual plant.

18. What are the main problems to consider in pruning an American elm?

Problems

1. Make accurately labeled diagrammatic sketches showing the fruiting habit of the grape, peach and apple.

2. Give detailed instructions and explanations for pruning the following apple trees. A group of trees ten years of age has had about one-half its one-year-old shoots thinned out and the remaining one-year shoots headed back about 50 per cent each year. The terminal growths are long and vigorous each year, but the trees are not bearing any fruit, and no fruit spurs have been formed. In another group only the yearly terminal growth was headed back, but these trees are now becoming quite dense in the top while some of the lower branches are not making very vigorous growth and very little fruit has been produced. A third group of the same variety, but five years older, has been neglected ever since the first pruning. These latter trees are larger, even for their age, than those in the other two groups, are producing quite heavily and look quite bushy, but some of the smaller low branches are dying.

3. Give detailed instructions and explanations for pruning the following apple trees. A group of mature trees is well formed, although it has not been pruned for several years. The trees blossom abundantly each year, but practically all the blossoms fall or the fruit drops soon after forming. The terminal shoot growth averages 2 or 3 in. in length, and there is an average of three medium-sized leaves on each spur, whereas some of the spurs on the interior of the tree have small yellowish leaves and the spurs are dying.

4. An apple tree about twenty years old has been growing in sod the past five or six years and as a result has decreased in fruit production. When the tree was about fifteen years of age, it produced fair crops each year, but for the last two years it has blossomed but not produced fruit. Describe the appearance of the spur growth of this tree when the tree was fruiting and since it has quit fruiting. Illustrate by diagrammatic sketches and explain fully. What should you do to bring the tree into production?

Suggested Collateral Readings

1. Gardner, V. R., *et al.*, "Fundamentals of Fruit Production," pp. 432–521, McGraw-Hill Book Company, Inc., New York, 1939.

2. Gregg, O. I., Pruning and Care of Ornamental Woody Plants, *Mich. Ext. Bul.* 172: 1–18, 1936.

3. Ricks, G. L., and H. P. Gaston, The Thin Wood Method of Pruning, *Mich. Agr. Exp. Sta. Spec. Bul.* 265: 1–45, 1935.

4. Thompson, H. C., "Vegetable Crops," pp. 464–467, McGraw-Hill Book Company, Inc., New York, 1939.

5. Wyman, Donald, "Hedges, Screens and Windbreaks," pp. 39–47, McGraw-Hill Book Company, Inc., New York, 1939.

CHAPTER XV

PESTS OF HORTICULTURAL PLANTS

In order to survive, each plant is struggling with its environment, with the competing plants about it and with the plant and animal pests that prey upon it. Under natural conditions, where man does not attempt to alter conditions, these interacting factors result in the establishment of a state of equilibrium, or "natural balance," which obviously is not fixed but changes as conditions are favorable or unfavorable for that particular plant or kind of plant. Man, however, is often able to modify this natural balance in favor of the plants that he prefers as compared with the plants that he considers less desirable. He does this by providing a favorable environment, by the elimination of competing plants and by checking or destroying the pests that interfere with the optimum growth and development of the plants that he wishes to grow. Horticultural plants need to be protected from unfavorable conditions in their environment and from plant and animal pests that prey upon them and injure or destroy them. The root, trunk, branches, shoots, foliage, flowers, fruits and seeds of horticultural plants are attacked by pests that are found in a wide range of both the plant and the animal kingdoms.

IMPORTANCE OF HORTICULTURAL TROUBLES

It is impossible to obtain a definite figure of the damage done by various horticultural troubles, as they fluctuate so widely from year to year, from region to region and with the different kinds of plants. An unfavorable winter may result in the severe injury if not the total destruction of plants over a wide area. Well-known cases of this are the destruction of peach orchards in the commercial peach-producing areas and the occasional freezing of citrus trees in citrus-fruit areas. A severe and untimely frost during the blooming period has been known to wipe out the entire fruit crop for that year in certain areas. A greatly reduced set of fruit has resulted because of cool, damp weather during the blooming period. A dry summer followed by early fall rains may result in the cracking or splitting of fruit and may result in the malformation of Irish potato tubers. Such climatic conditions may bring about secondary growth in the fruit trees and vines, a growth that will be killed or severely injured by even a favorable winter that follows because the new wood was not matured properly. A hailstorm of short duration will ruin the crop and may injure the plants to such an extent that it will require several years for them to recover. Rains and floods will wipe out the entire crop in a short time or over a period of years may cause great and lasting damage by eroding the soil. Specific cases of many of the conditions mentioned will come to the minds of all, and many other general examples could be given; but these will be sufficient to show why it is impossible and impractical to attempt to give a statement of the amount of damage done to horticultural plants by adverse environmental conditions.

Losses of horticultural crops due to plant and animal pests are probably greater but generally not so conspicuous as losses due to unfavorable environmental conditions. A constant warfare is being waged against these pests, and it is only occasionally that the damage reaches epidemic proportions. In such cases the great increase in the damage was brought about by the introduction of a new pest or by climatic conditions peculiarly favorable to an established pest and probably unfavorable to its natural enemies which would normally keep it in check. Apple scab, severe damage by the late-brood codling moth, brown rot on stone fruits and potato blight are but a few of the common and well-known examples of cases where pests have attained epidemic proportions over greater or smaller areas in certain seasons. The operation of spraying apples to protect them from plant and animal pests is responsible for 20 per cent of the cost of producing a bushel of fruit. Whetzel presents the damage in the following interesting fashion: "The estimates of the plant disease survey of the U.S. Department of Agriculture indicate that approximately one bean in every dozen, one peach in every eight, one bushel of Irish potatoes in every twelve . . . are destroyed annually by diseases in these crops." This does not include the cost of protecting the crops but just the quantity lost in spite of protection.

The introduction of a new pest into a locality may result in an immense loss until a suitable means of control is found, and it may result in the elimination of certain crops from the infested area. The introduction of the San José scale into California from China, the Japanese beetle into New Jersey from Japan, the citrus canker into Texas from Japan and the phylloxera into the grape vineyards of Europe from the United States are classic examples of cases in which introduced pests nearly destroyed established industries and are still taking a heavy annual toll from horticultural crops. The Colorado potato beetle spread rapidly eastward and is now invading the Irish potato fields of Europe. It found the introduced Irish potato a much more favorable food supply than the native plants upon which it fed previous to the introduction of the cultivated potato. This new and abundant food supply enabled this insect to multiply prodigiously and spread rapidly eastward, until today it is found universally in the potato fields of North America, and its control is considered a regular routine part of the cultural program.

Many of the virus diseases, as peach yellows, curly top of sugar beets and the mosaics of various plants, have made it impossible for years to grow certain plants in certain sections. The destruction of the white birch by the bronze birch borer, the chestnut by the chestnut blight and the white pine by the white-pine blister rust and the present threat of the Dutch elm disease in our American elms cannot be evaluated in dollars and cents. These few specific examples will suffice to show the impossibility of arriving at a satisfactory estimate of the immense financial and social damage done to our horticultural plants by the various and numerous troubles that interfere with their most satisfactory development.

CLASSIFICATION OF HORTICULTURAL TROUBLES

Many and diverse classifications have been given for horticultural plant troubles. The classifications differ in form rather than in content, the particular form, or basis, being influenced or determined largely by the particular purpose for which the classification was to be used.

In general, the troubles that beset horticultural plants may be separated into those due to environmental conditions and those due to animal and plant pests that prey upon and do injury to horticultural plants. The troubles due to the environmental conditions were discussed in some detail in the chapters devoted to the relations of temperature, moisture, light and soil and need not be discussed further at this time. The particular animal or plant pest responsible for the damage is often identified by the characteristic injury that it does to a part or parts of the plant. Brief descriptions are presented of a few of the more common, generally distributed, representative types of pests that are injurious to horticultural plants.

Animal Pests of Horticultural Plants

All members of the animal kingdom are dependent on the plant kingdom, directly or indirectly, for their food supply. When the animals become more injurious than beneficial to horticultural plants, the horticulturist considers them as pests.

Higher Animals as Horticultural Pests.—Occasionally a member of a higher group of animals does considerable damage to horticultural plants. One will occasionally hear of cases during severe winters where birds have eaten many of the buds from fruit trees, but these are isolated and infrequent. Rodents such as rabbits, mice and moles, when particularly abundant or when the food supply is scant, may frequently cause considerable damage in limited areas.

Rabbits.—In severe cases rabbits may gnaw a ring of "bark" from the trunk of the younger trees. When this occurs it is often necessary to bridge graft over the injured area to save the life of the tree. Although various repellent paints or washes have given some success as preventatives where rabbits were not especially numerous and the food supply was fairly abundant, the most satisfactory method of control is to wrap the trunks of young trees. The most common substances used for wrapping are paper, burlap, wood veneer and ¼-in. mesh galvanized-hardware cloth. The burlap and paper wrappings are wound tightly about the tree trunk beginning at the ground line and extending to a height of about 2 ft. and tied securely with soft wire or binder twine. The wood-veneer wrappers are bent into a cylindrical shape, placed around the trunk of the tree with the creosoted end down, pushed into the soil an inch or two and tied. A satisfactory wrap that can be left about the tree for eight or ten years or until danger of girdling by rabbits is practically over can be made of ¼-in. mesh hardware cloth. One-inch mesh poultry netting can be used for rabbits, but the ¼-in. hardware cloth will be effective against field mice as well as rabbits. Cut the hardware cloth in 24-in. squares. If the 24-in. width of cloth is used, a bound edge will be at both top and bottom. Bend the wire into a loose cylinder about the trunk of the tree, force it into the soil for an inch or two and tie with light wire. A very efficient method of fastening the wire into a cylinder about the trunk of the tree is by using three hog rings.

The wraps should be placed around the tree trunks about the time the leaves fall, and all wraps, except the wire guard, should be removed when growth starts in the spring. All the attention required by the wire guards is to keep them clean of debris and properly placed until the tree no longer requires such protection.

Mice.—There are two kinds of mice that may do damage to orchard trees, and they are destructive to trees of all sizes. One is known as the "field mouse," and the other as the "pine mouse." The field mouse lives chiefly on top of the soil. Its runways are made through the grass and vegetative ground cover so that it is especially troublesome in orchards that are in sod or in those in which a cover crop is used. It feeds on the bark of the trunk of the trees at the ground line, working away under the protection of the ground cover. The pine mouse eats the bark from the roots of the tree; and since it works entirely underground, its presence is not so easily detected, and a large amount of damage may be done without one's suspecting its presence.

Since the field mice live aboveground, the removal of all weeds and grass in the fall from an area 4 or 5 ft. in diameter about the trunk of the tree is very beneficial. Mounding and smoothing the soil about the trunk of the tree to a height of 4 to 6 in. would be a feasible practice if one had but a small number of trees. The ¼-in. hardware-cloth cylinder, as used against rabbits, may be placed about the trunk of the tree and forced into the ground as a protection from field mice but will be of no value as a protection against pine mice, since they feed on the roots underground. The most satisfactory means of control is the use of poisoned baits. Several formulas for poisoned baits have been developed and recommended by the Bureau of Biological Survey of the U.S. Department of Agriculture. Poisoned-grain baits have proved very satisfactory. For the control of field mice locate a feeding station under each tree, and place within it 2 tbsp. of the poisoned bait; if necessary, replenish the bait every three or four weeks. These feeding stations should be so constructed and located that only mice will have free access, and the bait will be kept dry and retained within the feeding station. The poisoned bait for the pine mice must be placed directly in their runways.

A formula for rolled-oats bait is as follows: Mix 1 oz. of powdered strychnine and 1 oz. of baking soda. Sift this mixture over 8 qt. of dry rolled oats, and mix it thoroughly with the oats. Warm the poisoned oats in an oven, and pour over and mix thoroughly with them 1¼ pt. of a warm fat-paraffin mixture. This mixture is made of 3 parts melted beef fat and 1 part melted paraffin. In order to obtain a suitable coating it is necessary that both the poisoned oats and the beef-paraffin mixture be warm and stirred constantly while being mixed.

Recently zinc phosphide has been found very effective in the control of mice. Pieces of apple are dusted with the poison and placed in the runways.

Moles.—The characteristic tortuous ridges in the lawn over the tunnels of the mole are readily identified. The smoothness of the turf is destroyed, and the grass on the raised soil may die because of excessive drying. These pests can be controlled by trapping, poisoning and asphyxiating. Different types of traps are available on the market, and these have proved to be very satisfactory. Calcium cyanide, carbon bisulfide and paradichlorobenzene have all been used successfully. In using calcium cyanide make openings in the tunnels every 5 or 6 ft., place a teaspoonful of the cyanide in each opening and close the opening without pressing down the burrow. Carbon bisulfide is applied in a similar fashion, pouring about a teaspoonful in each opening and closing it carefully. A teaspoonful of paradichlorobenzene placed into the burrows at intervals of 5 or 6 ft. and the openings covered as before will be found effective. If these chemicals are used in too large quantities, considerable damage to the grass may result.

Lower Animals as Horticultural Pests.—Members of the lower groups of animals are very serious pests to horticultural plants. Considerable injury may be done by particular members, as nematodes, but the greatest group of animal pests is popularly known as "insects."

Insect Pests.—The term "insects" in its popular use includes not only the true insects but also the "red spiders" and other mites that are destructive to horticultural plants. There are more species of insects than all other animals and plants combined, but fortunately the horticulturist does not need to know all the 625,000. In order to understand partially the control practice it is advisable to have a general knowledge of the main features of the life cycles of representative insect pests of horticultural plants.

Most of the insects have four distinct stages in their life cycle: the adult; the egg; the larva, or worm; and the pupa, or cocoon, stage which changes to the adult. This cycle is known as the "complete metamorphosis." Some insects have an incomplete metamorphosis, as the adult, the egg and the nymph stages, with the nymph finally developing into the adult form. Some, as the aphids, are both oviparous and viviparous; that is, they lay eggs and also give birth to living young.

Various systems of classification are used for the insects but a satisfactory one for the horticulturist is that based upon the feeding habit of the insect. With few exceptions the horticulturist is concerned only with two classes: (1) the biting, or chewing, insects; (2) the piercing, or sucking, insects.

INSECTS WITH BITING OR CHEWING MOUTH PARTS.—The biting insects are those which bite out and take into their digestive tract portions of the plant tissues. When such insects are feeding inside the plant tissue, they are often termed "boring insects." This class of insect generally does the largest amount of damage to horticultural plants, and its presence and the injury done is more noticeable than the presence of and damage done by insects with sucking mouth parts.

The damage is most generally done by the insects when they are in the larval and nymph stages, and the chief control measure is directed against the pests while they are in those stages. There are some exceptions in which the primary control is directed at the pest while it is in one of the other stages of its life cycle, but the secondary control measures are usually directed at it at such times. A few of the most common and most generally destructive examples of this group will be presented. Some are important in one locality, and some in others;

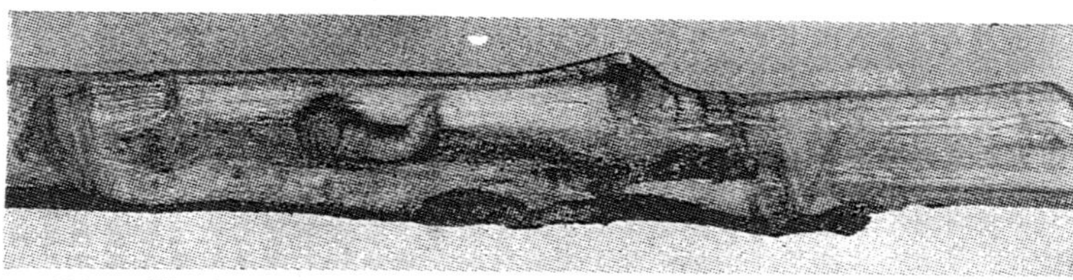

FIG. 121*a*.—Flat-headed apple tree borer hibernating in trunk of young apple tree. (*Missouri Agricultural Experiment Station.*)

FIG. 121*b*.—Round-headed apple tree borer; half-grown larvas in base of young apple tree. (*Missouri Agricultural Experiment Station.*)

some are destructive one year, and some another year in the same locality; and it is probable that some that are not mentioned are more important in certain localities than any that are discussed.

Codling moth (*Carpocapsa pomonella*, L.): The codling moth is undoubtedly the most destructive insect that attacks apples. It is destructive also to the pear, quince and hawthorn and has been reported on the plum, cherry and English walnut. It is native to southeastern Europe and is present wherever pomaceous fruits are grown. It does its damage either by tunneling about within the fruit, usually about the core, or by making "stings" or feeding punctures on the apple.

There may be one, two or more generations per year in a given locality. The number of generations will vary in the same locality in different years, influenced greatly by environmental conditions. This irregularity complicates the control measures. The insect winters over in the larval form in cocoons under the rough bark of the trees, in the grass about the trees, in favorable crevices in the lug boxes, packing houses, etc. With the advent of warm weather in the spring the larva enters the pupal stage. The duration of this stage varies considerably with local climatic conditions but is usually completed within 30 days. The moths emerge and soon begin flying about in the twilight hours laying the white flaky eggs on the very small apples or the leaves adjacent to them. During her life of about 10 days the female moth lays about 80 eggs. The eggs hatch within 6 or 7 days,

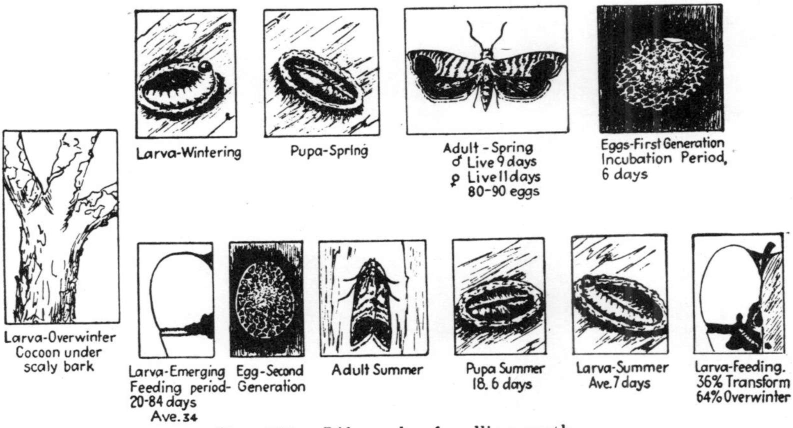

FIG. 122.—Life cycle of codling moth.

and the young larvae begin feeding on the young leaves and fruits and finally enter the young apple chiefly through the calyx, even though the latter has been closed for some time. They feed about within the apple for about a month, when they become mature and emerge from the fruit and form their pupal cocoon. Many of these larvae will not pupate until the following spring. Under favorable climatic conditions some will pupate and change to adults which emerge as moths about 50 days after the first brood of eggs was laid. These moths lay eggs for the second generation. This generation repeats the cycle of the first generation, and in some instances there appear to be three generations per year. The "stings" on the apples late in the season are caused by the feeding punctures of the second and later generation larvae. The wintering larvae will have representatives of all generations.

The codling moth is generally controlled by a stomach poison. Arsenate of lead is the one used most generally with the greatest degree of satisfaction. The abundance of the insects and the problem of undesirable spray residue on the fruit have led to a search for other suitable means of control. A promising substitute for arsenate of lead, especially after the first cover application, appears to be a form of nicotine, with or without oil. Supplementary means of codling-moth control are scraping and burning the loose scaly bark from the trunk and larger limbs by blooming time, banding the scraped trees

FIG. 123.—Diagrammatic sketch showing the injury to foliage by the canker worm.

with chemically treated bands of corrugated paper before the first-generation larvae leave the fruit to pupate and general sanitary measures, as the disposal of all cull fruits and the destruction of over-wintering larvae or the adult moths in the packing houses.

Spring cankerworm (*Paleacrita vernata*, Beck.): Because of their characteristic habit of locomotion the larvae of the cankerworm are known as "measuring worms," or "loopers." These insects feed largely on the apple and to a lesser extent on the stone fruits. They may prove very destructive on the elm and some other shade trees. They are native to North America. In some seasons when the insects are very abundant the damage may be severe in neglected orchards and unprotected shade trees. It is done by the larvae feeding on the foliage where they eat everything but the main ribs. This insect can often be seen suspended from the tree on the end of a silken cord.

There is but one generation a year. The insect winters over in the pupal stage a short distance below the surface of the soil. The adults appear early in the spring. As the females have only rudimentary wings which are useless for flying, it is necessary for them to crawl up the trees where they lay their eggs in groups in crevices on the trunk and larger limbs. The eggs hatch about the time the apple flowers are in the "pink" stage, and the larvae begin to feed. The larvae mature in late spring and spin silken threads as they lower themselves to the ground. They burrow under the surface a short distance and pupate to emerge as adults the following spring.

These leaf-eating insects are readily controlled by a stomach poison, as arsenate of lead. Since it is necessary for the female to crawl up the trunk of the tree to deposit her eggs, bands of various adhesive materials placed about the trunk have been found effective in catching the females as they ascend the tree.

Plum curculio (*Conotrachelus nenuphar*, Herbsti): The adult plum curculio is a brownish-black, rough-coated, snout-nosed beetle about ⅕ in. long that is a very troublesome pest. It would be recognized in the larval form as the yellowish-white, noticeably curved grub about ½ in. long near the pit of stone fruits. The beetle attacks all pomaceous and drupaceous fruits and is especially destructive to the plum, peach and cherry in the eastern part of North America and often causes considerable damage to the apple. The plum curculio is native to the eastern part of North America. This insect does its damage by the feeding and egg-laying punctures in the fruit and by the burrowing of the larvae within the fruit. The fruit may often be misshapen as it develops, and exudations of "gum" may appear on the plum and peach.

There may be one or two generations per year. This seems to be determined largely by the climatic conditions prevailing in the particular locality. The insect passes the winter in adult form, hibernating in protected places furnished by dead grass, brush piles and debris in and adjacent to the orchard. The beetles emerge from their hibernating quarters in the spring about the time the apples' flowers are in the prepink stage or at the time the shucks are falling from the peaches. They may feed for a short time on the young leaves, but as soon as the fruit is large enough the beetles begin feeding on and laying eggs in the fruits. The feeding puncture is a small, shallow, rather superficial cavity which heals over leaving a slight depression or a corky scar. The egg puncture is similar except that a crescent-shaped cut is made about one side of the puncture. This injury heals over and leaves the characteristic crescent scar with the egg puncture in the center. The egg hatches in about 7 days, and the creamy colored larva feeds within the fruit until it is a full-grown yellowish-white legless worm about ½ in. long that is slightly curved toward the underside. Practically all fruits in which larvae are developing fall to the ground. When mature, the larvae leave the fruit and burrow into the ground for a depth of about 2 in. where they pupate in earthen cells. The length of time in the soil will vary considerably with conditions, but 30 days would be a fair average. The adult beetles emerge from middle July through September. They fly to the fruit and begin to feed. In regions where there is but one brood, no eggs are laid at this time and only feeding punctures will be found. Upon the arrival of cool weather the adults migrate to hibernating places for the winter from which they emerge the following spring to attack the new fruit crop.

The curculio is a rather difficult insect to control by spraying, but a stomach poison properly applied will usually be effective, especially when combined with supplementary means as the elimination of suitable hibernating places, the cultivation of the soil during the pupating stage and the destruction of the "drops" by hand-picking or pasturing the orchard with hogs during that period when the infested apples are dropping. Upon being disturbed the adults will drop to the ground. Advantage is taken of this characteristic, and the ground under the tree is covered with canvas, the tree is jarred and the insects collected and destroyed.

Peach-tree borer (*Sanninoidia exitiosa*, Say): The peach-tree borer is most generally known and recognized in the larval form as the cream-colored, dark-headed worm about 1 in. long tunneling about underneath the bark of the peach tree a short distance above or below the ground line. The insect is indigenous to eastern North America where it apparently lived upon the native stone fruits until the introduction of the peach. The peach-tree borer is probably the most important insect pest of the peach in the commercial growing areas of North America east of the Rocky Mountains. It causes a large amount of injury to peach trees by actually destroying the trees or weakening them to such an extent that they are more readily attacked by other insects and diseases or are destroyed by unfavorable climatic conditions that would have had but little injurious effect upon them if their vigor had not been lessened by the action of the peach-tree borer. The insect does its damage in the larval stage by feeding and tunneling about in the cambial region of the trunk of the peach tree near the ground line. This results in a partial girdling of the tree with the characteristic effects of such injury. Gum will exude from the injured areas, and this mixed with the frass, or "sawdust," is an indication of the insect's presence.

This insect has a complete metamorphosis. Generally there is one generation per year, but some of the insects of the late-hatched larvae or those which developed slowly because of unfavorable low temperature or inadequate food supply may require more than one year to complete their cycle. This accounts in part for the variations in the size of the larvae found at the same time.

The adult peach borer is a dark steel-blue, clear-winged moth about ½ in. long and with a wing expanse of about 1¼ in. There are some differences in the color of the male and female, and the male is the smaller of the two sexes. In various parts of the country the adults appear at various times, but in all localities they will be emerging over a period of three months or longer. This will usually be from about the middle of June to the middle of September. The adults live from 4 to 6 days. During their short life the female will lay from 300 to 600 or more eggs. The eggs may occur singly but are usually in clusters. The majority of the eggs are deposited on the trunk of the peach tree near the ground line, but some will be deposited farther up the trunk and even on the main branches. The incubation period is influenced greatly by temperature, ranging from 5 days in the warm summer days if the eggs are exposed to the sun to 15 days in the cooler days of autumn or if the eggs are in shade during a considerable portion of the day. The eggs usually hatch sometime during the night or early morning.

The newly hatched larva makes its way down the trunk of the tree and burrows into the tree just below the ground line. A majority of the larvae will be within the 2-in. zone below the soil level, although a few may be found on the roots. They feed in the cambial area of this region until the soil drops to a temperature of 40 to 50°F., after which they remain inactive until the soil warms up to about 50°F. in the spring.

The full-grown larva forms a loose cocoon in which to pupate. By far the greatest number of the cocoons will be found near the surface of the soil adjacent to or within 2 in. from the trunk of the tree. The length of time that elapses from the spinning of the cocoon to the emergence of the adult will vary from 15 to 30 days according to environmental conditions.

Many different means have been used to combat the peach-tree borer, but until the discovery of paradichlorobenzene (P.D.B.) the most satisfactory method of control was by digging the larvae out of the trunks of the trees by hand. This is still the best method for trees under four years of age. The use of paradichlorobenzene is now the universal method employed to control the peach-tree borer in trees four years of age or over. This is a white crystalline substance at ordinary temperatures which changes to a gas when exposed to the air. The amount used per tree will vary according to the diameter of the tree trunk. The usual recommendations are ¾ to 1 oz. for trees from four to six years of age and 1 to 1¼ oz. for trees over six years old. Since most of the insects are in the larval stage early in the fall and the soil temperature is favorable, the trees are generally treated during that season of the year. The favorable soil temperature is 60 to 70°F., for at this range the gas is evolved sufficiently rapidly to kill the insects and is less injurious to the tree.

In preparing a tree for treatment clear and smooth the soil about its base for a distance of about 12 in. If a conspicuous amount of gum has exuded from the tree trunk, it will be well to remove it, as it appears to protect the insects somewhat from the gas. Distribute the required amount of paradichlorobenzene in a continuous circular band about the tree at a distance of about 2 in. from the trunk. Gently shake some fine soil over this band, and then add three or four shovelfuls of fine soil on top. Pack firmly with the back of the shovel. This should be sufficient to control the peach-tree borer.

Strawberry leaf roller (*Ancylis comptana*, Frohl): The presence of the strawberry leaf roller is most generally recognized by the folding and rolling of the leaflets of the strawberry plant. Upon unfolding the leaflet the small, cylindrical, greenish larvae will be found. This insect is common in Europe and in North America wherever strawberries are grown. In the United States it appears to be most abundant in the Mississippi Valley. In cases of severe infestation it causes considerable damage, as much of the foliage may be destroyed. The insect does its damage in the larval stage. Generally the larva begins feeding on the underside of a young leaflet and finally folds the leaflet along the midrib. As the larva becomes larger the leaflet will be rolled.

This insect has a complete metamorphosis. The number of generations seems to vary somewhat with the climatic conditions of the locality, but in most localities where the insect attains destructive numbers there appear to be three generations. There may be but two generations in the northern part of the United States and a partial fourth generation in the southern part. The adult reddish-brown moth has a wing expanse of ½ in. The moth will live for about 10 days during which the female will deposit about 100 eggs. The small, oval, pale yellowish-green, translucent eggs are deposited at night chiefly on the underside of the strawberry leaflet. The length of the incubation period is influenced greatly by temperature. The first-generation eggs which hatch in the cool spring (May) require 10 or more days, whereas the eggs of the second and third generations which hatch during much warmer periods require about 5 days.

The newly hatched larva begins to feed near the midrib and larger veins on the underside of the young leaflets. It soon crawls to the upper side where it begins to draw the two halves of the leaflet together by folding along the midrib. It feeds and attains full size in this folded, or rolled, leaf where it pupates. The duration of the larval stage, like that of the egg stage, is shortened by higher temperatures and ranges from 20 to 30 days. The pale-brown, ovate pupa is found only in the rolled leaf. As with the egg and larva stages the pupal stage is shortened by higher temperatures, being 7 or 8 days in the warm part of the year and twice that long in the cooler season. From 40 to 50 days is required for the complete cycle from egg to adult. The generations, however, will overlap.

The strawberry leaf roller is a biting insect and can be controlled with a stomach poison. Since the larva folds the leaflet and feeds from the inside, it is essential that the spray be applied at the proper time. Since there are usually three broods of larvae, the poison can be applied three times. Of the three the first is the most important and generally the only one required. This is just when the earliest blossoms appear. The other applications will be made according to the life cycle of the insect in the particular locality but will generally be applied just after harvest and about six or seven weeks later. Mowing the leaves after harvest and raking and burning them is an effective supplementary if not primary means of controlling the pest.

Colorado potato beetle (*Leptinotarsa decemlineata*, Say): The Colorado potato beetle is the most important insect pest of the Irish potato. It will be recognized as the humpbacked, soft, red-bodied, dark-headed larva or grub or the nearly semispherical yellow adult with the black bars running lengthwise on the wing covers that is found feeding on the foliage of the Irish potato. This insect was native to a section east of the Rocky Mountains and originally fed upon the native plants there that are fairly closely related to the Irish potato. With the introduction of the potato the insect found a satisfactory food supply. It moved rapidly eastward into the potato fields and today is found in

North America wherever the Irish potato is grown and has been introduced into Europe. The insect does its damage in the larval and adult stages, by eating the foliage of the plants.

This biting insect has a complete metamorphosis with from one to three, usually two, generations per year. The number of generations is influenced by the location of the potato-growing region, the number being larger in the southern regions. The adult beetle hibernates in the soil at a depth of 4 to 6 in. As the ground warms in the spring, the adults emerge and go to the potato or similar plant for food and a suitable place to deposit eggs. The adult beetle is about 3/8 in. long but somewhat narrower than long and decidedly rounded on top. The ground color is yellow, and the wing covers are marked by 10 longitudinal black lines. During the first month after emerging each female lays from 500 to 1,500 or more eggs. The adults will live for two months or longer and do considerable damage feeding on the young tender potato foliage.

The orange-colored cylindrical eggs are attached by one end to the underside of the leaf. They occur in rather compact groups of 35 to 45 near the tips of the young leaves. The eggs hatch in about 7 days. The larva is a humpbacked, fleshy, soft-bodied insect, red in color with a dark head and two rows of black spots on either side. It is a voracious feeder on the youngest leaves of the potato plant. It grows rapidly for about three weeks and then burrows into the ground for about 3 in. and forms an earthen cell and within two days forms the pupa.

The pupa is colored red like the larva. After a period of about seven days, depending upon the temperature, the pupa changes to the adult beetle which emerges and begins a new cycle. At the close of the season the adult beetle burrows into the ground for a depth of 4 in. or more and remains there until warm weather the following spring. A period of approximately 30 days is required to pass from the egg to the adult form.

The Colorado potato beetle is a biting insect and consumes large quantities of potato foliage. Since it is susceptible to stomach poisons, it can readily be controlled by such materials. Paris green was the old-time favorite but has been replaced largely by arsenate of lead and calcium arsenate.

INSECTS WITH PIERCING, OR SUCKING, MOUTH PARTS.—The sucking insects are those which obtain their food by inserting their beaks into the plant and sucking out the plant juices. This parasitism causes a weakening of the plant and, in severe cases, may result in its death. This type of insect is instrumental in transmitting various virus diseases.

San José scale (*Aspidiotus perniciosus,* Comst.): The San José scale is generally recognized by the shieldlike dark-gray covering with the yellowish or dark-colored raised portion, or nipple. Severely infested stems will have a grayish appearance. After settling down to feed, the insect secretes this shield cover which is entirely separate from its body. The shield covering the female scale is flat, circular, about 1/12 in. in diameter and with the raised nipple occupying the central portion. The shield covering the male scale is smaller and is elongated, being about twice as long as wide, and has the raised nipple toward one end. The insect was first observed in this country at San José, Calif. There is considerable uncertainty relative to the place from which it was introduced into this country, but China is believed to be its native home. It is now found quite generally throughout the fruit-growing regions of North America where climatic conditions are favorable for its development. The insect is a serious pest on most of our deciduous fruits and many ornamental trees and shrubs. The San José scale does its damage by sucking the plant juices from the stems, foliage and fruits. The feeding punctures cause conspicuous reddish circular spots on the fruit. In severe cases the young plants and even the older ones may be destroyed or so severely weakened that they are easily killed by other troubles which would not be serious to strongly growing plants.

This sucking insect has an incomplete metamorphosis, there being no egg stage, as the young are born alive. There are three or more generations per year depending upon the climatic conditions of the locality. The San José scales pass the winter as partially grown insects underneath their protective scales, which at this stage are black. In early spring the hibernating individuals that survived the winter continue their growth to maturity. When mature the male emerges from underneath his waxy scale as a small two-winged insect, but the legless, wingless, eyeless, saclike female remains under her scale where she is fertilized by the male which then dies. She remains alive under her scale for a period of about six weeks during which time she gives birth to 200 or more living young. It is possible for an overwintering female to give rise to over 2,000,000,000 young in a single favorable season. The young, nearly microscopic, orange-yellowish nymphs or larvae crawl about for a few hours and then insert their beaks into the plant and begin to feed. These crawling young may be transported considerable distances on birds, on insects or by wind and thus spread over large areas. Soon after settling down, the scale-like covering begins to be formed over the insect. All the young scales look alike; but as they develop, the two sexes take on their characteristic forms. In about 30 days from birth the insects are mature, the male emerges and the females are ready to be fertilized. Within three or four days she gives birth to living young, and the

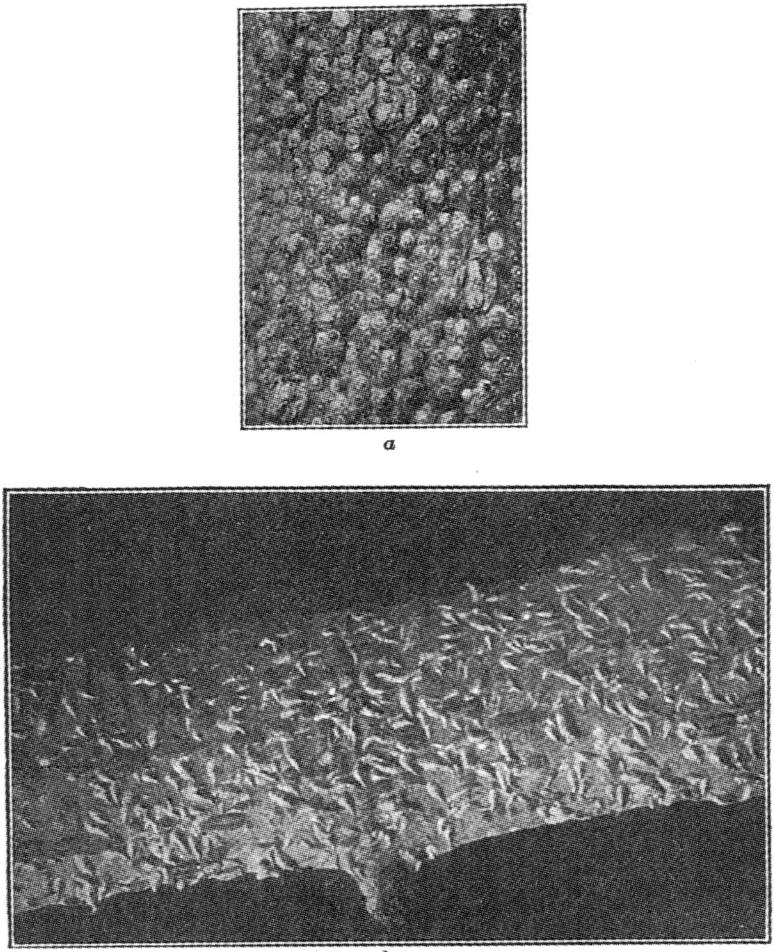

a

b

FIG. 124.—Scale insects: (a) San José; (b) Oyster shell. (*Gardner, Bradford and Hooker.*)

cycle is continued. Under favorable conditions, therefore, a new generation appears about every five weeks until the advent of cooler weather. Because the females give birth to living young for a period of six weeks, there is a general overlapping of generations so that in the fall practically all stages of development are found simultaneously.

The insects hibernate under their protective scales until favorable conditions for continued growth occur the next spring. This difference in the stage of development results in high mortality, and those insects which were about half developed at the time of hibernating are the chief ones to survive.

When first found in this country the San José scale was greatly feared. Although it is an expensive pest to combat, it can be kept under control by its natural enemies and an application of an oil spray or a strong lime-sulfur spray just before the infested plants start growth in the spring.

Oyster-shell scale (*Lepidosaphes ulmi*, L.): The oyster-shell scale is easily recognized by the scale that covers it and that looks like a minature elongated oyster shell. The striated shield is noticeably pointed at one end and gradually increases in width to the other broad and rounded end. The scale covering the fully developed female insect is about ⅛ in. long which is about twice as long as that covering the male. With the exceptions of tropical and arctic regions this insect has a world-wide distribution, and there is considerable uncertainty relative to its original home. It apparently was introduced into the New England colonies at an early date and from there has spread over a large part of North America. The oyster-shell scale has a wide range of host plants and is found on fruit plants and many ornamental trees and shrubs to which it often does more damage than it does to the fruit plants. Like the San José scale it does its damage by sucking the plant juices from the stems, foliage and fruits. The insects occur principally on the woody portions of the plant and, although not so destructive as the San José scale, cause the plant to become weakened and may cause the death of some branches or, in severe prolonged attacks, the death of the plant.

This sucking insect has a complete metamorphosis. In the cooler parts of the country there is but one brood per year, but in the warmer sections two broods occur yearly. The oyster-shell scale passes the winter in the egg stage underneath the scale that covers the shriveled, dead female insect. There will be from 50 to 100 small, oval, glistening white eggs under each scale. With the advent of warm weather the eggs begin to hatch and will continue hatching for a period of about a week. The length of the period is influenced by temperature, being shortened by higher temperatures. Because of temperature effects, the time at which hatching begins will vary with the locality, but for any locality hatching of the overwintering eggs will begin about the time the petals have fallen from the apple blossoms. After hatching from the egg the light cream-colored larva emerges from underneath the scale and moves about for an hour or two. After it has found a suitable location it inserts the hairlike sucking tube into the plant and begins to feed on the plant juices. The newly hatched insects may be carried some distance on birds and other insects or may be blown by the wind. Within three or four hours after settling down to feed, a white, glistening, cottony secretion begins to cover the insect. The larva continues to grow and mature and is protected by a scale that keeps pace with the growth of the insect. The scale becomes brown and shows the characteristic concentric markings of the scale covering the mature insect. The female remains under the scale permanently; but the males, of which there seems to be a very small number, emerge as small two-winged adults, fertilize the females before they are fully developed and die. The body of the female becomes distended with eggs and fills the entire area underneath the scale. When mature the female begins laying eggs, which operation extends over a period of about a month. As the eggs are laid, the body of the insect becomes smaller and smaller until all the eggs are laid when the female dies.

In regions where there is but one brood yearly these eggs do not hatch until the following spring. Where there are two broods yearly, all the stages are speeded up somewhat, and it is the second brood of eggs that winters over. In regions where there are two broods there is a period of about eight or ten weeks between hatching of the first and second brood.

The oyster-shell scale has a number of natural enemies which assist materially in keeping it within manageable limits. The most satisfactory control is obtained by spraying late in the dormant period with either lime-sulfur or oil at the strengths recommended for a dormant application.

Aphid species: Aphids, or plant lice, are common on a large number of horticultural plants. Different species attack different crop plants, and there is practically no crop that is entirely free from the attack of one or more species. Each species occurs in several different forms, but there is a general similarity in their habits and life cycles. They are most generally known and recognized as the small, soft-bodied, wingless insects of varying sizes that are found near the tips of shoots or on the under surfaces of leaves that are curled to a greater or lesser degree. Aphids are world-wide in their distribution. Various species are native to different countries, but now they are widely disseminated from one country to another. There is a large number of different kinds of aphid, but the number of different kinds of host plants attacked by each species is limited to one or a very small number of closely related forms. The aphids are sucking insects and do their damage by sucking the plant juices from the leaves, stems, roots or fruits. They feed principally on the younger, tender growing parts of the plant, checking the growth of the plant and fruits, causing the leaves to become distorted and curl downward and producing a malformation on the roots attacked. The presence of aphids is often indicated by the presence of ants which are attacted by the "honeydew" excreted by the aphids.

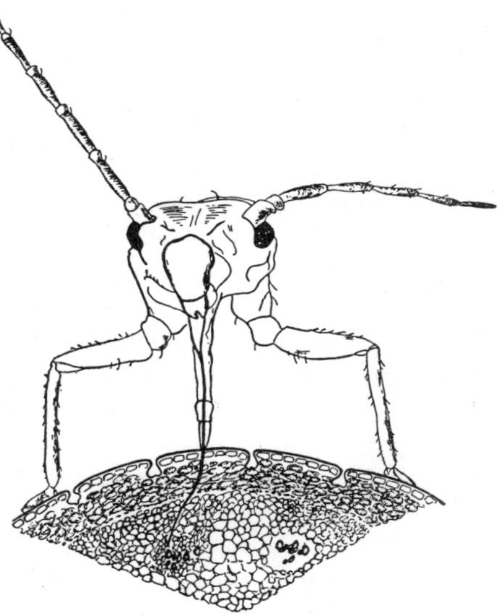

FIG. 125.—Diagrammatic sketch of aphid penetrating leaf.

The aphids have a remarkable and varied life cycle. The same species may exhibit several forms. The number of generations per year varies greatly. Aphids pass the winter in the egg stage. The eggs are small and of some species slightly oval and develop a shiny black color by spring. As the weather warms up, they hatch and give rise to wingless females which are termed "stem mothers." Upon hatching, the young nymphs move about quite actively and feed by sucking the plant juices from the succulent growths. After passing through several molts the stem mother becomes mature and without being fertilized begins to give birth to living young. In the case of the green-apple aphis (*Aphis pomi*) this period of development requires about three weeks. The stem mother may live for a period of about four weeks and produce 40 or more winged or wingless females. In this fashion a succession of generations may be produced until the approach of cool fall weather when the male and female forms are produced. In some of these generations winged forms are produced which in some types provide only a means of distribution, whereas in other types they serve also as a means of changing to different kinds of host plants. The egg-laying females are wingless and much smaller and more active than the females that give birth to living young. In some species the males are winged, and in other wingless. There is a very small percentage of males, and it is probable that many of the females are never fertilized. The length of life of the female and the number of eggs deposited will vary, but in *A. pomi* she will live about 7 days and lay about three eggs. The eggs that are not destroyed hatch the following spring, producing stem mothers, and the yearly cycle is repeated. This description of the yearly cycle of the aphids applies to all the different types in a general way, but the yearly life cycle will vary with the particular species of aphis.

Aphids have several natural enemies which, together with unfavorable climatic conditions, may often make additional control measures unnecessary. As aphids are sucking insects, they can be controlled by contact insecticides. The substance most generally used is a form of nicotine. Proprietary compounds containing rotenone and some oil sprays are used also. The spray should be applied just as the eggs have hatched in the spring, using the dilution recommended by the manufacturer of the product being used. Later applications may be necessary.

Miscellaneous Lower Animals.—A few of the lower animals are considered as pests to horticultural plants. Of these ants often prove troublesome to lawns but nematodes are the most destructive.

ANTS.—Ants are so common and so universally distributed that they are well known. There are many different kinds most of which are beneficial to man, but some become troublesome by building their nests in garden or lawn.

The ant has a complete metamorphosis: the egg, larva, pupa and adult. There are three classes or groups of adults: the workers, the queens and the males. Even within the same species considerable variation occurs in these forms. Except for the newly emerged males and females, ants are wingless. The males die soon after mating, and the females start a new colony and lose their wings. This life cycle is important in determining the efficiency of control measures. Ants are generally controlled by repellents, contact insecticides, stomach poisons and fumigants. Contact insecticides, as pyrethrum and nicotine, are used but are usually not effective, since the nest, or colony, is not destroyed. Various stomach poisons combined with sugar, honey or grease may be used. There are a number of commercial preparations of this nature. As this poison must be carried to the nest in order to destroy the brood and queen, it is not usually so effective as using a fumigant on the nest such as carbon disulfide or calcium cyanide. Using a sharp-pointed stick about the diameter of the little finger, make holes in the ant nest about 8 to 12 in. apart to a depth of 6 or 8 in. Pour a tablespoonful of carbon disulfide into each hole, and close the hole by pressing the soil together. When all holes have been treated, a wet burlap or canvas covering should be placed over the nest to retard the escape of the gas. Carbon disulfide should be handled with care, as it is a highly volatile, inflammable liquid and upon vaporizing and mixing with air is highly explosive. Calcium cyanide powder, a powerful and dangerous poison, may be used in a similar fashion.

NEMATODES.—Nematodes (eelworms, or roundworms), which are much lower in the animal kingdom than insects, are very small, slender, round, threadlike worms which live in and on plants throughout the year. There is a very large number of different kinds, many of which apparently are not injurious to economic plants. They are abundant and widely distributed but are best adapted to regions with a mild climate. They are destructive to greenhouse plants everywhere and to plants outdoors in the South Atlantic states, the Gulf states and parts of California. They do considerable damage to many fruits and ornamentals and to most vegetables by causing poor stands, reduced yields and weakened vitality with its accompanying dangers. Some species feed on but one or a few closely related plants; others feed indiscriminately on a wide assortment of plants. They work on all parts of the plant, but one widely distributed and probably the most injurious species (*Heterodera radicicola*, Greef.) feeds on the roots of a large variety of plants, causing characteristic swellings or galls which have received the descriptive name of "root knot."

This nematode has a life cycle somewhat like that of an insect with a complete metamorphosis. In the warmer areas there may be as many as 12 generations per year, but in the cool northern limits of its habitat there may be but 3 or 4 generations annually. The nematode passes the winter in the egg stage which is very resistent to unfavorable environmental conditions. The colorless, transparent, oval eggs are microscopic in size, each being about $\frac{1}{250}$ in. long. The fine hairlike, microscopic to barely visible worms hatch from these eggs and make their way into the soil to the young roots of a plant. They bore their way into the root and feed and develop to maturity. Their presence in the root causes the characteristic root-knot distortion. After a time, the length of which is determined largely by environmental conditions, the males mature and fertilize the still immature females. The females continue to grow, becoming about 1/25 in. long and changing from the cylindrical worm form to a pear shape. They begin laying eggs at the rate of 10 to 15 per day and continue to do so until they have laid from 300 to 500 eggs. The eggs will hatch in 2 to 6 days or may remain dormant for a long period of time. The period of time from egg to egg of the different generations during the summer varies from three to eight weeks.

Since the pest does its damage to the roots of the plant and remains either in the soil or in the roots within the soil, it is difficult to control. The destruction of root-infected residues, practicing a two- to four-year crop rotation and using plants that are resistant to attack, is beneficial. The most satisfactory means of control is soil sterilization, but this cannot always be practiced. Hot water, steam and carbon disulfide are all used as soil-sterilizing agents. Hot water is not satisfactory for any but small quantities of soil. If a pot or seed flat of soil is immersed in boiling water for 5 min., the nematodes will be destroyed. Soil to be treated with steam should be dry and loose when a treatment at 140 to 150°F. for 2 hr. or more will be effective in killing the nematodes to depths at which the high temperature prevails. Carbon disulfide has given satisfactory results in greenhouse soils. One gallon of a stock solution containing 68 per cent carbon disulfide, 26 per cent water and 6 per cent rosin fish-oil soap diluted to 50 gal. with water and added to the soil at the rate of 1 gal. per square foot has proved effective. All stages of the nematodes are destroyed by drying. Nematodes will be killed in soil spread out in a 2-in. layer in a hot dry greenhouse for a period of about 8 days. Recently, chloropicrin, one of the tear gases, has been used successfully to destroy the nematodes in seed beds.

Plant Pests of Horticultural Plants

There are many different kinds of plants in both the higher and the lower forms that are troublesome to horticultural plants.

Higher Plant Pests.—Although there are special cases, such as dodder, where a higher plant may be parasitic and be considered as a pest on other plants, the higher plants most troublesome to horticultural plants are those termed "weeds." In the cultivated areas of the garden and orchard these weeds can be kept under control by cultivation and cropping practices, and in the non-cultivated orchards by mowing and mulching. It is really only in the lawn and in the seed beds that the horticulturist must make special efforts to control weeds. Some weeds, as quack grass, are very difficult to control; whereas others, as dandelion, crab grass and plantain, are eradicated quite easily. The best control for any lawn weed is a well-grown, properly clipped, dense turf of grasses well suited to the soil, climate and location. Such turf will crowd out undesirable weeds and inhibit, if it does not prevent, the young weeds from becoming established. Weed seeds are destroyed in seed flats and beds by soil sterilization.

Dandelion (*Taraxacum officinale*, Weber).—The dandelion, with its beautiful yellow flowers that so quickly become unsightly fruit stalks and with its large coarse leaves, is a too common lawn weed. Hand digging is a slow and laborious process but if done properly is a satisfactory method of ridding the lawn of this weed provided suitable cultural practices are followed to maintain a satisfactory turf. A small amount of gasoline placed in the crown of the plant, by means of an oilcan, will kill the plant. The dandelion can be killed and the grass left uninjured by one or more applications of an iron sulfate spray. The spray is made by dissolving 1 lb. of iron sulfate in 1 gal. of water. This will cover about 500 sq. ft. of lawn. The first application should be made just before or just after the dandelions come into bloom. Two or three additional treatments at 10-day intervals may be necessary. Raking the leaves before spraying injures them and allows quicker and better action of the chemical. Iron sulfate should not be used during hot dry weather. This same treatment will kill white clover, the plantains and other wide-leaved weeds. Dandelions can also be destroyed by spraying the plants with kerosene. Spray 1 gal. on about 250 sq. ft. of turf. Too heavy an application will kill the grass as well as the weeds. Best results will be obtained by spraying in the fall, as the kerosene should be applied in cool weather and preferably in the evening or on cloudy days.

Lower Plant Pests.—Many diseases of horticultural plants are caused by members of the plant kingdom that belong to the lower groups. The two kinds most responsible for diseases of horticultural plants are bacteria and fungi. No attempt will be made to describe any large number of these plants that cause disease on other plants, but a few common representative examples will be given of those bacteria and fungi which attack horticultural plants.

Bacteria.—Technically bacteria are a subdivision of fungi. They are minute one-celled plants sometimes called "fission fungi" because of their common method of reproduction. Bacteria do not possess green coloring matter or chlorophyll.

FIRE BLIGHT (*Bacillus amylovorus*, Burrill).—The term "fire blight" is very descriptive of the appearance of the injury done to the foliage, shoots, spurs and flowers of apple and pear trees by this bacterium. The disease affects all parts of the plant both above- and belowground. The injury to the tips of the shoots and their foliage is the most conspicuous.

The disease is more prevalent during warm, humid weather on vigorously growing trees in fertile, moist soil. The bacterium is apparently native to eastern North America where it lived on the wild crab apple and a few other native plants. The planting of apple and pear orchards furnished a very satisfactory and more abundant host. The

FIG. 126.—Fire blight on apple. (*Cornell University Agricultural Experiment Station.*)

disease occurs generally in the apple- and pear-growing sections throughout North America. It is most destructive in the warmer area of the country and especially serious in the pear orchards of California. Apparently it has been introduced into Europe, as it has been reported in Italy. Its favorite host plant seems to be the cultivated pear, but it is found also on the apple and quince; some of the drupe fruits, as the plum and apricot; some of the ornamental trees, as hawthorn and mountain ash; and a few miscellaneous plants. Natural infection has been found only on plants of the Rose family.

Fire blight does the greatest amount of damage to the cultivated pear. Pear growing has been abandoned in certain areas owing largely to the damage done by this disease. Twig blight is the most conspicuous injury. Early in the growing season there is a rapid darkening and drooping of the leaves on the terminals of succulent growing shoots and the blackening of the shoots themselves. The leaves do not fall from the tree but remain attached even long after normal leaf fall in the autumn. Blossom blight is the most destructive injury to the present crop. It is the blighting of the flowers and foliage on the spurs similar to that on the shoots. Trunks or body blight is the most serious injury in the development of the cankered areas on the trunk and large limbs, as these furnish sources for new infection and, if not controlled, will result in the death of the plant.

This bacterium winters over in the cankered areas on the twigs, limbs and trunks of the infected plant. It survives mostly in the larger cankers, but in favored localities many of the twig cankers may be active. Both primary and secondary infection occur in the blossoms. Just before or at the time the blossoms are opening in the spring, gelatinous secretions, or "ooze," containing large numbers of bacteria appear about the edge of the cankered region where the diseased area is separated from the non-diseased. Heavy rains will wash this bacteria-containing secretion on to the flowers directly below the canker and cause them to become infected. A few insects, as certain flies, aphids and ants, will feed on this exudate and then infest the blossoms which are most susceptible to infection the first 2 days after opening. The bacteria enter the flowers through wounds or through natural openings and multiply rapidly under warm, humid conditions. Secondary infection of the flowers which takes place from blossom to blossom is carried on by insects—largely the honeybee. Rain is of but little if any importance as a direct agent of secondary blossom infection. The growing shoots may be infected any time during the growing season, as they are subject to both primary and secondary infection. The primary infection is usually slight, as the largest number of shoot infections occur after blossoming. Rain is of but slight importance in shoot infection, but aphids and other such insects that feed on the bacteria-containing exudate transfer the bacteria to the leaves and tips of tender rapid-growing shoots. The bacteria gain entrance through wounds or natural openings.

For many years the most common and effective means for controlling fire blight in apple and pear orchards has been cutting out and destroying the diseased parts. Since the disease moves downward, the affected shoot, twig or limb should be removed several inches below the apparently diseased area. On the larger limbs only the cankered area is removed. This is done by cutting away the diseased bark and some that is apparently healthy down to the wood. The holdover blight cankers are removed during the dormant season, and the infected shoots during the growing season. It is possible to control some of the early infections that do not show discoloration of the inner bark by removing only the outer bark. This work should be done by midsummer. All tools used and all wounds made in an effort to control the disease should be treated with a disinfectant at once. A pint of suitable disinfectant can be made by adding 2 oz. of glycerin to 14 oz. of water and then adding two each of the 7.3 gr. of mercuric chloride and mercuric cyanide tablets carried by druggists. This is a deadly poison. It should be kept only in glass or earthenware containers. Some success is being obtained by painting or drenching the cankered areas with zinc chloride solution, but this may result in injury to the tree. Spraying with a dilute, 1-3-50 Bordeaux mixture or dusting with 20-80 copper lime has been found effective in reducing blossom infection. If but one application is made, it is given during full bloom.

Crown Gall (*Pseudomonas tumefaciens*, (E. F. S. & Town) Duggar).—Crown gall is a bacterial disease caused by *P. tumefaciens*. The common symptom of the disease is the presence of wartlike growths of various sizes on the trunk and roots of the plant. The galls are found most frequently on the stem of the plant at or near the surface of the soil, but in some plants they are found on the roots at varying depths and at various distances from the trunk. Galls may be found on the trunk and main branches at some distance in the air. The bacterium that causes this disease is widely distributed in the fruit-growing districts of the United States. It attacks apples, peaches, raspberries, blackberries, dewberries, grapes, roses and many other kinds of plants, especially rosaceous plants. There appear to be different strains of the organism that attack specific kinds of plants and cause characteristic symptoms.

The organism gains entrance into the plant through wounds. With fruit trees it is chiefly a nursery disease, because infection takes place through the graft union at the time of propagating or shortly after planting, and the disease can be detected and the plant destroyed before the plant leaves the nursery. Not all the wartlike growths at the graft union are caused by this organism. Many of them are merely outgrowths of tissue due to defective graft unions. On the smaller plants the galls do considerable damage by distorting and restricting the vascular tissues, thus interfering with the normal flow of water and materials in the plant. The galls also use plant food for their own growth. These injuries result in a general weakening or drying up of the parts of the plant affected. The first slight swelling increases in size very rapidly, changing from a light-colored, soft tissue to a dark-brown, hard, woody, irregular outgrowth. The galls may decay, thus liberating immense numbers of bacteria which can survive in the soil apparently for at least one season.

The disease is usually controlled by planting disease-free stock. With tree fruits it is comparatively easy to detect it before the plants are set in the orchard, although callus outgrowths may be mistaken for crown-gall infection. With the smaller plants, as the brambles, that have been in the nursery but a comparatively short time the presence of the disease may not be detected so easily. In this case the plant should come from healthy plants growing on disease-free soil. They should be planted on land that has been cropped for two or three years to corn or other crop that is not attacked by the disease. After planting, care should be exercised to avoid injury to the roots and stems by cultural operations, as the organism gains entrance into the plant through wounds.

Fungi.—Fungi have more than one cell, are much larger than bacteria and reproduce in different ways. Fungi lack green coloring matter so necessarily are dependent on other plants or animals for their food. If they obtain their food from dead plant or animal tissue, they are termed "saprophytes"; but if they obtain their food from living plants or animals, they are termed "parasites." In obtaining their food and living on and within the host they bring about the diseased condition of the plant or animal host.

APPLE SCAB (*Venturia inaequalis*, Cke.).—Apple scab, the most universal and destructive fungous disease of the apple, is most generally recognized by the characteristic blemishes that it causes on the fruit. The spots are small at first but enlarge slowly from the edge. The center becomes dark brown, and the edge black. A whitish band of loose cuticle often surrounds the black margin. Because of drying of the apple tissue in the diseased area, cracks may occur in the centers of the larger spots. Apple scab apparently is present in every country where apples are grown, but as with most fungous troubles, it is most destructive in warm humid areas. The disease occurs on many species of the apple. It attacks the leaves, flowers, fruits and to some extent the twigs. It will be observed first as darker areas on the underside of the young leaves. Diseased areas will be found on the shoots and on the stems of the young fruits. Later the characteristic diseased areas of various sizes and degrees of development will be found on the fruit. In severe cases the diseased fruit is misshapen and smaller in size than the non-diseased fruit.

In the fall the diseased leaves drop to the ground, and the overwintering stage develops within them. In the spring this phase of the life cycle is completed within the fallen leaves with the development of spores known as "ascospores." These spores mature about the time the apple flowers show pink. As the leaves become moistened by the spring rains, the spores are discharged into the air. This discharge from the various leaves may continue for a period of 8 to 30 days

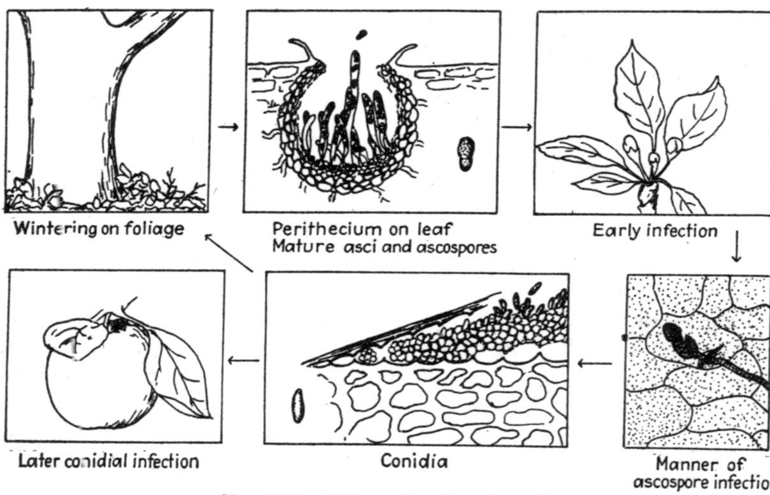

FIG. 127.—Life cycle of apple scab.

depending upon the abundance of diseased foliage and the frequency and extent of rains. The wind carries these small spores to the leaves where, in the presence of moisture, they germinate and penetrate directly. This is known as "primary infection." Growth of the fungus continues, and within a period of 8 to 17 days, depending largely upon the temperature, the conidium, a new type of spore, develops in this infected area. The conidia are discharged through the ruptured leaf covering. These spores give rise to new sources of infection on shoots, leaves and fruit known as "secondary infection." These infected areas in turn produce conidia which are spread to new areas. Under moist warm climatic conditions this production and dissemination of spores continues throughout the growing season so that there is a constant source of infection. With the approach of cool weather the formation of the conidia cease, and after a time the diseased leaves fall to the ground and the annual cycle of the disease is completed.

The only satisfactory means of controlling apple scab is by spraying or dusting with a suitable fungicide, as some form of sulfur or copper. If primary infection could be controlled completely, there would be no secondary infection; but since such control is not usually obtained, the fungicide must be applied throughout the growing season if the climatic conditions are favorable for the continued development of apple scab.

BROWN ROT [*Sclerotina fructicola* (Wint.) Rehm].—Brown rot is the most common fungous disease of the stone fruits. It is most generally recognized by its appearance on ripe or nearly ripe fruit where it produces a greyish-brown, moldy growth accompanied by the browning and decaying of the fruit underneath. It is this condition that gives the disease the common name of "brown rot." Some species of this fungus seem to be present over the entire world where drupe fruits are grown, but this particular species is prevalent in the United States and Canada. It is found especially on the peach, plum, cherry and apricot but may appear on the apple, pear and other members of the rose family. It is most prevalent and does its greatest damage in warm humid areas such as those in the Eastern and especially the Southeastern part of the United States where peaches are grown extensively. This fungus attacks the flowers, foliage, twigs and fruit but is most conspicuous and most destructive on the ripening fruit.

The blossoms are infected about the time they are opening. They turn brown, die and remain attached to the tree. Blossom blight in itself is usually not sufficiently abundant to be of primary importance, but it does provide numerous sources for later infestation. The fungus passes through the floral parts and even the fruits and attacks the twigs, causing twig blight, or canker. This, similar to the blossom blight, is not important in itself except that it furnishes additional sources for infection. Under optimum conditions for the growth of the fungus the leaves near the diseased flowers and fruits are attacked, thus providing additional sources for the production of spores that may infest the ripening fruit. This disease is essentially a fruit rot, and the most damage is done to the ripe or nearly ripe fruit. The young green fruit is fairly resistant, but the fungus can gain entrance through mechanical injuries, by direct contact with infested parts of the flower or through punctures in the skin made by insects of which the plum curculio is the chief offender. The partially ripe and ripe fruits are more readily infested than the green fruit, and the fungus grows much more rapidly in such fruits. Infection can take place through uninjured epidermis, but it occurs most commonly where the skin is bruised or broken. The tiny brown speck develops rapidly, and the disease penetrates deeply into the fruit. Under favorable conditions of high temperature and high humidity the fungus will form spores on the outside of the diseased fruit within 48 hr. after infection. These spores, like those from the flowers, leaves and cankered twigs, will be carried to other fruits where they may start new places of infection. Some of the diseased fruits may shrink into dry hard "mummies" which may remain attached to the tree until the following summer, and some may drop to the ground.

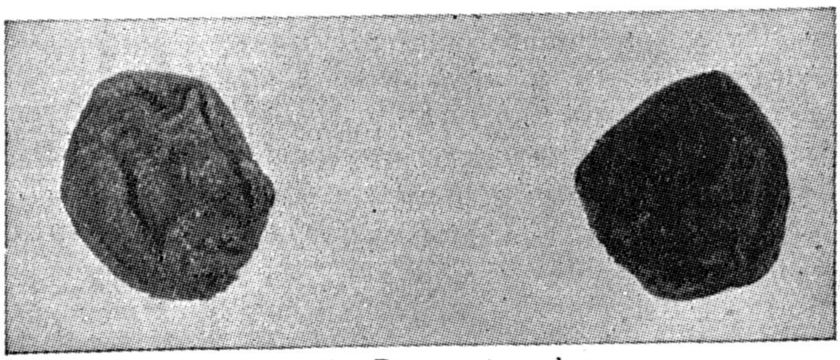

FIG. 128.—Brown rot on plum.

The fungus can winter over in three places: the persistent mummies on the tree, the mummies on the ground and, under favorable circumstances, the cankered areas on twigs. In the spring, about the time when the peach is in bloom, spore-bearing organs arise from the half-buried mummies on the ground, and ascospores are produced in large numbers during the blooming period of the peach. They are carried by the wind to the newly opened flowers. Under favorable conditions the spores will germinate in 4 to 6 hr. and produce the primary blossom infection. Primary infection of flowers or, more generally, young fruits can also take place from the spores (conidia) that either wintered over or are formed in the spring in the mummies attached to the tree. The primary infections soon give rise to millions of spores which are carried by wind, rain, insects and birds to new locations where they germinate quickly, and new sources of infestation are formed which repeat the cycle. These activities are occurring at the same time.

Although something will be gained by destroying the mummies, by picking them from the trees and ploughing under those on the ground and by pruning out the cankered areas, the only practicable means of control is by spraying or dusting with a suitable fungicide. Since the fruit becomes more susceptible as it becomes more mature, it should be thoroughly covered with a suitable sulfur fungicide while it is maturing.

Black Rot of Grape [*Guignardia bidwellii* (Ellis), Viala and Ravaz].—Black rot is the most serious fungous disease of the grape. Its most readily recognized symptom is the hard, shriveled and wrinkled berries ("mummies") many of which remain attached to the vine even during winter. This fungus was apparently indigenous to North America and lived on the native species of the grape. It was distributed from North America to Europe, and today probably no grape-growing region in the world is free from it excepting California where the climatic conditions appear to be unsatisfactory for its perpetuation. It attacks all members of the grape family. The European species of the grape are much less resistant than those which were native to North America. All green parts of the grape vine are attacked.

The disease appears on the leaves in the latter part of the growing season as small brown to black spots showing concentric rings of darker color. Spore-bearing bodies arise in these dark areas on the upper side of the leaf. The lesions on the shoot, tendrils and stems of the fruit cluster are not so noticeable. On the berries the disease first appears as a light circular spot with one or more encircling dark bands. Black spore-bearing specks appear in the lighter center, and the berry begins to shrivel. Shriveling and drying continue for about 10 days, when the berry has the characteristic mummy appearance.

The fungus winters over in mummied berries and in lesions on the canes, leaf petioles and fruit clusters. Absorption of water in the spring results in the discharge of the spores which, under moist conditions, germinate and penetrate any green part of the plant. Older leaves seem to be immune, but the berry is especially susceptible just as it loses the corolla cap and remains susceptible during the entire season. Within a period of 10 to 20 days these new lesions give rise to spores which in turn disseminate the disease to additional sources of infection. This discharge of spores continues until sometime in August when the fungus prepares for winter conditions.

The removal and destruction of fallen foliage, vine prunings and mummied berries and provisions for air drainage to facilitate the rapid drying of the vines after rains will all assist in controlling the fungus, but the disease can be controlled by the timely applications of a copper fungicide, as Bordeaux mixture. Since spores are discharged only after rains and germinate only in moisture, the disease will be more destructive and consequently will require more attention in seasons of high humidity.

Apple Blotch (*Phyllosticta solitaria* E. and E.).—This fungus attacks the fruits, leaves and twigs of its host the apple and is readily recognized by the stellar, or star-shaped, appearance of the fungus on the apple fruit. Raised, black spore-bearing bodies radiate in lines from a common center, giving the appearance of a bird's foot. This fungus apparently is indigenous to the United States and lives on the native susceptible crab apples. It is now common in the apple-growing sections of the United States east of the Rocky Mountains chiefly south of 40° north latitude. Apple blotch is restricted to certain species of the apple, and some of the commercial varieties of the apple are much more susceptible to the disease than others.

The fungus passes the winter in a comparatively inactive state in the cankered areas of the twigs. Growth is resumed in the early spring, and spores are produced and matured about the time the petals fall from the apple blossoms. In case of high humidity these spores are exuded from the diseased areas and are washed by rains to the foliage, fruit and shoots where they germinate, grow and produce the primary infection. These new lesions develop for a period of about three weeks when they give rise to spores which in turn cause the secondary infection that continues the life cycle of the fungus. The lesions on the fruit and foliage as well as those on the old cankered twigs will furnish spores for new infections until about August, at which time the foliage and fruit appear to develop a condition of immunity. The new infections on the shoots that give rise to the cankers are not apparent until about August. This infection appears first as small purplish cankers and is caused chiefly by the growth of the fungus through the leaf petiole into the twig, although infection does take place on the shoots directly through the bark. The fungus grows rapidly in these cankered areas until the coming of cool weather in the fall when it enters a resting stage. The cankered areas on twigs give rise to viable spores the following year and may continue as sources of infection for five or six years.

Since some varieties of apples are much more susceptible to injury from apple blotch than others, the more immune varieties can be selected. A certain amount of control can be obtained by pruning out the cankers at the regular pruning, but it is impracticable to attempt to control the disease in this fashion, because the cankers are too numerous. It can be controlled satisfactorily by applying a suitable fungicide at the proper times. The most important single application would be made about 10 days after petal fall, for it is at that time that primary infection is taking place from the holdover cankers.

ANTHRACNOSE (*Plectodiscella veneta*, Burk).—Anthracnose is probably the most destructive fungous disease of the black raspberry. It is most readily recognized by the small, eliptical, light-colored lesions with a dark margin that are found mostly near the base of the canes. The original home of the disease is undetermined, as it seems to be present in Europe, Australia and North America wherever its host plants grow. It attacks various species of Rubus, as the raspberries, blackberries and dewberries. It is most destructive on the black raspberry. The purple cane types are also susceptible, but the red raspberry is fairly resistant. The fungus attacks the canes, leaves, fruits and petioles, peduncles and pedicels.

The fungus winters over in the lesions on the canes. Early the next spring its growth is continued in these lesions, and spores (ascospores and conidia) are produced. About the time the new shoots are 6 to 8 in. high these spores are carried by the air currents or water to the new growth where they germinate, penetrate the host plant and produce new lesions. In a short time spores (conidia) are produced from these new lesions and are carried by water to new areas of the plant where they, in turn, attack the young tissues. This continues during spring and summer.

The disease first appears in the spring as small, purplish, slightly raised spots a short distance behind the tips of the new shoots when they are about 6 in. high. As the shoot grows and the disease develops, the spots enlarge, becoming sunken and light colored in the center with a raised dark margin about the edge. The elliptical lesions have their long axes up and down the shoot; but after a time several lesions may coalesce, forming irregular lesions which may encircle the shoot. As the shoot matures into a cane, longitudinal cracks may appear in the diseased areas. About the same time or a little after the lesions appear on the shoots, small purplish spots with light-colored centers will be found scattered over the upper surface of the leaf. Later in the season some of these diseased spots may drop out, giving a "shot-hole" effect. The disease may attack individual drupelets or cause the entire fruit to become brown and woody. The lesions on the petioles, peduncles and the pedicels are similar to but smaller than those on the shoots. These diseased conditions on the leaves and shoots weaken the plant, thus decreasing the yield. Disease of the peduncles and pedicels prevents the proper development of the fruit and may cause it to dry up before ripening.

Although the removal of all portions of the old cane at the time of setting a new planting, the selection of resistant varieties and good cultural practices to promote satisfactory growth and the elimination of weeds will assist in keeping the disease in check, this fungous disease can be controlled satisfactorily by spraying with Bordeaux mixture or lime-sulfur. Two applications of lime-sulfur will usually prove adequate. The first one is a 1–10 concentration given just as a few of the leaves unfold in the spring, and the second a 1–40 concentration given about one week before blossoming.

Apple Rust (*Gymnosporangium juniperi-virginianae*, Schw.).—Under favorable climatic conditions apple rust proves to be a destructive fungous disease on susceptible varieties of the cultivated apple. The disease is interesting in having alternate hosts. It spends one phase of its life cycle on the red cedar and its closely related cedars and the other stage on the various species of the apple. It is readily recognized on the apple leaf by the bright orange spots on the upper surface and the yellowish-brown, elongated, curved appendages arising from this diseased area on the undersurface of the leaf. On the red cedar and its close relatives, the disease is recognized by the reddish-brown galls ("cedar apples"), which in the spring exude bright orange-colored gelatinous masses.

This disease is common throughout the central part of the United States and throughout the eastern United States and Canada. It is present wherever the apple and red cedar are grown in proximity to one another. It is most destructive on the commercial apple, but some varieties are much more susceptible than others. Cedar rust causes injuries to apples, which necessitates their being discarded at time of packing; it causes a decrease in size of fruits and, by its injury to the foliage, weakens the vigor of the tree.

The fungus winters over in the reddish-brown, corky cedar apples on the red cedars. During the warm spring rains numerous orange-colored, gelatinous, finger-like projections develop from these cedar apples. Upon drying, the spores produced in this gelatinous mass are carried by the wind to the foliage and fruit of the apple. Those which light on young fruits and young leaves germinate under favorable conditions of moisture and temperature and produce the conspicuous, somewhat circular, yellow- to orange-colored spots on the upper side of the apple leaf which may attain a diameter of a ½ in. Late in the summer elongated, slightly curved, yellowish-brown appendages arise from these diseased areas on the underside of the leaf. On the apple fruit the disease produces yellowish-orange, more or less circular spots which vary in size from about ¼ in. to the entire side of the fruit. Small, circular, dark, raised areas appear in the diseased area. The infection from the red cedars may extend over a period of about six weeks. About two months after infection has taken place on the apple, spores are produced which during the remainder of the season are carried by the wind to the red cedar where they cause reinfection on the young growth. No noticeable indication of infection on the cedar is apparent for nearly a year, and it requires two years for the full development of the gall.

The disease can be controlled by planting resistant varieties of apples and destroying all red cedars and their close relatives within a radius of ½ mile of the apples. It can be controlled by the use of sulfur or copper sprays similar to those used for the control of apple scab.

Virus Diseases of Horticultural Plants.—A group of characteristic diseases of horticultural plants is caused by substances known as "viruses." The diseased conditions were recognized for a long time before the cause was known. They were believed to be physiological because something was obviously interfering with the normal functioning of the plant, but no causal organism could be detected. Even yet very little is known of the exact nature of a virus except that it can be carried in various ways from a diseased to a healthy plant and that the healthy plant will soon exhibit the characteristic symptoms of the diseased plant. Insects are common carriers of the disease from one plant to another, but in many cases the carrying agent is as yet undetermined. All we can say definitely is that a virus is a substance that, under suitable conditions, causes characteristic diseased symptoms in a plant. The general control measures for this group of diseases are prevention of mechanical transmission, spraying to control insect carriers, elimination of weed hosts and diseased plants, rotation of crops and attempts to develop resistant varieties.

Peach Yellows.—Peach yellows is a very destructive virus disease of the peach. It not only reduces the yield on diseased trees but necessitates their destruction to prevent spreading of the disease. The symptoms are readily recognized. Bunches of fine wiry shoots develop from the upper surfaces of branches. Diseased branches start growth earlier in the spring than do normal branches. The leaves, though turgid, will droop, and the edges and tip will roll slightly inward. The fruit produced on diseased branches will mature a few days early and will be spotted and mottled with crimson and have more color than normal about the pit. Peach yellows seems to be confined to the peach-growing regions east of the Rocky Mountains in North America.

The only satisfactory means of control is the prompt removal and destruction of the diseased tree when it shows its first symptoms.

Mosaic Diseases.—When attacked by a virus a large number of plants show a characteristic mottling of the foliage which has been given the general name of "mosaic disease" with the name of the plant preceding, as tobacco mosaic, potato mosaic, raspberry mosaic. The particular virus causing the disease appears to be quite specific for the particular kind of plant, and a number of different viruses causing different types of mosaic may attack the same kind of plant. A lengthy discussion of these diseases would be inadvisable, but the raspberry mosaic will be used as representative.

There appear to be several different viruses, each of which produces a mosaic condition in the raspberry plant. Mosaic diseases as a group are the most destructive of all raspberry troubles. The prevalence of this disease has caused the abandonment of commercial raspberry

FIG. 129.—Mosaic on leaf of Latham raspberry. (*Michigan Agricultural Experiment Station.*)

culture in many areas. Raspberry mosaic, in some form, appears to be distributed universally in the raspberry-growing sections of North America. It is very destructive on red raspberries, black raspberries and purple cane raspberries but less so on dewberries and blackberries.

Symptoms range from a mild mottling on a few of the leaves to dwarfing and to very pronounced mottling in many leaves accompanied by irregular, brown, dead areas in the foliage and a production of rosettes. The mottling is a yellowing due to the absence of chlorophyll. It is probable that these different symptoms and varying degrees of the same symptom are caused by more than one virus.

The virus is transmitted by species of aphids. Satisfactory control measures are obtaining disease-free plants, selecting resistant varieties and removing and destroying plants as soon as symptoms appear.

Review Questions

1. How can one promote a favorable "natural balance" for the plants that he is attempting to grow?

2. What various parts of horticultural plants are attacked by pests?

3. Name some of the environmental conditions that are detrimental to horticultural plants.

4. Why is the loss likely to be unusually heavy shortly after a new pest has been introduced into a locality?

5. Classify the animal pests that are injurious to horticultural plants.

6. Give an example of each type of animal pest that is injurious to horticultural plants.

7. Give the life cycle of an insect that has a complete metamorphosis.

8. Classify the plant pests that are injurious to horticultural plants.

9. Give an example of each type of plant pest that is injurious to horticultural plants.

10. Give the life cycle of a fungus that causes a diseased condition on a specific horticultural plant.

11. Are viruses considered to be plants or animals?

12. What is a general symptom of plants affected with a virus?

Problems

1. A diseased horticultural plant or plant product has been brought to you for diagnosis. After careful examination, you stated that the trouble was ——. State the specific factors that convinced you of the identity of the trouble.

2. You have a commercial apple orchard in a region where the codling moth is a serious problem. State the procedure that you would follow in determining the times at which you would apply sprays to combat the pest.

3. You have been spraying your apple trees each year for the control of apple scab. You have sprayed on the same dates during the past two years and have used the same crew machinery and materials. You had excellent control one year but very poor control the next year. Explain.

Suggested Collateral Readings

1. Melhus, I. E., and G. C. Kent, "Elements of Plant Pathology," The Macmillan Company, New York, 1939.

2. Metcalf, C. L., and W. P. Flint, "Destructive and Useful Insects," McGraw-Hill Book Company, Inc., New York, 1928.

3. Pirone, P. P., Diseases of Ornamental Plants, *N. J. Agr. Exp. Sta. Cir.*, 385: 1–80, 1939.

4. Thompson, H. C., "Vegetable Crops," pp. 144–157, McGraw-Hill Book Company, Inc., New York, 1939.

5. White, E. A., "The Florist Business," pp. 202–219, 220–235, The Macmillan Company, New York, 1933.

CHAPTER XVI

CONTROLLING PESTS OF HORTICULTURAL PLANTS

Ever since man first selected certain plants that he wished to preserve, he has attempted to protect them from their various pests. His first conscious effort was probably in selecting the plants that seemed, under natural conditions, to withstand best the attacks of the pests. His next step was the use of purely physical means of protection by the manual destruction or exclusion of the pest. Finally as his knowledge of the pest and of chemicals increased, the latter were used to protect the plants. The means now employed to protect horticultural plants may be considered conveniently under the two methods of natural means of control and artificial means of control.

NATURAL MEANS OF CONTROLLING HORTICULTURAL PESTS

Natural means of control of horticultural pests are those factors which are operative without the assistance of man. These would include the inherent resistance, or immunity, of the plant; environmental conditions, as unfavorable temperature, moisture and light; and the natural plant and animal enemies of the horticultural pest. The last are often spoken of as "biological control."

Resistance and Immunity

It is a commonly known fact that among different varieties of the same kind of plants various degrees of resistance, even to the state of immunity, will be exhibited toward adverse environmental conditions and to plant and animal pests. Different degrees of resistance will be exhibited by the same kind of plant at different ages, by the same plant at different seasons of the year and by different parts of the same plant in different stages of growth or development of the plant or plant part. The exact cause or causes of resistance and immunity are not known. Certainly some of them are hereditary. They may be either physical or chemical or both physical and chemical. The thickness of the cuticle and cell walls, the size and abundance of stomates and lenticles, the presence and abundance of pubescense and other physical structures of the plant and many other factors may contribute to resistance, or immunity, to attacks by plant and animal pests. The chemical composition of various parts of the plant may serve as a barrier to the germination and growth of fungi and the development of bacterial and virus disease or may be distasteful, repellent or even toxic to insect pests. Since both physical and chemical conditions are influenced by nutritional conditions of the plant, the growers can exercise an influence on the resistance and immunity by suitable cultural practices. As a general rule a vigorous plant is more resistant than a feeble one. As some of the factors favorable for resistance or immunity are hereditary, progress will be made in improving the resistance of plants to pests by breeding and selecting and by fortuitous mutations.

Environmental Conditions

The different plant and animal pests of horticultural plants have various conditions of temperature, moisture and light that are favorable for their growth and development. One can select kinds and varieties of plants and grow them in environments less favorable than others for the survival of their pests. This matter was discussed in Chap. III.

Natural Enemies of Horticultural Pests

The plant and animal pests that prey upon horticultural plants in their turn have plants and animals that prey upon them, so a natural system of checks and balances exists. Animals and insects have been introduced into new territories and have become very destructive before their natural enemies were present in sufficient quantities to keep the introduced animal or insect under satisfactory control. Fungal and bacterial diseases also have competing fungi and bacteria. The horticulturist encourages the perpetuation and multiplication of birds, insects, fungi and bacteria that aid in keeping the plant and animal pests of horticultural plants under control. A classic example of one insect used to control another is that of the ladybird which feeds on certain scale insects that attack citrus trees. The ladybirds are raised in large numbers and distributed through the citrus orchards where they destroy the scale insects. A continuous search goes on for beneficial bacteria, fungi, insects and higher animals that can be used to assist in the control of plant and animal pests of horticultural crops.

ARTIFICIAL MEANS OF CONTROLLING HORTICULTURAL PESTS

Under undisturbed natural conditions a "natural balance" will be maintained among the different plants and animals of that community. The horticulturist, however, does not grow plants under natural unrestricted conditions. He will grow large areas of a single kind of plant and often a plant that would not be produced, much less maintained and perpetuated, under natural conditions. Since the horticulturist wishes to grow specific plants for a definite purpose in large quantities under conditions that would not exist in undisturbed natural conditions, he must resort to artificial means to control the plant and animal pests harmful to the plants that he wishes to grow. The chief artificial control measures can be conveniently grouped as physical, chemical and legal.

Physical Means of Controlling Horticultural Pests

Probably the first conscious effort that man made to destroy the pests on the plants that he wished for his own use was the physical destruction of the pest causing the trouble or of the diseased plant or plant part. This method is still used extensively and in some cases is the only effective means known for certain pests. Other physical means of control are the use of mechanical guards, sanitation, cultivation of the soil, rotation of different crops on the land, pruning, purifying and grading seeds and sterilizing seeds and soil with heat.

Mechanical Guards.—Wooden, paper or wire guards are placed about tree trunks to protect them from injury by mice and rabbits. Metal bands and bands of various adhesive substances are placed around the trunks of shade trees to prevent certain types of crawling insects from climbing the trees. A number of years ago the apple orchardist put bands of cloth around the trunks of the trees, and the codling moth hibernated under these bands and were destroyed by the orchardist. Cloth bands have been superseded by corrugated paper which is impregnated with a chemical that kills the hibernating codling moths. Many orchards store their harvesting equipment in the fruit-packing house and screen all the openings. The following spring the codling moths that hibernated in the harvesting equipment are trapped by the screens and are unable to return to the orchard. Paper collars are forced a couple of inches into the ground about the stems of tomato and other young vegetable plants to protect them from cutworms. Traps are used commonly to catch moles in lawns.

Sanitation.—Inasmuch as many of the pests of horticultural plants harbor or winter over in plant refuse, as dropped leaves and fruits, weeds and other plants about the trees and in fence rows, its destruction and the elimination of weeds and other harboring places are valuable aids in keeping the pests to controllable numbers.

Soil Cultivation.—Many of the horticultural insect pests spend one phase of their life cycle in the soil. During this period they can often be destroyed by cultivating the soil, as this exposes them to their enemies and to unfavorable climatic conditions. Certain pests can be destroyed by plowing under or disking the infested fruits and foliage that have fallen to the ground.

Crop Rotation.—Certain insects and diseases live over from season to season in the soil or the diseased-plant refuse. Since many of the pests are quite specific in the plants that they attack, it is possible to destroy them by growing different plants on the area for two or three seasons. This is a well-recognized practice with the vegetable grower.

Pruning.—Some diseases and some insects are controlled entirely, or their numbers greatly reduced, by pruning out and destroying the injured part of the plant. Proper pruning will permit the better penetration of sunlight into the interior of the plant, and many of the pests, especially the fungous diseases, are destroyed by sunlight. Such pruning also permits the better movement of air through the plant which facilitates the drying of the foliage and fruit after rains and in the mornings so that the fungi do not have satisfactory conditions for germination and growth. Properly pruned plants permit better penetration and coverage of sprays and dusts that are applied to control the pests.

Purifying and Grading Seeds.—For many years seeds have been cleaned of weed seeds and graded by means of various-sized screens. More recently this has been done by a current of air which sorts the seeds according to their difference in weight or specific gravity. By such purifying one is enabled to avoid planting the weed seeds, and by grading the seeds the possibility of obtaining more vigorous seedlings is increased.

Chemical Means of Controlling Horticultural Pests

As man's knowledge of the insects and diseases that were injurious to plants increased, and as his understanding of chemistry became greater, he began to use chemicals to repel, control or destroy these pests and to protect his plants from injury. Unfortunately the chemicals are themselves injurious to the host as well as to the pests, so that we have not, at the present time, a chemical that is entirely non-injurious to the host plant and at the same time suitably toxic to the plant pest.

It has been noted previously that most of the insects that are injurious to horticultural plants have either biting or sucking mouth parts. They either bite out a portion of the plant tissue and take it into their digestive tract or insert a tube into the plant and suck out the juices from the interior of the plant. The first group of insects can be controlled by the application of a stomach poison to the plant tissues that the insect eats. The eating is done mostly during the larval stage of the life cycle. Since poison cannot be applied to the food of the second group of insects, the chemicals used for their control must come into contact with the insect and consequently are known as "contact insecticides." Some of the contact insecticides will also control biting insects and will destroy the eggs of many insects. Another group of chemical insecticides which is used for both chewing and sucking insects is known as "fumigants." This group is volatile, or changes into a gas, and kills the insect by contact. Some of the common contact insecticides really change into a gas before they are effective. The lines of demarcation for the different types of insecticides are not sharply drawn, but the grouping into stomach poisons, contact insecticides and fumigants is convenient.

The chemicals generally used to combat fungi and bacteria are a form or compound of sulfur or copper, although some miscellaneous chemicals are used successfully. Since bacteria usually live within the plant tissues, chemical means of control are usually not effective, but spraying is sometimes satisfactory, as in the case of Bordeaux mixture used for the direct control of fire blight.

There are many proprietary products on the market that are usually combinations of substances that will combat biting and sucking insects and fungous diseases. These products, when used according to directions, are satisfactory for the control of the pests for which they are designed. They are used extensively, chiefly by those who need but a small quantity of material.

Efforts are being made continuously to obtain new compounds that are toxic to the pest, are non-injurious to the host, do not injure the product for human consumption, are economical to use and can be applied satisfactorily. Some progress is being made in this search, and eventually some of the chemicals now used may be replaced by others that are more satisfactory.

Insecticides.—An insecticide may be considered as a chemical that is used to control insects. Many different chemicals have been and are being used to combat insect pests of horticultural plants. These insecticides are often classified as stomach poisons, which are used to combat the insects with biting mouth parts, and contact insecticides, which are used chiefly against insects with sucking mouth parts. Some insecticides are effective against both types of insect.

Stomach Poisons.—Many substances have been used as stomach poisons for biting insects, but only a comparatively small number has given sufficient satisfaction to be widely adopted. The most widely used chemical for biting insects is a form of arsenic. Since certain quantities of this chemical, as well as of others, are injurious to man as well as to insects, laws have been passed limiting the amounts that can be left upon products intended for human consumption. At the present time the arsenicals are the most satisfactory chemicals to use as stomach poisons for biting insect pests of horticultural plants. It would be impracticable to discuss the various arsenicals used, but brief mention will be made of a few of the most important ones. The relative efficiency of an arsenical depends primarily upon the amount of arsenic present on a given surface.

Paris Green.—Paris green, a complex compound of copper and arsenic, is one of the oldest arsenicals used to combat biting insects. It is a beautiful light-green powder composed of crystals of microscopic fineness. It has a high arsenical content and a quick killing capacity. It does not mix readily with water, settles rapidly, does not adhere well to the plant but washes off easily in rains and is likely to cause injury to the plant. It has been used extensively on vegetable crops, especially on the potato for the control of the Colorado potato beetle, but in recent years is being replaced by more satisfactory materials.

Calcium Arsenate.—Calcium arsenate is a compound similar to lead arsenate except that calcium has replaced lead. It contains a higher percentage of arsenic than lead arsenate and consequently is more toxic. It is a fine powder and is excellent to apply as a dust. It causes some injury to fruits but is used extensively, especially as a dust, on vegetable crops. The calcium arsenate comprises about 10 per cent of the dust mixture.

Lead Arsenate.—Lead arsenate is a compound of lead and arsenic. There are several forms, but the usual one, acid arsenate of lead, is the most widely used and, at present, the most satisfactory stomach poison for biting insects on horticultural plants—especially fruits. Arsenate of lead is an amorphous fluffy white powder. It is fairly toxic, moderately rapid in its killing properties, only slightly soluble in water, remains in suspension well, adheres well and may cause some injury to the host plant. It is compatible with many insecticides and fungicides. It is commonly used at the rate of 3 lb. in 100 gal. of solution with an equal quantity of hydrated lime. When used as a dust about 10 per cent of the dusting mixture is lead arsenate. Mixtures of lead arsenate with petroleum-oil emulsions and lead arsenate with nicotine sulfate have recently been applied successfully for the late codling-moth attacks.

Contact Insecticides.—Contact insecticides, or contact poisons, comprise those substances which kill the insect by coming into contact with it. Such insecticides are usually used for the control of sucking insects, but they are also effective against some biting insects and against insect eggs. The kinds of contact insecticides vary considerably, and the exact manner in which they cause the death of the insect or the destruction of the egg is unknown.

NICOTINE.—Nicotine, an extract of tobacco, is a water-soluble, volatile alkaloid which is a valuable insecticide. It is classed as a fumigant as well as a contact insecticide. It appears on the market under several trade names. Nicotine sulfate is a brownish-black, disagreeable-smelling liquid which contains various percentages of actual nicotine. At a concentration of 1 pt. per 100 gal. (1 to 800) of a 40 per cent solution it is a standard spray for aphids and other soft-bodied sucking insects. It is a valuable ovicide, as it destroys the eggs of certain insects. Recently it has been used satisfactorily with oil as a stomach poison for the late-brood codling moth. Nicotine dusts are prepared by impregnating fine powders with nicotine or nicotine sulfate. Paper containing nicotine is prepared in a similar fashion.

PYRETHRUM.—Pyrethrum is a contact insecticide made from the pulverized flower heads of certain species of pyrethrum. It is noninjurious to man and kills insects by contact only. It is used to a certain extent by greenhouse operators and is an important constituent of household insecticides and fly sprays.

ROTENONE.—Rotenone is a colorless crystalline substance obtained from certain plants, chiefly tropical, which functions as a contact insecticide and as a stomach poison for leaf-eating insects and destroys certain mites and ticks on plants and animals. Its specific mode of action is as yet undetermined. The plants (*Derris spp.* and *Lonchocarpus spp.*) from which rotenone is obtained are indigenous to such regions as the Malay Archipelago and Latin America where they have long been used by the natives to kill fish. Crushed roots placed in the water kill the fish which then float to the surface where they can be collected. Rotenone is, at present, generally applied to the plant as the finely ground root of the plant mixed with other powder as a carrier. It is the active constituent of many proprietary insecticides.

OILS.—The use of oils as contact insecticides has increased greatly in recent years. Two general classes of oil spray are used for spraying. An oil emulsion is a mixture of oil, water and an emulsifying agent that has been treated mechanically so that the oil globules have been broken up into very fine globules and remain suspended in the water. Oil emulsions can be purchased, but large users often make the emulsion as needed, as emulsions are not very stable. A miscible oil is one in which the emulsifier is dissolved in the oil. It will mix readily with water, forming a milky white spray. Miscible oils are commercial preparations and are quite stable. The dilution adopted will be influenced by the oil used, the pest to be controlled and the season of the year. In preparing and diluting the oils one should follow carefully the instructions relative to the particular oil or product being used as the different oils vary considerably.

SOAPS.—Soaps possess contact insecticidal properties. Many of the sucking insects on house plants can be controlled by washing the plants with soapy water.

SULFUR.—Sulfur, in the form of elemental sulfur and as lime-sulfur, is used effectively as a contact insecticide. In this case the red spider mites are considered as an insect. Liquid lime-sulfur at the rate of 1 gal. to 9 gal. of water is used late in the dormant season for the control of San José scale. Oil sprays are replacing the lime-sulfur as a contact insecticide.

Fumigants.—Fumigation is merely the application of the toxic chemical in a volatile or gaseous form. It is probable that some of the insecticides applied as liquids or solids really are volatilized before they become effective and therefore might justly be considered as fumigants. Most of the fumigants are applied as liquids or solids and change to the gaseous form. Fumigation is confined to greenhouses, to plants that can be inclosed temporarily, to the treatment of seeds and to the treatment of soil where the too rapid escape of the gas can be prevented. Only a few of the fumigants will be considered.

HYDROCYANIC ACID.—Hydrocyanic acid is used for the control of insects in greenhouses and scale insects on citrus trees. In the latter case each tree is enclosed in a cloth tent while being treated. The required dosage is determined according to the insects to be controlled and the volume of the greenhouse or tent to be treated. The hydrocyanic acid is then liberated from potassium cyanide or sodium cyanide by treating with specific amounts of sulfuric acid and water. Hydrocyanic acid is also liberated by the action of moisture on calcium cyanide and bicarbonate. The calculated amount of one or the other of these powders is sprinkled on the wet paths in the greenhouse. The anhydrous liquid gas is used for the treatment of citrus trees. This is a dangerous substance, and one should know how to use it with the necessary precautions.

CARBON DISULFIDE.—Carbon disulfide is a heavy colorless liquid which changes to a gas upon exposure to the air. Since the gas is heavier than air, this substance is used to destroy insects in the soil. It is used commonly to destroy ants in the lawn. For the best use of carbon disulfide it should be applied when the soil is fairly dry, and the treated area covered with soil, straw or other suitable material for several hours after treating. Borers have been destroyed in trees by injecting carbon disulfide into their tunnels by means of an oilcan and then plugging the opening.

PARADICHLOROBENZENE (P.D.B.).—Paradichlorobenzene is a white crystalline chemical which has proved very effective in the control of the peach borer in trees three years old and older. The crystals volatilize, and the gas penetrates through the soil and into the tree where it kills the insect. This chemical is also being used in the form of a paint to destroy borers in shade trees.

Fungicides.—A fungicide may be considered as a chemical that is used for the control of a fungus. Some of these substances are proving of value also in controlling certain bacteria. Many different chemicals have been tried, but sulfur, copper and mercury have proved the most satisfactory. They are used in the dry form as dusts and in the liquid form as sprays or paints.

Sulfur.—Sulfur, either as elemental sulfur or combined with another substance forming a chemical compound, has been used for a long time to control both insects and diseases of plants.

LIME-SULFUR.—Lime-sulfur is a chemical compound formed when elemental sulfur is boiled in water with lime for a suitable time. It is an amber-colored, foul-smelling liquid slightly heavier than water and is used at various strengths and seasons of the year to destroy sucking insects and fungi on horticultural plants. Until a few years ago it was always used in the liquid form, but a process of manufacture has been perfected by which it is possible to evaporate the water from liquid lime-sulfur and obtain a dry, powdered lime-sulfur. In preparing for spraying, 1 lb. of dry lime-sulfur is equivalent, in fungicidal properties, to 1 qt. of the concentrated liquid lime-sulfur. For the control of scale insects this liquid form is used at a concentration of 1 gal. of the concentrated liquid to 9 gal. of water. For later applications the concentration of the lime-sulfur is decreased depending upon the plant to be sprayed, the pest to be combated and the climatic conditions. It is most injurious during hot dry weather. Even with these precautions the plant is injured by lime-sulfur, and elemental sulfur in various forms is gaining in use for the control of fungi, especially for the cover-spray applications.

DRY-MIX SULFUR-LIME.—Sulfur does not mix or go into suspension readily in water. Lime-sulfur is injurious to the host plant; and self-boiled lime and sulfur, which is really a physical mixture of lime with sulfur, is quite variable and troublesome to prepare. A method was devised to mix finely pulverized sulfur and hydrated lime by the addition of a wetting agent. This is commonly prepared by mixing thoroughly 8 lb. of sulfur, 4 lb. of hydrated lime and ½ lb. of calcium caseinate or other commercial wetting agent. This 12½ lb. of powder mixture is then added to 50 gal. of water to make the spray.

ELEMENTAL SULFUR.—Elemental, or uncombined, sulfur is pure sulfur which in its unchanged physical form is dusted or sprayed on the plant. It is not so effective a fungicide as lime-sulfur but is practically non-injurious to the host plant and gives satisfactory commercial control in many cases.

DUSTING SULFUR.—Dusting sulfur is a sulfur so finely ground that it will pass through a 300-mesh screen. It is used extensively on peaches, plums and sweet cherries and to a much more limited extent on apples and pears. The common dust mixture is 80-10-10 sulfur, lime and arsenate of lead; or if either the lime or the arsenate of lead is omitted, it becomes a 90-10 dust mixture with 90 per cent sulfur and 10 per cent either lime or arsenate of lead.

WETTABLE SULFUR.—A number of proprietary wettable sulfurs are on the market. They are more convenient than dry-mix sulfur-lime and are replacing it in many places. They are also being used instead of lime-sulfur for some of the later cover sprays of apples and pears. They are made of finely ground sulfur to which a wetting agent has been added so that the sulfur will mix readily with water. Some of these sulfurs are so finely ground that they are practically colloidal when added to liquids. They are used at the concentration recommended by the manufacturer.

Copper.—Copper in some form is the basis of a large and important group of fungicides. It has a stronger fungicidal action than has sulfur. It is quite injurious to certain kinds of plants and more injurious to special varieties. It is most injurious during cool damp weather.

BORDEAUX MIXTURE.—Bordeaux mixture, or a modification thereof, is the most important fungicide containing copper. It is prepared from copper sulfate and lime. The first mixtures were quite concentrated, containing a large amount of both copper sulfate and lime. More recent investigations have shown that weaker concentrations are just as effective in controlling disease and less injurious to the host. A formula now commonly used is 1 lb. of copper sulfate and 2 lb. of fresh hydrated lime in 50 gal. of water. Recently insoluble or "fixed copper" sprays are being used successfully on plants susceptible to injury by Bordeaux mixture.

COPPER DUSTS.—Many seeds are treated with a form of copper dusts for seed-borne diseases or for diseases that harbor over in the soil. Copper carbonates and copper oxides are used for this purpose. The seed is shaken up with the required amount of the chemical which adheres to the seed coat in sufficient quantities to furnish protection until the seedling no longer needs protection.

Mercury.—Mercury forms the basis of many of the preparations used to disinfect seeds. It is also used to sterilize soil and is added to some of the paints and preparations used as wound dressings. It is a powerful poison.

Legal Means of Controlling Horticultural Pests

We have noted that the pests that prey upon horticultural plants are kept in control to some extent by their natural enemies and also by physical and chemical means employed by man. This group of protective measures would not be satisfactory or effective without legal control.

Laws Governing the Inspection and Transportation of Horticultural Plants.—In order to prevent or at least to delay the introduction of a new pest into a non-infested area all imported plants and plant parts are examined carefully, and all possible precautions taken to prevent the entrance of such pests. Even a casual study of the loss due to the gypsy moth, San José scale, Japanese beetle and citrus canker will convince the most skeptical of the value of such regulations. All nursery stock intended for interstate shipment is inspected for the presence of pests. Many states and certain sections have special restrictions. It has been found advisable in some instances to place a quarantine on horticultural plants or plant parts produced in certain countries or certain parts of the United States.

Laws Governing the Manufacture and Sale of Chemicals.—Laws governing the manufacture and sale of chemicals applied to horticultural plants and plant products are necessary to protect both the producer and the consumer.

THE MECHANICS OF CONTROLLING HORTICULTURAL PESTS

Under this heading we shall discuss the equipment used, the preparation of the materials and the application of the materials.

Equipment Used in Controlling Hortlcultural Pests

Very great improvement has been made in the past twenty years in the equipment used for controlling horticultural pests. At the present time a great diversity exists in type and size of equipment.

Spraying Equipment.—The spraying equipment ranges from a whisk broom and a small pail to the stationary spray outfit that provides for 100 acres or more of tree fruits. It will vary according to the amount and kind of plants to be sprayed. For a few vegetables and low ornamental plants the whisk broom and the pail may be entirely satisfactory. A more efficient application could be made by one of

COMMON HORTICULTURAL PESTS AND THEIR CONTROL

Group of pest	Class of pest	Class control	Example of pest	Specific control for example cited
Insects	Biting or chewing	Stomach poison	Codling moth	(1) (2) Arsenate of lead, 1½ lb. 50-gal. spray
			Cankerworm	Arsenate of lead, 1½ lb. 50-gal. spray
			Curculio	Arsenate of lead, 1½ lb. 50-gal. spray
			Potato bug	Arsenate of lead, 1½ lb. 50-gal. spray
			Cabbage worm	Arsenate of lead, 1½ lb. 50-gal spray
			Rose slug	(3) Arsenate of lead, 1½ lb. 50-gal. spray
			Strawberry leaf roller	Arsenate of lead, 1½ lb. 50-gal. spray
	Sucking or piercing	Contact insecticides	Aphis	Nicotine sulfate ½ pt. 50-gal. spray
			San José scale	(4) Lime-sulfur, 1 gal. in 8-gal. spray
			Oyster-shell scale	(4) Lime-sulfur, 1 gal. in 8-gal. spray
	Boring	Mechanical means	Peach borer	Paradichlorobenzene
			Flat-headed apple-tree borer	Dig out or inject carbon disulfide and plug hole. Keep plant vigorous
			Elm borer	Same as above
Diseases	Fungous	Fungicides	Apple scab	(5) Lime sulfur, 5 qt. 50-gal. spray
			Apple blotch	(6) Bordeaux mixture, 2-4-50
			Black rot (grape)	Bordeaux mixture, 2-4-50
			Brown rot (peach)	(7) Wettable sulfur
			Mildew (rose)	Wettable sulfur or sulfur dust
			Anthracnose	Lime-sulfur
	Bacterial	Mechanical means	Fire blight	(8) Cut off and burn diseased parts
			Crown gall	Destroy diseased plants in severe cases
Virus	Virus	Mechanical means	Mosaics Yellows	Destroy diseased plants; control aphis; use resistant varieties
Rodents	Rodents	Mechanical means	Rabbits	Wire guards
		Stomach poisons	Mice	Poisoned bran

(1) Other arsenicals may be used.
(2) Nicotine and oils are being used in commercial orchards.
(3) Many proprietary products on the market.
(4) Oil sprays at makers' recommendations.
(5) Dry lime-sulfur and commercial sulfurs and Bordeaux mixture can be substituted. Follow recommendations of manufacturer for preparation of spray.
(6) Use same materials as for apple scab.
(7) Use according to manufacturer's recommendations.
(8) Use Bordeaux mixture for blossom stage.

the pint or quart hand sprayers. For the same, or a little larger, area one might select a small tank or bucket sprayer which is quite satisfactory for all low-growing plants and even a few trees. One progresses through the wheelbarrow sprayer, the 50-gal. barrel sprayer, the small power sprayer, the large power sprayer, finally to the large stationary spray outfit with pipes that convey the material to all parts of the orchard.

Dusting Equipment.—The dusting equipment varies as greatly as the spraying equipment, ranging from a folded piece of cheesecloth to

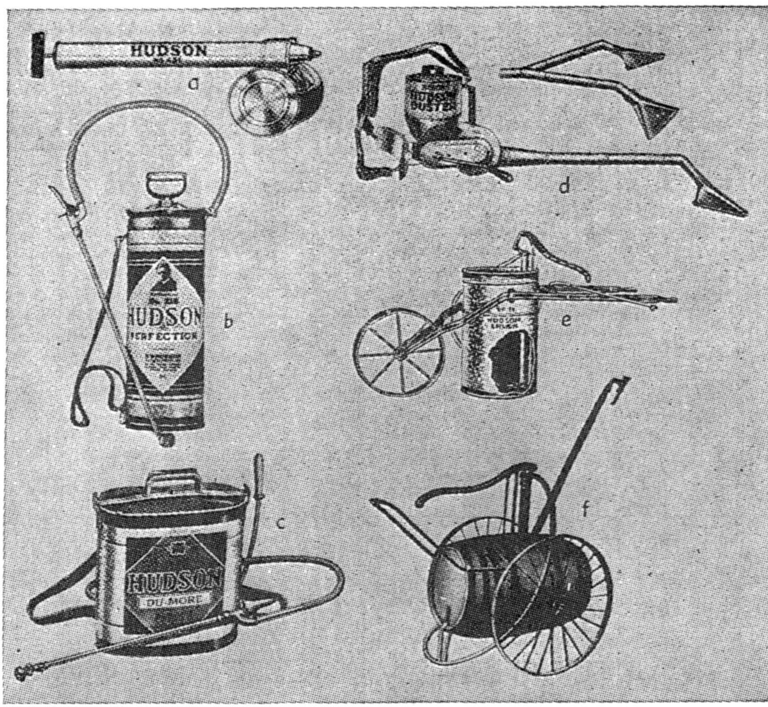

FIG. 130.—Representative pest control equipment for small plantings: (*a*) hand sprayer, (*b*) compressed-air sprayer, (*c*) knapsack sprayer, (*d*) hand duster, (*e*) wheelbarrow sprayer, (*f*) barrel sprayer. (*Courtesy of Hudson and Bean Co.*)

the airplane. The folded cheesecloth is adequate for dusting a few vegetables and flowers, the hand duster can be used satisfactorily for rather large areas of vegetable crops, the power duster is used for orchard crops and large acreages of truck crops and the airplane is used for dusting in some of the specialized potato- and other truck-crop sections.

Sterilization.—For soil sterilization in nursery seed beds and greenhouse benches, chemicals are applied either as solids or as liquids. The universal method of sterilizing soil is by the use of steam. Some greenhouse benches are equipped so that the soil in them can be treated with steam. In other cases and in nursery seed beds steam is liberated from pipes in the soil that is first covered with specially constructed galvanized pans. Seed flats are often prepared for planting and placed in autoclaves for sterilization. One of the types of tear gas is being used for sterilizing soil.

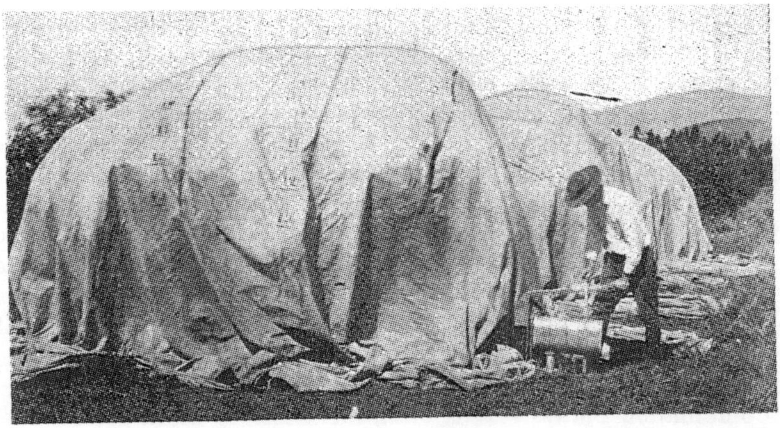

FIG. 131.—Fumigating citrus trees in California. (*Courtesy of Owl Fumigating Corp.*)

Fumigation.—Fumigation varies from burning a sulfur candle, placing cyanide in diluted sulfuric acid and sprinkling calcium cyanide on moist areas or discharging it into a humid atmosphere to liberating liquid hydrocyanic acid gas in confined areas.

PREPARATION OF MATERIALS USED IN CONTROLLING HORTICULTURAL PESTS

With the improvements in materials and equipment the manner of making or preparing the chemicals for application has been greatly simplified. For the proprietary compounds one should follow the directions given for the particular product. Sprays and dusts are prepared for the control of biting insects, sucking insects and fungous diseases. It is possible and commonly advisable to combine two or even three materials and apply as one. All sprays should be strained into the sprayer through a fine screen provided for that purpose.

Arsenate of Lead.—Lead arsenate is insoluble in water but mixes readily with it, forming a fine suspension. Unless thorough agitation is possible, as in a power sprayer, it is advisable to add the required amount of powdered arsenate of lead to a small quantity of water while stirring rapidly. This is then added to the required amount of water or spray solution for the total dilution. In case lime is not present in the spray, it will be advisable to add a quantity of hydrated lime equal to the quantity of lead arsenate.

Bordeaux Mixture.—Bordeaux mixture is made of copper sulfate, hydrated lime and water. Before the introduction of the quickly soluble pulverized copper sulfate it was customary to make a stock solution of copper sulfate by dissolving it in water in a wooden vessel. One gallon of the stock solution contained 1 lb. of dissolved copper sulfate. A similar stock mixture was made using 1 lb. of hydrated lime per gallon of water. The sprayer was nearly filled with water; and with the agitator operating, the required amount of lime mixture was put through the strainer; then the required amount of the copper sulfate stock solution was added. Sufficient water was then added to bring the spray to the required volume. At present the quickly soluble form of copper sulfate is dissolved in water in the sprayer or in a wooden container. This solution is diluted with water to near the required volume when the hydrated lime that has been made into a "milk" by being added to water is strained into the sprayer while the agitator is running. Water is then added to bring the spray to the required amount.

Combination Sprays.—It is often advisable to use a combination spray composed of lime-sulfur and arsenate of lead. With this combination it is advisable to add a quantity of fresh hydrated lime equal to the amount of arsenate of lead used in the spray. For this combination it would be well to add the various ingredients to separate pails of water to get the chemicals well dissolved or in suitable suspension before combining them in the spray tank. In case dry lime-sulfur is being used, it should be dissolved in a pail of water. The hydrated lime should be added to water in another pail; and the arsenate of lead, to water in a third pail. After the spray tank is half or more full of water, start the agitator, and add the lime-sulfur through the strainer. Then add the water mixture of hydrated lime, then the arsenate of lead, and then fill the tank to the required amount. Since chemical action takes place between the lime-sulfur and the arsenate of lead, this combination spray should be used at once, as it deteriorates and will loose a considerable part of its value after a couple of hours.

If desirable, nicotine sulfate may be added to the foregoing combination spray.

Dusts.—Dusting mixtures are purchased already prepared or are made by mixing thoroughly the proper amount of the different ingredients.

Applying the Materials Used in Controlling Horticultural Pests

Other than using a suitable substance for the particular pest or pests to be controlled, the effectiveness of the chemical used will be determined by the time of application, the amount that is applied and the manner in which it is applied.

Time of Spraying.—The chemicals applied to horticultural plants for the control of pests are chiefly to prevent the damage rather than to cure the injury. They are applied as an insurance against possible loss. Consequently, the time of application is determined by the vulnerable stage in the life cycle of the pest and the susceptibility of the host to injury from the treatment. In certain stages of its life cycle the pest is practically immune to control measures; in others the chemicals used must be applied during the dormant season when the host plant will be least injured by them. Spray calendars therefore will vary according to the pests to be controlled and the locality in which they are intended to be used. Even in the same locality they will vary from season to season because of climatic differences. Spray calendars, therefore, are to be considered as valuable suggestive guides and not as sets of rules. Since the times designated for the early applications are given according to the appearance or state of growth of the host plant, they are more accurate than the times indicated for the later sprays. This is true because various pests respond to environmental conditions and have generally reached a certain stage in their life cycle when the host plant in that locality has reached a certain stage in its yearly growth. The host plant is used as an index of the stage of growth or development of the pest. Subsequent growth of the host and pest may vary considerably in the same environment.

Manner of Spraying.—Any method of applying the chemical that will give a thorough, even coverage with the minimum amount of material will be the most satisfactory, as it will control the pests, result in the minimum amount of injury to the host and leave the least amount of spray residue. The large power-spray outfits that use pressures ranging from 400 to 600 lb. produce a fine mist spray which penetrates to all parts of the plant and covers all exposed areas with a thin film. The high-power dusters also envelope the plant in a cloud of dust which settles on all exposed parts.

There must be enough material on the plant at all times to prove toxic to the pest. This is not an easy condition to obtain. The plant parts are increasing in size so rapidly that new unprotected areas are being exposed constantly; the materials are being washed off by rains and are undergoing chemical changes that make them less toxic to the pest. Consequently the frequency of the applications may vary considerably. Furthermore, the amount required to cover the plant effectively will vary with the efficiency of the machine. One will notice, therefore, that the amount of spray material used to cover the same sized tree once in different parts of the United States will vary from 5 to 30 gal. or more.

FIG. 132.—Large sprayer with 600-gallon tank. (*Courtesy of Friend Company.*)

Amount of Spray.—The necessity of spraying frequently and in considerable quantities even after the fruits have attained a good size led to the passage of a law that requires that spray residues above certain specified amounts must be removed from fruits intended for human consumption that enter into interstate commerce. If the amount of residue is but slightly over the specified amount, the fruit may be brought within the tolerance by brushing, but brushing has not proved to be so satisfactory or reliable as washing.

In the eastern part of the United States where the applications of arsenate of lead are less frequent, where the quantity of material is less per tree, where summer oils are not used so frequently and where the natural rainfall washes off some of the spray residue, the apples can be brought within the legal requirements by washing for 1½ min.

Apple-spray Schedule

Name of spray	Pests to control	Materials to use	Remarks
Dormant	Scale insects	Liquid lime-sulfur, 1 gal. in 8 Oil emulsion—3 per cent oil Miscible oil—manufacturers' specifications	Apply late in dormant season, but complete before buds show green tips
Prepink	Apple scab Cankerworm Curculio	Liquid lime-sulfur, 2 gal. or Dry lime-sulfur, 8 lb. and Lead arsenate, 3 lb. Water to make 100 gal.	If aphis is still present, add 1 pt. nicotine sulfate
Pink	Apple scab Curculio	Liquid lime-sulfur, 2 gal. or Dry-lime-sulfur, 8 lb. and Lead arsenate, 3 lb. Water to make 100 gal.	Begin spraying as flower buds separate and show petal color, and complete spraying before blossoms open. To decrease lime-sulfur injury add 8 lb. hydrated lime
Calyx-cup or petal fall	Apple scab Curculio Codling moth	Liquid lime-sulfur, 5 qt. or Dry lime-sulfur, 5 lb. and Lead arsenate, 3 lb. Hydrated lime, 5 lb. Water to make 100 gal.	Begin spraying when most of the petals are off
First cover or first brood cover[1]	Apple scab Curculio Codling moth	Liquid lime-sulfur, 5 qt. or Dry lime-sulfur, 5 lb. or Wettable sulfur-manufacturers' specifications and Lead arsenate, 3 lb. Hydrated lime, 5 lb. Water to make 100 gal.	Apply 10 days after petals fall. Wettable sulfurs can be substituted for lime-sulfur and reduce danger of spray injury
Second cover	Curculio Codling moth	Same as for first cover; as the season advances, the strength of lime-sulfur should be decreased to lessen danger of burning	Apply 10 to 14 days after first cover; if apple blotch is a problem, use Bordeaux instead of lime-sulfur
Third cover	Apple scab Codling moth	Same as for second cover	Apply 10 to 14 days after second cover; if apple scab has been controlled and blotch is not present, the fungicide may be omitted. Do not apply on summer varieties
Second brood, fourth cover	Apple scab Codling moth Apple maggot	Same as for third cover	Apply 9 to 10 weeks after petals fall

[1] The number, frequency and composition of the cover sprays will vary with conditions.

PEACH-SPRAY SCHEDULE

Name of spray	Pests to control	Materials to use	Remarks
Dormant	Scale	Liquid lime-sulfur, 1 gal. in 8 or Oil emulsion—3 per cent or Miscible oil—manufacturers' specifications	For leaf curl alone use liquid lime-sulfur 5 gal. in 100 gal. Apply before buds swell in spring
Shuck fall	Curculio	Lead arsenate, 2 lb. Zinc sulfate, 4 lb. Hydrated lime, 4 lb. Water to make 100 gal. or Dust, lime-lead 90-10	Apply as shucks or floral parts are splitting and falling from plant
First cover	Curculio Brown rot Scab	Lead arsenate, 2 lb. Iron or zinc sulfate, 4 lb. Hydrated lime, 4 lb. and Wettable sulfur—manufacturers' specifications or Dry-mix sulfur-lime 16-8-1 and Water to make 100 gal. or Dust-sulfur 80, lead arsenate 10, hydrated lime 10	About two weeks after shucks fall
One month before harvest	Brown rot Scab	Wettable sulfur or Dry-mix sulfur-lime or Dust sulfur 80, lime 20	Do not use lead arsenate at this stage if it can be avoided
Two weeks before harvest[1]	Brown rot Scab	Wettable sulfur or Dust sulfur 80, lime 20	

[1] Additional applications for the control of brown rot may be necessray after this one.

POTATO-SPRAY SCHEDULE

Time to spray	Pests to control	Materials to use	Remarks
Plants 6 to 8 in. tall	Colorado potato beetle Flea beetle Leaf hopper	Calcium arsenate, 4 lb. or Lead arsenate, 4 lb. and Bordeaux mixture, 8-6-100 gal.	Dusts can be used instead of liquid sprays. Use 20 lb. copper dust, 5 lb. arsenicals, 75 lb. hydrated lime
7 to 14 days later	Colorado potato beetle Flea beetle Leaf hopper	Calcium arsenate, 4 lb. or Lead arsenate, 4 lb. and Bordeaux mixture, 8-6-100 gal.	If Colorado potato beetle and flea beetle persist, use the arsenical in later sprays
7 to 14 days later	Leaf hopper Blight	Bordeaux mixture, 8-6-100 gal. or Copper dust, 20 lb.; hydrated lime 80, lb.	The frequency and number of subsequent applications will depend upon the presence of the pests
7 to 14 days later	Leaf hopper Blight	Same as above	Late blight will be more prevalent in seasons that are cool and damp. Under such conditions more frequent and more numerous applications may be necessary
7 to 14 days later	Leaf hopper Blight	Same as above	

with a 1.5 per cent solution of hydrochloric acid. Apples produced in the irrigated sections of the country where the lead residue is high or summer oil sprays have been used may have to be given a weak alkaline bath, then the acid bath. All fruits are rinsed thoroughly after being washed.

Review Questions

1. What was probably man's first conscious effort to control pests that attacked his horticultural plants?

2. Classify the natural means of controlling pests of horticultural plants.

3. Give an example of each natural means used in controlling pests of horticultural plants.

4. Classify the artificial means of controlling pests of horticultural plants.

5. Name the various physical means used to combat the two principal classes of insects.

6. What is an insecticide?

7. Give an example of a specific chemical used to combat each of the two principal classes of insects.

8. What is a fungicide?

9. What are the two most common chemicals used to combat diseases caused by fungi?

10. What legal means are used to control horticultural pests?

11. What is the simplest spray equipment?

12. What is the most advanced spray equipment now in use?

13. What is the simplest dusting equipment?

14. What is the most advanced dusting equipment?

15. What is meant by a combination spray?

16. If the same kinds of pests are present, are the same sprays applied at the same dates to the same trees in different years?

Problems

1. You have a mature apple orchard of 1,000 trees which you intend to spray in the "pink stage" for the control of apple scab. State the materials that you would use, the quantity of each and the manner of preparing the spray.

2. Specify the horticultural plantings about your home, and describe the type of spray outfit that you would purchase that would be adequate to use under such conditions.

Suggested Collateral Readings

1. AUCHTER, E. C., and H. B. KNAPP, "Orchard and Small Fruit Culture," pp. 253–350, John Wiley & Sons, Inc., New York, 1937.

2. LAURIE, ALEX, and G. H. POESCH, "Commercial Flower Forcing," pp. 258–310, P. Blakiston's Son & Company, Philadelphia, 1939.

3. MASON, A. F., "Spraying, Dusting, Fumigation of Plants," The Macmillan Company, New York, 1928.

4. TALBERT, T. J., and A. E. MURNEEK, "Fruit Crops," pp. 187–207, Lea & Febiger, Philadelphia, 1939.

5. WATTS, R. L., and G. S. WATTS, "The Vegetable Growing Business," pp. 159–182, Orange Judd Publishing Company, New York, 1939.

CHAPTER XVII

HARVESTING AND STORING THE PRODUCTS OF HORTICULTURAL PLANTS

Harvesting and storing the products of horticultural plants are important practices. Harvesting consists of removing the product from the plant, and storing consists of placing it under environmental conditions that retard the rate of the life processes.

FIG. 133.—Harvesting potatoes with a two-row, tractor-drawn digger. (*Courtesy of W. E. Schrumpf.*)

The many commercial problems of grading, packing, transporting and storing that are associated with harvesting and storing horticultural products will not be discussed, but only these problems in their relationship to the amateur or home gardener.

HARVESTING

The time and method of harvesting are the two principal problems involved in the removal of the product from the plant.

TIME

The best time for harvesting various horticultural products is usually determined by the degree of maturity of the particular kind and variety of product. The proper degree of maturity for harvesting, in turn, depends upon the keeping qualities of the particular kind and variety of product, the particular method of disposing of the product and the method of storage.

The proper degree of maturity for harvesting often depends upon the particular kind of product and occasionally on a particular variety. Apples and pears are practically the only fruits stored by the average home gardener as fresh products. The other kinds of fruits are too highly perishable to be stored with success in common storage. There is a wide variation in the storage quality of the different varieties of apples and even of the same variety grown in different regions. For example, the McIntosh grown in the Middle Atlantic states is practically an early fall apple suitable for only a few days' storage, whereas the same variety grown in northern New York or New England can be held in storage satisfactorily for as long as four months. Summer apples, which are generally used for immediate consumption, soften so rapidly, even when picked green, that picking before they are fully ripe is often necessary. Most pears should never be allowed to ripen on the trees. Their fine eating quality develops only when they are picked in an early stage of maturity and ripened away from direct sunlight. If picked too early, they shrivel and wilt before ripening. The relative maturity of apples and pears can be judged by the changing of the ground color from green to yellow-green and by the ease with which the stem separates from the tree. Peaches and most varieties of red raspberries and cherries develop their best flavor if allowed to ripen on the plant.

The proper degree of maturity for harvesting also differs with the kind and variety of vegetable. Parsnip, parsley and horseradish may be kept all winter in the ground where they are grown if proper care is taken to prevent alternate freezing and thawing. Kohlrabi, beets, turnips and rutabagas may become tough and woody if allowed to reach too large a size before harvest. On the other hand, the development of the hard outer shell of squash and the protecting surface skin of the Irish potato are developed only on the matured product, and this protective covering aids in successful storage. Sweet potatoes must be mature for successful storage, but cauliflower and celery should be harvested before fully mature when placed in storage. Varieties of the same crop often vary in the time they are ready for harvesting. Onions for home storage should be of the American type, because white Bermuda onions start to grow in common storage. Late varieties of cabbage keep far better in storage than do the early varieties.

The best degree of maturity for harvesting such cut flowers as China aster, chrysanthemum, calendula, gaillardia and zinnia is after the blossoms are opened fully. On the other hand, the best time for harvesting roses, peonies and poppies is when the flower is in an early opening bud stage. Daffodils and other bulbous flowers should be harvested when the flowers are half open.

The best time for harvesting is influenced greatly by the disposition that is to be made of the particular product. When fruits or vegetables are to be consumed immediately or are to be frozen and kept for a period of time, they should be harvested when they are in that degree of maturity associated with the highest eating quality. Products destined for common storage, however, should be picked earlier than those used at once. Apples that are to be consumed soon after harvesting should ripen on the tree until softening begins, but apples destined for storage must be picked at an earlier stage of maturity. The longer the storage period the less should be the degree of maturity when harvested. Irish potatoes may be harvested from the time they are of sufficient size until the vines have fully ripened, but Irish potatoes that are to be stored should be harvested after the vines are fully withered.

The method of storage is also an influencing factor in determining the degree of maturity at which the particular product should be harvested. Cold storage, where temperatures are kept constantly low by mechanical refrigeration, makes it possible to store a more mature product or a product of any degree of maturity for a longer length of time than is possible in cool storage or a house cellar, where temperatures are usually higher and more irregular. Some fresh fruits and vegetables may be frozen and kept in refrigerated storage lockers indefinitely.

Method

The removal of the product from the plant and the necessary handling during harvest generally bruises and wounds some of the tissues. This wounding or bruising permits the entrance of decay organisms and leads to the more rapid decay of the product. Consequently, all harvesting operations should be done carefully, and handling of the products should be reduced to the minimum. Commercial growers use padded receptacles for harvesting easily injured products and exercise every precaution to reduce injury by handling.

Much handwork is required in harvesting horticultural products. Most fruit and many vegetable crops are harvested entirely by hand. Instruments such as sharp knives, spading forks and special diggers are used to facilitate harvesting in the case of certain crops. A sharp knife is the best instrument to use in harvesting cut flowers; spading forks are often employed for digging Irish potatoes; plows with special attachments are used for sweet potatoes; and the commercial Irish potato crop is dug mainly with a special type of digger which digs one to four rows at a time.

STORING

Storing at low temperatures makes it possible to retard the rate of the plant's life processes, thus lengthening the season during which horticultural products can be obtained in the fresh state. The rather pronounced seasonal distribution of perishable horticultural products has been largely eliminated because of satisfactory means of storage.

Factors Influencing Storage

The successful storage of horticultural products requires a consideration of the temperature, humidity, air and light conditions in the place of storage.

The proper temperature for storing horticultural products varies; but unless 40°F. or less can be maintained for most of the storage period, it is not advisable to store many vegetables or fruits. Some few vegetables, such as sweet potatoes, require a temperature of 55°F. The rate of chemical change within the tissues of the stored products is affected materially by temperature. The starch of Irish potatoes begins to change to sugar if the tubers are held below 38°F. Later if the tubers are stored for a time at 45 to 50°F., much of the sugar is changed back to starch. Kieffer pears may be stored at 32°F.; but for the attainment of maximum eating quality they should be ripened at a temperature of 60 to 65°F. Root crops stored at high temperature lose weight because of the oxidization of dry matter and the loss of moisture. Squash stored at 50°F. or above undergo rapid destruction of carbohydrates. Apples are commonly stored at 32°F., but they keep equally well with less evidence of certain storage troubles at a temperature of 34°F.

The relative humidity of the air in the storage rooms has a direct relation to the keeping qualities of the products held in them. If it is too low, wilting occurs in most fruits and vegetables; and if it is too high, it favors the development of decay, especially in rooms where the temperature is high and variable. The proper humidity of the storage room varies with the kind of product stored. For most fruits a relative humidity of 80 to 85 per cent is best. Leafy vegetables and root crops require a relative humidity of 90 to 95 per cent; and other vegetables, with a few exceptions, do well at a humidity of 85 to 90 per cent. A satisfactory atmospheric humidity approximates the percentage of water in the product stored.

Ventilation, in the sense of providing pure air, is rarely necessary but air circulation is useful mainly for cooling and preventing moisture from collecting on the stored product. What little ventilation is necessary is provided by the opening and closing of doors in conducting the necessary work in the storage room and in common storage by the unavoidable leaks around doors and air inlets and outlets.

Some products, like Irish potatoes and onions, become green if exposed to much light while in storage, and celery and endive fail to blanch. Generally darkness is maintained in storage rooms except when light is necessary for the performing of necessary operations.

Types of Storage

Both cold storage and common storage are in general use for horticultural products. Both types are successful if in favorable locations and suitably managed.

Cold storage is that type in which the desired temperature is obtained and maintained by artificial means of refrigeration. The common mechanical home refrigerator is a small cold-storage plant. Cold storages have been used commercially for a long time, but they have become important in the storing of fruits and vegetables for farm and urban families only rather recently with the introduction of refrigerated lockers. At the present time many fruits and vegetables that are produced by the home gardener are frozen and stored in refrigerated lockers where they may be kept indefinitely.

Common storage is that type in which the natural outdoor temperature is depended upon to establish and maintain the desired storage temperature. House cellars, caves, pits and specially constructed buildings are used by the home gardeners as common storages. A simple common storage may be built around a north window of the basement. The window may act as an inlet for cold air and be constructed so that it can be held open at various widths. An air outlet over the door of the storage room will allow much of the warm air to be removed with the upward drafts from the basement. No furnace flues or steam or hot-water pipes should go through the storage room.

Caves are often used by farm families as a place for storing fruits and vegetables. Permanent caves may be made of concrete, clay blocks or bricks. Pits, trenches and mounds of various styles can be used successfully. Shallow pits or a barrel or box sunk into the ground may be used during the winter for vegetables that are not injured by being held in a frozen condition. The soil should be mounded around the pit, barrel or box to prevent surface water from entering. Some of the vegetables adapted for storing in this way are cabbage, parsnip, parsley, horseradish and salsify. These vegetables may be harvested after freezing weather begins, stored in convenient piles and covered to protect them from alternate freezing and thawing.

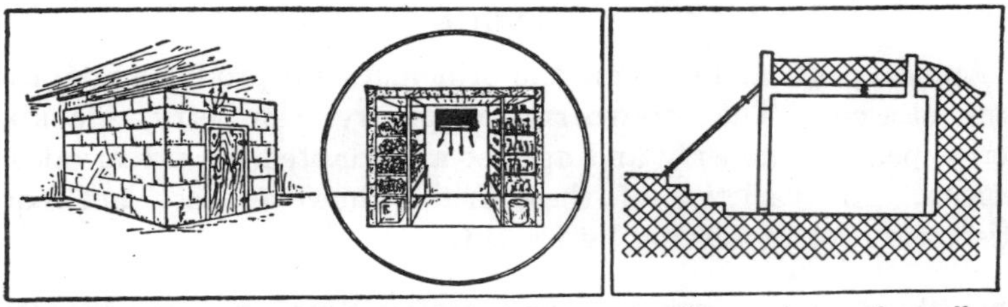

FIG. 134.—Left, a storage room inside a basement. It may be much smaller and simpler than this one. Air is let in through a window and out above the door. Center, the inside of this storage room with stored products at the side so that freezing air won't drop onto them directly. Right, the simplest form of outdoor cave. (*Iowa Extension Service.*)

FIG. 135.—A permanent storage cellar built in side hill. (*Michigan Extension Service.*)

They should be placed in pits as soon as frozen. The box or barrel containing the vegetables should be covered first with a piece of burlap or carpet, then with a mouseproof board cover and finally with straw or similar material. When they are taken from the pit, they may be thawed out overnight in cold water, after which they may be kept in common storage for a considerable length of time.

Rooms in the house that are kept close to freezing may serve as places for storing onions, cabbage, potatoes, apples and nuts. Rooms kept dry and at a temperature of 50 to 55°F. are also good storage places for squash, pumpkins and sweet potatoes.

HARVESTING AND STORING SELECTED PRODUCTS

The basic principles influencing the harvesting and storing of horticultural products are the same, but the specific practices will vary somewhat with the particular plant or crop.

Fruits

Apples, pears and nuts are the principal fruit products adapted for home storage. The strawberry, raspberry, dewberry, blackberry, cherry, peach, blueberry and apricot are adapted for preservation by freezing. All varieties of these fruits, however, are not equally satisfactory for preserving by freezing.

TABLE 35.—A FEW VARIETIES OF THE PRINCIPAL KINDS OF FRUITS ADAPTED FOR FREEZING PRESERVATION IN LOCKER STORAGES

Strawberry	Raspberry	Cherry	Plum	Peach	Apricot
Senator Dunlap	Red:	Montmorency	Red Wing	Elberta	Blenheim
Beaver	Latham	Early Richmond	Monitor	J. H. Hale	Moorpark
Blakemore	Chief	English Morello	Damson		
Premier	Black:		Elliott		
Wayzata (everbearing)	Cumberland				
	Black Pearl				

Apples and Pears.—Apples to be placed in common storage should be picked when in the hard, ripe stage of maturity. Red varieties will gain in color as long as they remain on the tree; but if they become too mature, their storage period will be greatly shortened. Only sound fruit is suitable for storage. The fruit should be picked carefully, and every precaution taken to avoid bruising. After picking, it should be sorted for condition, and the sound fruit placed in clean boxes, baskets or crates. The containers of fruit are stacked in the storage room, where a temperature of 34 to 36°F. is maintained. They should never be so full as to permit bruising.

Most varieties of apples develop scald in storage. This disease causes the skin of the apple to turn brown and later permits rot to start. Scald can be partially prevented by wrapping each fruit in specially prepared oilpaper or by mixing about ½ lb. of shredded oilpaper through each bushel of fruit. Early winter varieties, including Grimes Golden, Jonathan, Delicious, Northwestern Greening and McIntosh, may be kept in common storage until the middle of December in northern latitudes. Later varieties including Winesap, Stayman Winesap, Ben Davis, Willowtwig, Ralls and Mammoth Black Twig may be stored satisfactorily until February and March. Fall varieties such as Wealthy should not be stored longer than February.

Pears are harvested and stored in much the same way as that suggested for apples, the most notable exception being in the degree of maturity at time of harvest. Most pears develop their fine eating quality when they are picked green and ripened away from direct sunlight. The proper degree of maturity for harvesting is determined to a great extent by the appearance of the minute pores, or lenticels, that cover the surface of the fruit. These lenticels are first visible as small light-colored spots. These spots eventually become brown; when this occurs, the openings have been covered by the development of a layer of cork, and there is little danger of the fruit's shriveling after removal from the tree. Some varieties of pear, like Kieffer, if left too long on the trees become gritty because of the development of stone cells.

Vegetables

Many kinds of vegetables, including Irish potatoes, cabbage, onions, beets, turnips, carrots, squash and sweet potatoes, can be kept successfully in common storage. The most satisfactory kinds for preserva-

Table 36.—A Few Varieties of the Principal Kinds of Vegetables Adapted for Freezing Preservation in Locker Storages

Snap beans	Sweet corn	Lima beans	Peas	Spinach	Broccoli	Squash	Asparagus	Greens
Green pod: Burpee's Stringless Bountiful Giant Stringless Asgrow Stringless Wax pod: Pencil Pod Black Wax Round Pod Kidney Wax Keeny's Stringless Kidney Wax Pole bean: Kentucky Wonder	Golden Sunshine Golden Bantam Golden Cross Bantam Improved Golden Bantam Tender-gold Country Gentle-man Stowell's Ever-green	Hender-son's Bush Burpee Improved	Little Marvel Thomas Laxton Telephone	King of Denmark Long Standing Blooms-dale Nobel (Giant Nobel)	Italian Green Sprouting (Cala-brese)	Golden Hubbard Golden Delicious Green Hubbard	Mary Washing-ton	Swiss Chard, Lucullus Kale, Dwarf Scotch Mustard Giant Curled Beet tops Other greens

tion by freezing and storage in refrigerated lockers are peas, lima beans, corn cut off the cob, snap beans, asparagus, spinach, broccoli, Swiss chard, kale, cauliflower, carrots and squash. All varieties of these vegetables, however, are not equally satisfactory for preservation by freezing.

Irish Potatoes.—Irish potatoes are harvested from the time they are of sufficient size to use satisfactorily until a considerable time after the tops have died. The time of digging depends upon whether the crop is for immediate consumption or for storage. Potatoes that have been allowed to remain in the ground until injured by frost should never be stored, because it is almost impossible to sort frozen potatoes from sound ones.

Fig. 136.—Large commercial potato-storage houses at Caribou, Me.

Although power-drawn diggers are used on large commerical acreages, hand tools such as a spading fork or a potatoe hook are used by the home gardener. The latter should be used carefully in order to avoid bruising the crop. Early morning and late evening are good times for digging potatoes in hot weather. Potatoes dug in the even-[illegible] may be left to cool all night on the ground and be picked up and [illegible]d early the following morning. In large commercial plantings they are picked up immediately after digging, placed in suitable containers and stored as soon as possible.

Irish potatoes may be stored satisfactorily in open-mesh sacks which are stacked in tiers with 4-in. air spaces between them. The lower tier of sacks should rest upon a slatted or false floor of wood, because burlap next to a concrete or earthen floor will rot. The best temperature for storing is just under 40°F.

Cabbage.—Cabbage should be harvested when the head is firm. Late varieties are the best for storage and should be harvested before the head is too mature. The head is severed from the stem with a sharp knife or hatchet. A good moist potato storage is also a good cabbage storage. Cabbages may be stored in crates, in baskets or on slatted shelves. The temperature of the storage should be 32°F., and the humidity about 90 per cent.

Onions.—Onions should be harvested before prolonged rains in the fall and thoroughly cured or dried at once. Bermuda onions are harvested as soon as large enough for market, but varieties to be stored must be more mature. When about 60 per cent of the tops have ripened at the neck and fallen over, the bulbs may be pulled from the row. A well-cured onion should be firm and not readily dented at the base when held in the hand and pressed with the thumb. The bulb should show no sprouting and no new roots. Onions are best for storage when they are topped about 1½ in. above the bulb. They should be stored in a cold, dry storage because high temperatures start sprouting, and dampness starts new roots.

Root Crops.—Early root crops, as carrots, beets and turnips, are usually harvested when they reach edible size. Early beets are usually pulled when they reach 1½ in. in diameter, and turnips are harvested when they attain a diameter of 2 in. The late crop should be harvested before the roots become too mature and woody. The best storage temperature for these crops is 32°F., and the humidity should be 90 to 95 per cent. These crops may be stored for home use with the roots placed in moist sand or earth.

Sweet potatoes may be harvested when they reach marketable size. The main crop, which is to be stored, should be allowed to ripen fully, as indicated by the yellowing of the foliage. The commercial crop is ordinarily dug with a plow with a sharp rolling colter and a small moleboard with rods attached. The rolling colter on the beam cuts the vines ahead of the plow, and the iron rods projecting from the moleboard free the potatoes from the soil and vines.

Successful storage depends upon proper curing. This usually requires a period of ten days to three weeks when the potatoes are exposed to a temperature of 80 to 85°F. After curing, the potatoes are stored at a temperature of 55°F. and a humidity of 60 to 70 per cent.

Ornamentals

Cut Flowers.—The best time for harvesting cut flowers is when the plant is turgid, usually before ten o'clock in the morning. The maturity of the bloom at cutting time is another important consideration, because many kinds of flowers are not at their best when fully mature. No definite rule can be given as to the proper stage of maturity, but experimental data indicate that the composites, such as China aster, chrysanthemum, calendula, gaillardia and zinnia, may be allowed to open fully before cutting. Commercial florists often cut the aforementioned kinds of flowers a trifle in advance of the mature stage. Roses, peonies and poppies will open their blooms when their stems are placed in water, and it is best to cut them when the flower is in an early opening-bud stage. Daffodils and most bulbous flowers are gathered commercially when the flowers are about half open, because they develop quickly after they are removed from the plant.

The best instrument for removing flowers from a plant is a sharp knife. It causes less injury to the tissues of the stem than do scissors or the method of breaking off of the flowers. A long, clean, slanting cut made between the nodes allows for more surface in the water after the flowers are placed in containers. A cut at right angles to the stem often leaves a small surface which comes in contact with dirt in the bottom of the container and shortens the time of keeping the flowers in good condition. After cutting, flower stems should be placed in water as soon as possible. In fact, it is a good plan to place the stems in a pail partially filled with water immediately after they are removed from the plant.

Most cut flowers are benefited by a freshening period of a few hours in water at 50°F. immediately after being cut. After flowers are removed from the garden and placed in a roomy receptacle filled with cool water, they should be left in a cool moist room for at least 2 hr. Various kinds of cut flowers vary in keeping qualities regardless of the method of preservation. Under the average room temperature of 70 to 75°F. one may expect such cut flowers as sweet alyssum, columbine and lupine to keep from 2 to 5 days, whereas bachelor's-button, chrysanthemum, coreopsis, gaillardia, pentstemon and veronica will keep from 6 to 10 days. Flowers will keep better and longer at room temperature if a small portion of the base of the stem is removed each day when they are given fresh water.

Rhizomes, Tubers, Corms, Bulbs.—Many hardy herbaceous perennials, as peony, iris, crocus, narcissus and hyacinth, are allowed to remain in the ground over winter. Some of the tender herbaceous perennials, however, as dahlia, canna, gladiolus and often certain varieties of tulip, must be dug and stored during the winter months in order to avoid damage by low temperatures. Dahlias and cannas are usually dug after the tops are killed back by frost. Gladiolus should be dug before the ground is frozen. Tulips are dug before the tops are completely dried, as the top assists in the operation of harvesting.

Many of the hardy herbaceous perennials are usually stored immediately after digging. Gladiolus and often tulips are generally cured in a well-ventilated shed or room for a period of three to six weeks before they are placed in storage. Cannas and dahlias are stored in shallow trays containing dry soil or sawdust. Gladiolus should be stored dry in shallow trays with ample ventilation.

Review Questions

1. What are the two principal factors considered in harvesting horticultural products?

2. What determines the time for harvesting a particular horticultural product?

3. Why is careful handling so important in harvesting horticultural crops?

4. What is the function of storage?

5. What factors influence the successful storage of horticultural products?

6. What are the two principal types of storage?

7. What is meant by cold storage?

8. What is meant by common storage?

Problems

1. Select a location, and make diagrammatic sketches showing the construction of a common storage suitable for the crop from the home apple orchard and potato patch.

Suggested Collateral Readings

1. AUCHTER, E. C., and H. B. KNAPP, "Orchard and Small Fruit Culture," pp. 1–117, John Wiley & Sons, Inc., New York, 1937.

2. CARRICK, D. B., and A. M. GOODMAN, The Common or Air-cooled Apple Storage and Its Management, *N. Y. Agr. Col. (Cornell) Ext. Bul.*, 286: 1–29, 1934.

3. COMIN, DONALD, The Common Storage, Its Construction and Management, *Ohio Agr. Exp. Sta. Bul.* 573: 1–49, 1936.

4. ROSE, D. H., *et al.*, The Commercial Storage of Fruits, Vegetables and Florists' Stocks, *U.S. Dept. Agr. Cir.* 278: 1–39, 1938.

5. THOMPSON, H. C., "Vegetable Crops," pp. 170–176, McGraw-Hill Book Company, Inc., New York, 1939.

6. WATTS, R. L., and G. S. WATTS, "The Vegetable Growing Business," pp. 204–211, Orange Judd Publishing Company, New York, 1939.

INDEX

A

B

M

N

O